DNA Repair: Novel Insights

DNA Repair: Novel Insights

Edited by **Nas Wilson**

New York

Published by Callisto Reference,
106 Park Avenue, Suite 200,
New York, NY 10016, USA
www.callistoreference.com

DNA Repair: Novel Insights
Edited by Nas Wilson

International Standard Book Number: 978-1-63239-151-3 (Hardback)

Contents

Preface

In my initial years as a student, I used to run to the library at every possible instance to grab a book and learn something new. Books were my primary source of knowledge and I would not have come such a long way without all that I learnt from them. Thus, when I was approached to edit this book; I became understandably nostalgic. It was an absolute honor to be considered worthy of guiding the current generation as well as those to come. I put all my knowledge and hard work into making this book most beneficial for its readers.

This book provides novel insights into DNA and is designed to serve as a useful resource for scientists as well as students associated with the field of DNA repair. Topics described in this book for the illustration of clinical translational efforts based on paradigms established in DNA repair such as diagnostic, therapeutic and analytical. The aim of this book is to serve as a valuable source of reference in seminars as well as courses and for biologists interested in the field of DNA repair.

I wish to thank my publisher for supporting me at every step. I would also like to thank all the authors who have contributed their researches in this book. I hope this book will be a valuable contribution to the progress of the field.

Editor

Interface with Clinical Medicine

Genetic Polymorphisms of DNA Repair Genes and DNA Repair Capacity Related to Aflatoxin B1 (AFB1)-Induced DNA Damages

Qiang Xia, Xiao-Ying Huang, Feng Xue,
Jian-Jun Zhang, Bo Zhai, De-Chun Kong,
Chao Wang, Zhao-Quan Huang and Xi-Dai Long

Additional information is available at the end of the chapter

1. Introduction

Aflatoxin B1 (AFB1) is an important aflatoxin produced by some strains of the moulds Aspergillus parasiticus and Aspergillus flavus [1-3]. This aflatoxin was discovered as a contaminant of human and animal food, especially peanuts (ground nuts), core, soya sauce, and fermented soy beans in tropical areas such as the Southeastern China as a result of fungal contamination during growth and after harvest which under hot and humid conditions in the late 1950s and early 1960s [1-4]. Increasing evidences have shown that AFB1 exposure levels are consistent with hepatocellular carcinoma (HCC) risk values [1, 2, 4-7]. DNA damage by AFB1 plays the central role of carcinogenesis of HCC-related to this toxin in the toxic studies [2, 8-10]. Today, AFB1 has been classified as a known human carcinogen by the International Agency for Research on Cancer [1, 2, 5, 10, 11]. However, more and more epidemiological evidence has exhibited that although many people are exposed to the same levels of AFB1, only a relatively small proportion of exposure person develop HCC [6, 12-23]. This indicates individual DNA repair capacity related to AFB1-induced DNA damage might be associated with HCC carcinogenesis [4].

This study attempts to briefly review currently available data on genetic polymorphisms of DNA repair genes and DNA repair capacity related to AFB1-induced DNA Damages, with emphasis on: (1) DNA damage types, (2) DNA repair pathways, (3) the role of DNA repair genetic polymorphisms in the repair process of DNA damage by AFB1, and (4) the elucidation of corresponding DNA repair capacity. Additionally, we summarized the association between genetic polymorphisms of DNA repair genes and AFB1-related DNA repair capacity via a meta-analysis based on published data.

2. AFB1's chemistry

In 1963, Asao *et al.* accomplished the structural elucidation of AFB1 and found AFB1 was a member of aflatoxins family (AFF) highly substituted coumarins containing a fused dihydrofurofuran moiety [24]. AFF consists of four members: AFB1, aflatoxin B2 (AFB2), aflatoxin G1 (AFG1), and aflatoxin G2 (AFG2). Among of these members, AFB1 is the most important toxin and structurally is characterized by fusion of a cyclopentenone ring to the lactone ring of the coumarin moiety [24]. AFB1 is so named because of its strong blue fluorescence in ultraviolet light. These properties facilitated the very rapid development in the early 1960s of methods for monitoring peanuts, cores, grains, and other food commodities for the presence of the toxins (Fig. 1) [1]. This type AFF possesses an unsaturated bond at the 8,9 position on the terminal furan ring, and subsequent studies have demonstrated that AFB1 may be metabolized by cytochrome P450 (CYP) enzymes to its reactive form at this position, also called AFB1-8,9-epoxide (AFB1-epoxide) [2, 10]. AFB1-epoxide can covalently bind to DNA and induce DNA damage, thus this epoxidation at at the 8,9 position is critical for AFB1's DNA genotoxic and carcinogenic potency [2]. Noticeably, another important chemiatric feature of AFB1 is the attraction of liver organ, possibly because the metabolic enzymes CYPs are mainly produced by liver [2, 10, 25].

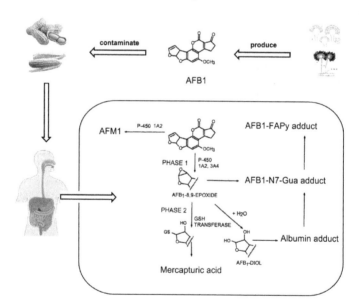

Figure 1. Biotransformation pathways for AFB1. AFB1, mainly produced by the moulds Aspergillus parasiticus (right upper figure) and Aspergillus flavus (right under figure), is metabolized by cytochrome P450 enzymes to its reactive form, AFB1-8,9-epoxide (AFB1-epoxide). AFB1-epoxide covalently binds to DNA strands and results in the formation of AFB1-DNA adducts (including AFB1-N7-Gua adduct and AFB1-FAPy adduct).

3. DNA damage by AFB1

Several previous reviews have significantly summarized the DNA toxicity of AFB1 [1, 2, 8]. Generally, the severity of DNA toxic effects in human or animals vary with exposure levels, exposure years and nutritional status [1, 2, 26]. For large doses of exposure, this agent can induce acute damage of DNA such as inhabiting DNA synthesis, decreasing DNA-dependent RNA polymerase activity, and restraining messenger RNA (mRNA) and protein synthesis, and subsequently resulting in the lethal changes of liver cells: hepatocellular severe degeneration and necrosis [1, 2].

For long-times and low-levels exposure mainly induces chronic DNA damage [1, 2]. This damage can result in neoplasia, primarily HCC, in many animals or human. Chronic DNA damages induced by AFB1 include AFB1-DNA adducts, oxidative DNA damage, DNA strand break damage, and gene mutation [1, 2, 4].

3.1. AFB1-DNA adducts

AFB1-DNA adducts, including 8,9-di-hydro-8-(N7-guanyl)-9-hydroxy–AFB1 (AFB1-N7-Gua) adduct and formamidopyridine AFB1 (AFB1-FAPy) adduct (Fig. 1), is the main type of AFB1-induced DNA damage [1-4, 25-39]. Among these AFB1-DNA adducts, AFB1-N7-Gua adduct is the most common type identified and confirmed in vivo researches [2, 25]. This type adduct is formed from two pathways: (1) Binding reaction of AFB1-epoxide with DNA; and (2) enzymatic oxidation of AFP1, AFM1, and others with unsaturated in the 8,9-position [2, 25]. In the first pathway, the formations of AFB1-N7-Gua adduct proceeds by a precovalent intercalation complex between double-stranded DNA and the highly electrophilic, unstable AFB1-epoxide isomer. After that, the induction of a positive charge on the imidazole portion of the formed AFB1-N^7-Gua adduct gives rise to another important a DNA adduct, a ring-opened AFB1-FAPy adduct. Accumulation of AFB1-FAPy adduct is characterized by time-dependence, non-enzyme, and may be of biological basis of genes mutation because of its apparent persistence in DNA. Another pathway only gives rise to minor AFB1-DNA adducts [1, 2, 25]. Additionally, some other DNA-adducts types, ex. covalent binding of AFB1 to adenosine or cytosine in DNA, has also been reported, however, needing more evidences to support this adducts [2].

Although AFB1-DNA adducts are mainly produced in liver cells, they are also found in the peripheral blood white cells [2]. Recent studies have shown that the levels of AFB1-DNA adduct of the peripheral blood white cells are positively and lineally correlated with that of liver cells, implying analysis of AFB1-DNA adducts in the peripheral blood white cells may substitute for the elucidation of tissular levels of adducts [40].

3.2. Oxidative DNA damage

In the process of agent AFB1 metabolism, this agent can induced reactive oxygen species (ROS) [2]. Especially, the metabolic particulate phases, including I and II phase involved by detoxicate enzymes such as CYP and glutathione S-transferase (GST), is postulated to contain long-lived ROS that can lead to oxidative DNA damage [2, 4]. Nowadays, ROS have also been suggested to be involved in the progression of chronic liver disease and the occurrence of HCC; whereas its' subsequent Oxidative DNA damage is generally regarded as a significant contributory cause of cancer from environmental exposures such as AFB1 exposure [41]. Of oxidative DNA damage, 8-oxodeoxyguanosine (8-oxodG), a kind of especial DNA adduct, is found as a sensitive marker of the DNA damage due to hydroxyl radical attack at the C8 of guanine [2, 4, 25, 42]. This adduct, different from the aforementioned AFB1-DNA adducts, is the most abundant endogenous DNA lesion caused by ROS, and has been classified as a biomarker of oxidative DNA damage [2, 10, 43, 44].

Previous studies have shown that in vitro treatment of hepatocytes with AFB1 resulted in a dose-dependent increase in ROS formation [45]; whereas exposure of rats to AFB1 produced a time- and dose-dependent increase in 8-oxodG in hepatic DNA [46, 47]. In 2007, Wu, et al. investigated the association between AFB1 exposure levels and oxidative damage levels in high AFB1 areas from Taiwan, China [48]. In this case–control study nested within a community-based cohort (74 HCC cases and 290), researchers tested the levels of urinary excretion of 8-oxodG, a biomarker of oxidative DNA damage and urinary AFB1 metabolites, a biomarker of AFB1 exposure, through enzyme-linked immunosorbent assays (ELISA). Results showed 8-oxodG levels were significantly positive correlated with AFB1 exposure, suggesting AFB1 exposure should induce oxidative DNA damage [48]. Together, these data suggest that AFB1-induced oxidative DNA damage may constitute an important pathway in AFB1 toxicity.

3.3. DNA strand break damage

Previous reviewed adducts are capable of forming subsequent repair-resistant adducts, depurination, or lead to error-prone DNA repair resulting in single-strand breaks (SSBs) and double-strand breaks (DSBs). SSBs and DSBs are two kinds of important DNA damage types by AFB1 exposure. For SSBs, there are three pathways to produce this type DNA damage under the AFB1 exposure conditions: direct attack by ROS, through base hydrolysis, and enzymatic consequence of the repair of spontaneous base damage and base loss (such as resulting from abasic AP. sites arising spontaneously or from the action of glycosylases in the process of BER pathway) [49-51]. As the most abundant lesion occurring in cellular DNA, SSBs can play havoc with replication and transcription if not efficiently eliminated. However, they might cause other DNA damage such as genic mutations, DSBs, or carcinogenesis of cells [51, 52]. While DSBs is rare and severe DNA damage type among DNA damage induced by AFB1 exposure [25], mainly produced under the high-dose AFB1 exposure conditions. This damage can lead to chromosomal rearrangements at the first mitosis after exposure to the DNA strand-breaking agent [53].

3.4. Gene mutations

For genes mutations induced by AFB1 exposure, the experimental and theoretical researches are briefly on the p53 gene [54-56]. Reaction with DNA at the N^7 position of guanine preferentially causes a G:C > T:A mutation in codon 249 of this gene, leading to an amino acid substitution of arginine to serine [54-56]. In high AFB1-exposure areas, this mutation is present in more than 40% of HCC and can be detected in serum DNA of patients with preneoplastic lesions and HCC. While codon 249 transversion mutations are either very rare or absent in low or no AFB1-exposure areas [4]. Using the human p53 gene in an in vitro assay, codon 249 has been exhibited to be a preferential site for formation of AFB1-N^7-Gua adducts evidence consistent with a role for AFB1 in the mutations observed in HCC [57-65]. Therefore, the codon 249 mutation of p53 gene has been defined as the hot-spot mutation of p53 gene (TP53M) resulting from AFB1 and has become the molecular symbol of HCC induced by AFB1 exposure. The frequency of TP53M is also regarded as the molecular biomarker of AFB1-related DNA repair capacity [4].

4. DNA repair pathways of AFB1-related DNA damage

A wide diversity of DNA damage induced by AFB1 exposure, if not repaired, may cause chromosomal aberrations, micronuclei, sister chromatid exchange, unscheduled DNA synthesis, and chromosomal strand breaks, and can be converted into gene mutations and genomic instability, which in turn results in cellular malignant transformation [4]. Nevertheless, human cells have evolved surveillance mechanisms that monitor the integrity of genome to minimize the consequences of detrimental mutations [9]. AFB1-induced DNA damage can be repaired through the following pathways: nucleotide excision repair (NER), base excision repair (BER), single-strand break repair (SSBR), and double-strand break repair (DSBR) [4, 25].

4.1. NER pathway

NER pathway, a major DNA repair pathways in human cells featuring genomic DNA damage, can remove structurally such diverse lesions as pyrimidine dimers, irradiative damage, and bulky chemical adducts, and DNA damage from carcinogens and some chemotherapeutic drugs [66]. To date, the mechanism of this pathway is well understood and has been reconstituted in vitro. It consists of several sequential steps: lesion sensing, opening of a denaturation bubble, incision of the damaged strand, displacement of the lesion-containing oligonucleotide, gap filling, and ligation [66, 67]. In the fibroblast cells with the deficiency of xeroderma pigmentosum A (XPA) gene, conversion of the initial AFB1-N7-Gua adduct to the AFB1-FAPy adduct has been found to be more extensive. This suggests that NER should be a major mechanism for enzymatic repair of AFB1 adducts. Its defects lead to severe diseases related AFB1 exposure, including liver injury and HCC [4].

4.2. BER

Of the oxidative DNA damage resulting from AFB1 exposure, the formation of 8-oxodG is thought to be important due to being abundant and highly mutagenic and hepatocarcinogenesis [4, 25]. The 8-oxodG lesions are repaired primarily through the BER pathway. The BER pathway facilitates DNA repair through two general pathways: a. the short-patch BER pathway, leading to a repair tract of a single nucleotide; b. the long-patch BER pathway, producing a repair tract of at least two nucleotides [68, 69]. In these two repair sub-pathways, DNA glycosylases play a central role because they can recognize and catalyze the removal of damaged bases [68, 69]. This suggests that the defect of DNA glycosylases should be related to the decreasing capacity of the BER pathway and might increase the risk of such toxicity as AFB1 [4, 25].

4.3. SSBR

SSB is a relative severe type of DNA damage produced by AFB1 exposure. If not repaired, it can disrupt transcription and replication and can be converted into potentially clastogenic and/or lethal DSBs [51]. This DNA damage is repaired via SSBR pathway. SSBR pathway includes four basic steps: a. SSB detection and signaling, through poly (ADP-ribose) polymerase (PARP); b. DNA break end processing, through the role of polynucleotide kinase (PNK), AP endonuclease-1 (APE1), DNA polymerase β (Pol β), tyrosyl phosphodiesterase 1 (TDP1), and flap endonuclease-1 (FEN-1); c. gap filling, involving in multiple DNA polymerases; d. DNA ligation, involving in multiple DNA ligases [49, 50, 52]. This pathway mainly plays an important role in the repair process of SSBs induced AFB1.

4.4. DSBR

DSBs, although only make up a very small proportion of AFB1-induced DNA damage, are critical lesions that can result in cell death or a wide variety of genetic alterations including large- or small-scale deletions, loss of heterozygosity, translocations, and chromosome loss [70]. This type damage is repaired DSBR consisting of non-homologous end-joining (NHEJ) and homologous recombination (HR) [71, 72]. There are several decades DNA repair genes involve in DSBR pathway and the defects in these genes cause genome instability and promote tumorigenesis [71-77]. During the process of damage removed by aforementioned repair pathways, DNA repair genes play a central role, because their function determines DNA repair capacity [4]. It has been shown that reduction in DNA repair capacity related to DNA repair genes is associated with increasing frequency of genic mutation, levels of DNA adducts, and risk of cancers [8, 78]. Thus, genetic polymorphisms in DNA repair genes might be correlated with AFB1-related DNA repair capacity.

5. The elucidation of DNA repair capacity related to AFB1-induced DNA damage

As shown in the previous review, two main characteristics of AFB1-induced DNA damage are AFB1-DNA adducts and the hot-spot mutation of tumor suppressor gene p53 at codon 249 (TP53M) [4, 25]. Thus, DNA repair capacity related to this type DNA damage might be elucidated using the analysis of AFB1-DNA-adducts levels and TP53M frequency in the liver tissues or other tissues. For AFB1-DNA adducts, many researchers in the relative fields regard AFB1-FAPy adduct as a validated biomarker of AFB1 exposure based on as following reasons: (1) that AFB1-FAPy adduct is the imidazole ring-opened product of AFB1-N7-Gua adduct, also the stable of form of the later adduct, and may play an important role in the development of HCC. Moreover, the accumulation of this adduct is time-dependent and non-enzymatic, and may have potential biological importance because of its apparent persistence in DNA; (2) that AFB1-N7-Gua adduct is unstable and easily lost from DNA. Increasing evidences have exhibited that AFB1-FAPy-adducts levels in the liver or placenta tissues are lineally correlated with AFB1 exposure levels and HCC risk [79, 80], suggesting this adduct should be regarded as a biomarker for DNA repair capacity related to AFB1-induced DNA damage. Remarkably, the monoclonal antibodies recognizing AFB1-FAPy adduct have been developed by several research groups. These types of antibodies are not only used to orientationally and semi-quantitatively test AFB1-DNA adduct information in the tissue specimens through immunochemistry (IHC), but to quantitatively analyze the levels of this adduct using a competitive enzyme-linked immunosorbent assay (ELISA) in human liver and placenta tissue specimens. Additionally, a quantitative indirect immunofluorescence method using monoclonal antibody 6A10 has also been developed to measure AFB1-DNA adducts in liver tissues. In 2009, Long et al. evaluated the validation of AFB1-FAPy adduct in DNA samples from peripheral blood leukocytes representing AFB1 exposure levels [40]. Through linear regression analysis of the adduct levels in DNA samples from peripheral blood leukocytes and from liver tissue specimens, they found peripheral blood leukocytes' adduct levels were positively and linearly related to AFB1-DNA adduct levels of the HCC cancerous tissue. These data suggested that the levels of peripheral blood leukocytes' DNA adducts were representative of the tissues' DNA-adduct levels and might be regard as a biomarker for AFB1 exposure [4, 25, 40, 78]. Together, AFB1-FAPy adduct in DNA from such tissues as liver and placenta or from such as blood leukocytes should be potential biological importance in the elucidation of DNA repair capacity related to AFB1-caused DNA damage.

As regard of the mutations of p53 gene, because AFB1 exposure results in G to T transversion in both bacteria and human cells and AFB1 preferentially binds to codon 249 of p53 gene, as previous mentioned, AFB1 mainly induces the transversion of G → T in the third position at codon 249 of TP53M. The frequent value of TP53M is more persistent biomarker and more directly represents DNA repair capacity compared with AFB1-DNA adducts. In the studies from higher AFB1 exposure areas, researchers have found TP53M frequency associates with AFB1 exposure levels and HCC risk. Thus, this mutation is the selective elucidative marker for DNA repair capacity correlated with AFB1-induced DNA damage as well as AFB1-DNA adducts.

Additionally, HCC is the most common malignant tumors caused by AFB1 exposure. More and more epidemiological studies have shown AFB1-related HCC risk is related to different DNA repair capacity [4, 8, 15, 22, 40, 78, 81-90], suggesting that tumor risk value might be regard as a selective elucidative marker for DNA repair capacity correlated with AFB1-induced DNA damage.

6. Genetic polymorphisms of DNA repair genes involved in NER pathway for AFB1-related DNA damage repair

Accumulating evidences have implied that genetic polymorphisms in NER genes are associated with DNA repair capacity related to AFB1-induced DNA damage. Molecular epidemiology studies in this field are mainly from high AFB1 exposure areas such as in China. To date, two genes involved in NER pathway, namely xeroderma pigmentosum C (XPC) and xeroderma pigmentosum D (XPD), have been investigated in the DNA repair capacity analysis.

6.1. XPC

XPC gene (Genbank accession NO. AC090645), consisting of 16 exons and 15 introns, spans 33kb on chromosome 3p25. This gene encodes a 940-amino acid protein, an important DNA damage recognition molecule which plays an important role in NER pathway. XPC protein binds tightly with another important NER protein HR23B to form a stable XPC-HR23B complex, the first protein component that recognizes and binds to the DNA damage sites [91-98]. XPC-HR23B complex can recognize a variety of DNA adducts formed by exogenous carcinogens such as AFB1 and binds to the DNA damage sites [4, 91, 99]. Therefore, it may play a role in the process of DNA repair of DNA damage related to AFB1 exposure.

Some recent studies have showed that defects in XPC have been related to many types of malignant tumors [99-114]. Transgenic mice researches have also exhibited predisposition to many kinds of neoplasms in mice model with XPC gene knockout [115]. Moreover, pathological and cellular studies have shown that increasing expression of this gene is associated with hepatocarcinogenesis [116]. Together, these studies suggest the genetic polymorphisms localizing at conserved sites of XPC gene might modify the risk of HCC induced by AFB1 exposure and decrease DNA repair capacity related to AFB1-related DNA damage. Recently, four studies from high AFB1-exposure areas have supported abovementioned hypothesis [84, 89, 101, 117].

The first study conducted by Cai et al.[117] is from Shunde area, Guangdong Province which is characterized by high AFB1 exposure and high incidence rate of HCC. Researchers analyzed the association between two common polymorphisms—Ala499Val and Lys939Gln—of XPC gene and risk of HCC via an 1-1 case-control study (including 78 HCC patients and 78 age- and sex-matching controls) method, and found these two polymorphisms modified HCC risk [adjusted odds ratios (ORs) were 3.77 with 95% confidence interval (CI) 1.34-12.89 for Ala499Val and 6.78 with 95% CI 2.03-22.69]. Although they did not directly evaluated the effects of genetic polymorphisms of XPC gene and DNA repair capacity related to AFB1-caused DNA damage, study population in their study is from high AFB1 exposure areas and.

The other three studies, from Guangxi Zhuang Autonomous Region which is the most common of high AFB1 exposure area all over the world [4, 118], directly investigated the modifying effects of genetic polymorphisms XPC on AFB1-related DNA repair capacity and HCC risk based on hospitals via molecular epidemiological studies [84, 89, 101]. Their results showed XPC codon 939 Gln alleles increased about 2-times risk of HCC and decreased AFB1-related DNA repair capacity. Furthermore, Wu, et al.[89] and Long, et al. [84] quantitatively elucidated AFB-exposure time and levels and their interactive effects with the genetic polymorphisms of XPC gene and found some evidence of AFB1 exposure-risk genotypes of XPC codon 939 on AFB1-related DNA repair capacity (HCC risk: XPC risk genotypes and 18.38 > 1.11 × 4.62 for the interaction of AFB1-exposure levels and XPC risk genotypes; 22.33 > 1.88 × 8.69 for the interaction of AFB1-exposure time).

Additionally, Long, et al. [84] also observed that Gln alleles at codon 939 of XPC gene was associated with the decrease of XPC expression levels in cancerous tissues ($r = -0.369$, $P <$ 0.001) and with the poorer overall survival of HCC patients (the median survival times are 30, 25, and 19 months for patients with XPC gene codon 939 Lys/Lys, Lys/Gln, and Gln/Gln respectively). Interestingly, this decreasing 5-years survival rates would be noticeable under high AFB1 exposure conditions (the median survival times are 17 month for the joint of XPC gene codon 939 Gln/Gln and high AFB1-exposure level and 15 months for the joint of XPC gene codon 939 Gln/Gln and long-term AFB1-exposure time) [84].

As a result, these data suggest that genetic polymorphism at codon 939 of XPC gene is not only a genetic determinant in the DNA repair process of DNA damage induced by AFB1 exposure, and a risk and prognostic factor influencing HCC developing, but also is an independent genetic factor of evaluating DNA repair capacity related to AFB1-caused DNA damage. A possible reason is that this genetic polymorphism down-regulates XPC expression [84] and decrease the repair function of XPC protein [116].

However, Li et al. [101] reported that the proportional distribution of the Val/Val genotype at codon 499 of XPC gene did not differ between HCC cases and controls in Guangxi Zhuang Autonomous Region, China ($P > 0.05$), dissimilar to the data from another high AFB1 exposure area of China, Guangdong Province, suggesting this genetic polymorphism might not modify AFB1-related DNA repair capacity. Possible explanations for these inconsistent finding may be either due to unknown confounders or due to small sample size.

6.2. XPD

XPD protein, a DNA-dependent ATPase/helicase encoded by DNA repair gene XPD (also called excision repair cross-complementing rodent repair deficiency complementation group 2 (ERCC2), COFS2, EM9, or TTD.) (Genbank ID. 2068) which spans about 20 kb on chromosome 19q13.3 and contains 23 exons and 22 introns is one of seven central proteins in the NER pathway [119-122]. This protein is associated with the TFIIH transcription-factor complex, and plays a role in NER pathway [66, 67, 119-121, 123-125]. During NER, XPD participates in the opening of the DNA helix to allow the excision of the DNA fragment containing the damaged base [119-122].

There are four described polymorphisms that induce amino acid changes in the protein: at codons 199 (Ile to Met), at codon 201 (His to Tyr), at codon 312 (Asp to Asn) and at codon 751 (Lys to Gln) [123]. To date, the first two polymorphisms have not investigated because they are quite rare (~0.04%) in most population, whereas the latter two polymorphisms in conserved region of XPD have been extensively studied [123]. Several groups have done genotype-phenotype analyses with these two polymorphisms and have shown that the variant allele genotypes are associated with low DNA repair ability [126, 127]. Recent studies have showed the polymorphisms at codon 312 and 751 of XPD are correlated with DNA-adducts levels, p53 gene mutation, and cancers risk [86, 123, 128-131]. In a hospital-based case-control study conducted in a high AFB1 exposure area [40], Long, *et al.* found that the variant XPD codon 751 genotypes (namely Lys/Gln and Gln/Gln) detected by TaqMan-MGB PCR was significantly different between HCC cases (35.9% and 20.1% for Lys/Gln and Gln/Gln, respectively) and controls (26.3% for Lys/Gln and 8.6% for Gln/Gln, $P < 0.001$). Individuals having variant alleles had about 1.5- to 2.5-fold risk of developing the cancer (adjusted OR 1.75 and 95% CI 1.30-2.37 for Lys/Gln; adjusted OR 2.47 and 95% CI 1.62-3.76 for Gln/Gln). Based on relative large sample size (including 618 HCC cases and 712 controls), researchers stratified genotypes of XPD codon 751 according to matching factors and observed some evidence of interaction between XPD codon 751 Gln alleles and sex. These female with Gln alleles featured increasing HCC risk compared with those without these alleles. Moreover, the multiple interactive effects of between mutant genotypes of XPD gene codon 751 environment variant AFB1 or another NER gene XPC on HCC risk were also found, with interactive value 0.85, 1.04, and 1.71 for AFB1-exposure years, AFB1-exposure levels, and XPC gene codon 939 risk genotypes ($P_{interaction} < 0.05$).

Together, these results suggest the genetic polymorphisms at conserved sequence of XPD gene such as at codon 751 may have potential effect on AFB1-related HCC susceptibility. This supports different AFB1-related DNA repair capacity might be modified by genetic polymorphisms at codon 751 in DNA repair gene XPD. However, the study from AFB1-exposure areas shows that the genetic polymorphism at codon 312 in XPD polymorphism is not significantly correlated with DNA repair capacity related AFB1-induced DNA damage [4, 40].

7. Genetic polymorphisms of DNA repair genes involved in BER pathway for AFB1-related DNA damage repair

As previous described, DNA glycosylases play a central role in the BER pathway because they can recognize and catalyze the removal of damaged bases [68, 69]. Among having been reported genetic polymorphisms of DNA glycosylases, only human 8-oxoguanine DNA glycosylase (hOGG1) correlates with DNA repair capacity [132-143]. This gene (Genbank ID# 4968), also called HMMH, OGG1, MUTM, OGH1, 8-hydroxyguanine DNA glycosylase, AP lyase, DNA-apurinic or apyrimidinic site lyase, and N-glycosylase/DNA lyase, consisting of 7 exons and 6 introns, spans 17 kb on chromosome 3p26.2 (PubMed). This gene encodes a 546-amino acid protein, a specific DNA glycosylase that catalyzes the release of $8\text{-}_{oxod}G$ and the cleavage of DNA at the AP site [142]. Genetic structure study has shown the presence of several polymorphisms within hOGG1 locus [136]. Among these polymorphisms, the polymorphism at position 1245 in exon 7 causes an amino acid substitution (namely Ser to Cys) at codon 326, suggesting this polymorphism may glycosylase function and decrease DNA repair capacity [136].

In the past twenty years, increasing epidemiological evidences have validated aforementioned the hypothesis [132-144]. In 2003, Peng, *et al.* [138] analyzed the correlation among 8-$_{oxod}$G levels, hOGG1 expression, and hOGG1 Cys326Ser polymorphism in the high AFB1 exposure areas Guangxi Autonomous Region. They found that individuals having genotypes with hOGG1 codon 326 Cys alleles faced lower level of hOGG1 expression and higher 8-$_{oxod}$G levels. Supporting their results, Cheng, *et al.* [141] reported that hOGG1 expression was significantly linear correlated with HCC. Recently, using the molecular epidemiological methods, Zhang, *et al.* [134] found that the distribution of Cys alleles at codon 326 of hOGG1 in HCC cases (43.0%) significantly differed from in controls (33.1%). Logistic regression analysis next showed that the genotypes with Cys alleles, compared to without this alleles, increased HCC risk of Chinese population, with adjusted OR-value (95% CI) 1.5 (0.79-2.93) for Cys/Ser and 1.9 (0.83-4.55) for Cys/Cys. Similar results are also observed in the study from low AFB1 exposure areas of China [144]. A functional complementation activity assay exhibited that hOGG1 protein encoded by the 326 Cys allele had substantially lower DNA repair activity than that encoded by the 326 Ser allele [140]. Similar results were observed in human cells in vivo [137, 139]. Therefore, low capacity of 8-$_{oxod}$G repair resulting from hOGG1 326 Cys polymorphism might contribute to the persistence of 8-$_{oxod}$G in genomic DNA in vivo, which, in turn, could be associated with increased cancer risk [4, 137, 138].

As a result, these findings suggested the genetic polymorphism at codon 326 of DNA repair gene hOGG1 should modify AFB1-related DNA repair capacity. However, another case-control study from Japan shows this genetic polymorphism is not associated with HCC risk. This might result from lower AFB1 exposure in this area and not showing the relative low DNA repair capacity related to AFB1-induced DNA damage.

8. Genetic polymorphisms of DNA repair genes involved in SSBR pathway for AFB1-related DNA damage repair

SSBR pathway involves in several central DNA repair genes such as XRCC1, poly (ADP-ribose) polymerase-1 (PARP-1), APE (or DNA glycosylase), DNA ligase III, Pol β, and so on [49-51]. Of these DNA repair genes, only XRCC1 is investigated to correlate with AFB1-related DNA repair capacity. This gene, also called RCC, spans about 32 kb on chromosome 19q13.2 and contains 17 exons and 16 introns is one of three submits of DNA repair complex in the SSBR pathway (Gene dbase from PubMed). Its' encoding protein (633 amino acids), consists of three functional domains: N-terminal domain (NTD), central breast cancer susceptibility protein-1 homology C-terminal (BRCT I), and C-terminal breast cancer susceptibility protein-1 homology C-terminal (BRCT II) [4, 51, 145-151]. This protein is directly associated with Pol β, DNA ligase III, and PARP, via their three functional domains and is implicated in the core processes in SSBR and BER pathway [4, 51, 145, 150-152]. Mutant hamster ovary cell lines that lack XRCC1 genes are hypersensitive to DNA damage agents such as ionizing radiation, hydrogen peroxide, and alkylating agents [4, 51]. Furthermore, this kind of cells usually faces increasing frequency of spontaneous chromosome aberrations and deletions. Three single nucleotide polymorphisms in the coding region of XRCC1 gene that lead to amino acid substitution have been described and investigated [25]. Among these polymorphisms, the codon 399 polymorphism is of special concern, because this polymorphism resides in functionally significant regions (BRCT II) and may be related to decreasing DNA repair capacity [85, 153-179].

In 2008, Long, et al. [85] investigated the effects of genetic polymorphism at codon 399 in DNA repair gene XRCC1 based on the analysis of 501 AFB1-related HCC samples. They found that the HCC patients with XRCC1 genotypes with 399 Gln alleles (namely: XRCC1 codon 399 Arg/Gln or Gln/Gln) faced a significantly higher frequency of TP53M than those with the wild-type homozygote of XRCC1 [namely: XRCC1 codon 399 Arg/Arg, adjusted odds ratio (OR) = 6.13, 95% confidence interval (CI) = 3.87-9.72 for Arg/Gln; OR = 13.66, 95% CI = 4.44-42.08 for Gln/Gln, respectively]. Additionally, another study from high AFB1 areas Taiwan in China exposure showed the XRCC1 codon 399 Gln alleles were significantly associated with higher levels of AFB1-DNA adducts. Individuals with these alleles were at risk for detectable adducts (OR, 2.4; 95% CI, 1.1–5.4; $P = 0.03$) [80].

As regards of risk biomarker for DNA repair capacity namely AFB1-related HCC risk, a total of fourteen molecular epidemiological studies involving genetic polymorphism at codon 399 of DNA repair gene XRCC1 were found in PubMed database, Sprinker database, Ovid database, Wangfang Database, and Weipu database [22, 81, 83, 162, 164, 175, 180-186], summarized in Table 1. However, associations between this genetic polymorphism and DNA repair capacity have been reported in these case-control studies with the results being contradictory [172, 187]. Possible reasons are as follows: different study population, non-scientific design, the loss of matching methods or improper match, the loss of stratified analysis based on AFB1 exposure information, repeated data, and so on. To avoid above error and achieve more scientific results, we analyzed the possible causes of contradictory using meta-analysis method (Comprehensive Meta-Analysis Version 2, http://www.meta-analysis.com/). Fig. 2, 3, and 4 showed the meta-analysis results of the modifying effects of genetic polymorphism at codon 399 of XRCC1 gene on AFB1-related DNA repair capacity. Based on meta-analysis of overall studies including known published literature (Fig. 2), we found contradictive results; whereas we would observe significant modifying effects of genetic polymorphism at codon 399 of XRCC1 gene on DNA repair capacity related to AFB1-caused DNA damage if these possible repeated studies from the same researchers (Fig. 3) or adding these studies from low/no AFB1 exposure areas (Fig. 4) were excluded. Actually, although Yang, et al. [162] and Ren, et al. [173] did not observed significantly modifying effects of XRCC1 gene codon 399 polymorphism in crude logistic regression, they found Gln alleles would decrease DNA repair capacity in stratified analysis with susceptive environment variants. A individually matching case-controls demonstrated that subjects having codon 399 Gln alleles might feature remarkably increasing risk of HCC under longer-term AFB1-exposure years or higher AFB1-exposure levels conditions (adjusted OR > 10) [22]. This suggests that the genotypes with codon 399 Gln alleles of XRCC1 should be a risk biomarker of low DNA repair ability related DNA damage by AFB1 exposure.

NO.	Ref.	Year	Population	AFB1 exposure[a]	Methods	Matching Factor	Cases (n)	Controls (n)	Risk value[b] (OR)
1	Yu et al. (2003)	2003	Taiwanese	high	case-control	age, sex	577	389	1.54 (P = 0.129)
2	Han et al. (2004)	2004	Qidongese	high	case-control	age, sex	69	136	about 1 (P > 0.05)
3	Kirk et al. (2005)	2005	Gimbia	high	case-control	age, sex	149	294	2.66 (P < 0.05)
4	Long et al.(2005)	2005	Guangxiese	high	case-control	age, sex, HBV, HCV, race	140	536	2.18 (P = 0.0001)
5	Long et al.(2006)	2006	Guangxiese	high	case-control	age, sex, HBV, HCV, race	257	649	2.47 (P = 0.0001)
6	Ren et al. (2008)	2008	Beijingese	low	case-control	age, sex	50	92	0.49 (P < 0.05)
7	Borentai n et al. (2007)	2007	French	low	case-control	age, sex	56	61	1.84 (P = 0.015)
8	Kiran et al.(2009)	2009	Indian	low	case-control	no	63	142	0.33-0.63 (P < 0.05)
9	Kiran et al.(2009)	2009	Indian	low	case-control	no	63	142	0.33-0.63 (P < 0.05)
10	Su et al. (2008)	2008	Liaoningese	low	case-control	age, sex	100	111	2.95 (P < 0.001)
11	Yang et al.(2004)	2004	Qidongese	high	case-control	age, sex	69	136	about 1 (P > 0.05)
12	Pan et al. (2012)	2012	Shangdongese	medium	case-control	age, sex	202	236	1.35-1.55 (P > 0.05)
13	Li et al. (2012)	2012	Shangdongese	medium	case-control	age, sex	150	158	1.69-1.78 (P < 0.05)
14	Chen et al.(2005)	2005	Taiwanese	high	case-control	age, sex	577	389	1.57 (P > 0.05)

[a] Defined by means of Ref Henry, et al. (Science, 1999).

[b] AFB1-related DNA repair capacity is evaluated using risk biomarker AFB1-related HCC risk (see "DNA repair capacity elucidation related to AFB1-induced DNA damage" section). Based on this thesis, AFB1-related DNA repair capacity will decrease if OR > 1 and corresponding P-value < 0.05; will increase if OR < 1 and corresponding P-value < 0.05; and will not change if OR is about 1 and/or corresponding P-value > 0.05.

Table 1. Characteristics of studies about genetic polymorphism at codon 399 of DNA repair gene XRCC1 and risk biomarker for DNA repair capacity (namely HCC risk)

A

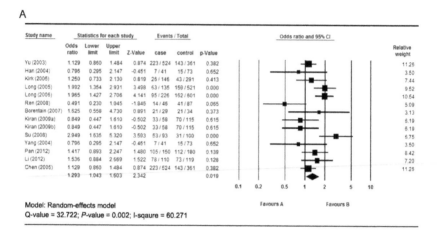

B

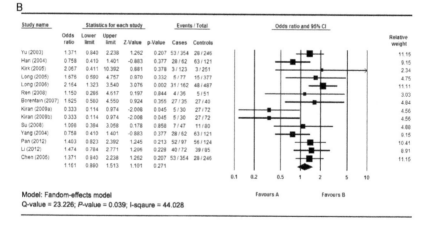

Figure 2. The meta-analysis of the relationship between genetic polymorphism at codon 399 (Arg/Gln) XRCC1 and AFB1-related HCC risk, a biomarker for DNA repair capacity correlated with AFB1-induced DNA damage, based on overall studies size. Compared with Arg/Arg genotype, Arg/Gln (A) genotype decreased AFB1-related DNA repair capacity. This effect was not observed in Gln/Gln genotype (B).

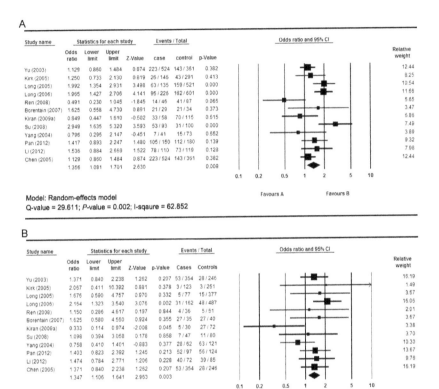

Figure 3. The meta-analysis of the relationship between genetic polymorphism at codon 399 (Arg/Gln) XRCC1 and AFB1-related HCC risk, a biomarker for DNA repair capacity correlated with AFB1-induced DNA damage, based on overall studies size excluded possible repeated studies. Compared with Arg/Arg genotype, Arg/Gln (A) and Gln/Gln (B) genotype decreased AFB1-related DNA repair capacity.

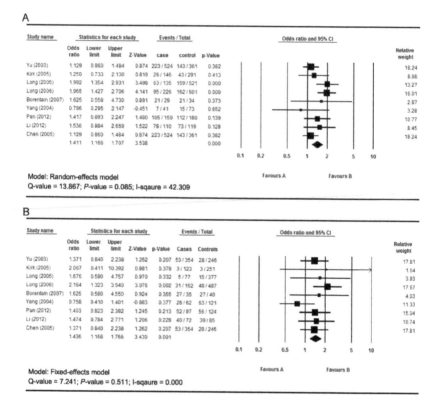

Figure 4. The meta-analysis of the relationship between genetic polymorphism at codon 399 (Arg/Gln) XRCC1 and AFB1-related HCC risk, a biomarker for DNA repair capacity correlated with AFB1-induced DNA damage, based on overall studies size excluded possible repeated studies and studies from low AFB1 exposure areas. Compared with Arg/Arg genotype, Arg/Gln (A) and Gln/Gln (B) genotype decreased AFB1-related DNA repair capacity.

These data support XRCC1 codon 399 Gln alleles decrease AFB1-related DNA repair capacity. Additionally, several studies have shown that the other two genetic polymorphisms (at codon 194 and codon 280) of XRCC1 also decrease DNA repair capacity related AFB1-induced DNA damage, with adjusted value 2.25-2.27 for codon 194 polymorphism and 4.95-6.27 for codon 280 polymorphism ($P < 0.05$) [175]. Furthermore, this decreasing DNA repair ability might more noticeable under the haplotypes with both codon 194 Arg alleles and codon codon 280 His alleles conditions [183].

9. Genetic polymorphisms of DNA repair genes involved in DSBR pathway for AFB1-related DNA damage repair

DSBR pathway involves a series of DNA repair genes. In published molecular epidemiological studies, only XRCC3 gene codon Thr241Met polymorphism and XRCC7 rs#7003908 polymorphism affect AFB1-related DNA repair capacity [8, 15, 78].

9.1. XRCC3

The product of the XRCC3 gene is one of identified paralogs of the strand-exchange protein RAD51 in human beings [188-192]. This protein correlates directly with DNA breaks and facilitates of the formation of the RAD51 nucleoprotein filament, which is crucial both for homologous recombination and HRR [188-192]. Previous studies have shown that a common polymorphism at codon 241 of XRCC3 gene (Thr to Met) modifies the function of this gene [193-205]. Two reports from high AFB1-exposure areas all of world supported above-mentioned conclusions [15, 90].

In the first frequent case-control study in Guangxiese [90], we observed that the genotypes with XRCC3 codon 241 Met alleles (namely Thr/Met and Met/Met) was significantly different between controls (33.01%) and HCC cases (61.48%, $P < 0.001$). Met alleles increases about 2- to 10-fold risk of HCC and this running-up risk is modulated by the number of Met alleles (adjusted OR 2.48 and 10.06 for one and two this alleles). Considering small sample size in this study, we recruited, in another independent frequent case-control study [15], a relatively larger sample size to compare the results. Subjects included in this study, 491 HCC cases and 862 age-, sex-, race, hepatitis virus infection information-matching controls, were permanent residents of Guangxi areas. Similar to the results of the first report, the distribution of XRCC3 codon 241 Met allele frequencies was found to be significantly different between cases (59.7%) and controls (32.1%). Individuals having the Thr/Met or Met/Met were at a 2.22-fold or 7.19 fold increased risk of developing HCC cancer. Above two studies showed this allele multiplicatively interacted with AFB1 exposure in the process of hepato-tumorigenesis. These results exhibits that the polymorphism at codon 241 of XRCC3 gene is a genetic determinant in AFB1-related DNA repair ability for DSBR pathway.

9.2. XRCC7

DNA repair gene XRCC7, called DNA-dependent protein kinase catalytic subunit (DNA-PKcs), DNAPK, DNPK1, HYRC, HYRC1, or p350) (Genbank ID. 5591), spans about 197 kb on chromosome 8q11 and contains 85 exons and 86 introns (Gene dBase in PubMed). This gene encodes DNA-PKcs that constitutes the large catalytic subunit of the DNA-PK complex. When DNA-PKcs is recruited to the site of DSBs by the Ku70/Ku80 heterodimer, DNA-PK complex changes into its active form and subsequently initiates the non-homologous end joining (NHEJ) repair, an important DSBR pathway [206-213]. Murine mutants defective in the XRCC7 have non-detectable DNA-PK activity, suggesting that XRCC7 is required for NHEJ pathway protein. More than 20 polymorphisms have been reported in the XRCC7 gene, some of which are correlated with malignant tumors such as bladder cancer (dbSNP in NCBI Database). Of these genetic polymorphisms in XRCC7 gene, only two loci (rs7003908 and rs10109984) are investigated their modifying effects on AFB1-related DNA repair capacity [8].

In this hospital-based case-control study conducted by Long, et al. [8], they found that these individuals with XRCC7 rs7003908 G alleles increased HCC risk compared the homozygote of XRCC7 rs7003908 T alleles (XRCC7-TT), with OR value 3.45 (2.40–4.94) for XRCC7-TG and 5.04 (3.28–7.76) for XRCC7-GG, respectively. Additionally, they also found this genetic polymorphism was correlated with higher the levels of AFB1-DNA adducts ($r = 0.142$, $P < 0.001$). However, another polymorphism rs10109984 did not modify AFB1-related HCC risk ($P > 0.05$). As a result, these data explore that genetic polymorphism of XRCC7 rs7003908 but not rs10109984 might decrease AFB1-related DSBR capacity, inquiring more studies to support this conclusion.

10. Future directions

Recently, great progress has been made in understanding the molecular mechanisms of the genetic susceptibility to DNA repair capacity related to AFB1-induced DNA damages. However, we are still far from a comprehensive view of the issue. The molecular mechanism about genetic polymorphisms in the DNA repair genes modifying DNA repair capacity related AFB1-induced DNA damages remains largely unknown. Although several reports have shown the spot mutation resulting from genetic polymorphisms may decrease DNA repair capacity via changing the structure of DNA repair proteins, downregulating expression of DNA repair genes or decreasing the function of DNA repair genes, more direct evidence is lost. Disclosing the roles of different genetic types of DNA repair genes in the different toxicity of AFB1 will greatly benefit our understanding of pathological mechanisms of the genetic polymorphisms in the DNA repair genes affecting DNA repair capacity related to AFB1-induced DNA damage, and will shed important light on the clinical therapy for these patients with risk types.

11. Summary

AFB1 is an important environment variation of DNA damage. This toxic variation is characterized by: (1) the attraction of specific organs, especially liver; (2) genotoxicity, mainly inducing the formation of AFB1-DNA adducts and the hot-spot mutation of p53 gene; and (3) carcinogenicity, primarily causing HCC. Among these chronic DNA damage characteristics, AFB1-DNA adducts play a central role because of their genotoxicity and interactions with genetic susceptive factors. In human, there are several repair pathways, including NER, BER, SSBR, and DSBR, is able to repair this type damage. Genetic polymorphisms in DNA repair genes might modify the expression and the functions of

DNA repair proteins encoded by the relative genes and decrease the AFB1-correlated DNA repair capacity. Based on this knowledge, DNA repair capacity related to AFB1-induced DNA damage can be elucidated via the following three methods: testing the levels of AFB1-DNA adducts (mainly AFB1-FAPy adduct), analyzing the frequency of TP53M, and evaluating the risk of HCC by AFB1 exposure.

Numerous studies reviewed in this paper have demonstrated that the hereditary variations in DNA repair genes are associated with DNA repair capacity of DNA damage induced by AFB1. These molecular epidemiological studies have significantly contributed to our knowledge of the importance of genetic polymorphisms in DNA repair genes in the individual's susceptibility to AFB1 exposure. It would be expected that genetic susceptibility factors involved in DNA repair genes for AFB1-induced DNA damage repair could serve as useful biomarkers for identifying at-low-DNA-repair-capacity individuals by AFB1 exposure and, therefore, targeting prevention of this toxicity-related malignant tumor.

However, there are several issues to be noted. The conclusions should first be drawn carefully, because of conflicting data existing in the same ethnic population in view of between some genotypes of DNA repair genes and the AFB1-related DNA damage repair capacity. Second, because of the fact that AFB1-related DNA repair is polygenic, no single genetic marker may sufficiently predict DNA repair capacity. Therefore, a panel of susceptive biomarkers is warranted to define individuals at low DNA repair capacity. Last, the corresponding molecular mechanisms of risk types modifying DNA repair capacity correlated with AFB1-induced DNA damages should be paid close attention.

Acknowledgments

We are grateful to Yuan-Feng Zhou for the collection and management of data. This study is partly supported by the National Nature & Science Foundation of China (NO. 81160255), the Science Foundation of Youjiang Medical College for Nationalities (NO. 2005 and 2008), and Guangxi Key Construction Project of Laboratory Room (NO. 2009).

Abbreviations

AFB1, Aflatoxin B1; AFB1-epoxide, AFB1-8,9-epoxide; AFB1-N^7-Gua, 8,9-di-hydro-8-(N^7-guanyl)-9-hydroxy–AFB1; AFB1-FAPy, ring-opened formamidopyridine AFB1; AFF, aflatoxins family; APE1, AP endonuclease-1; BER, base excision repair; CI, confidence interval; DNA-PKcs, DNA-dependent protein kinase catalytic subunit; DSB, double-strand break; DSBR, double-strand break repair; HBV, hepatitis virus B; HCV, hepatitis virus C; HCC, hepatocellular carcinoma; hOGG1, Human oxoguanine glycosylase 1; NER, nucleotide excision repair; OR, odds ratio; 8-$_{oxod}$G, 8-oxodeoxyguanosine; PARP, poly (ADP-ribose) polymerase; PLC, Primary liver cancer; PNK, polynucleotide kinase; Pol β, DNA polymerase β; ROS, reactive oxygen species; SSB, single-strand break; SSBR, single-strand break repair;

XPA, xeroderma pigmentosum A; XPC, xeroderma pigmentosum C; XPD, xeroderma pigmentosum D; XRCC1, x-ray repair cross complementary 1; XRCC3, x-ray repair cross complementary 3; XRCC4, x-ray repair cross complementary 4; XRCC7, x-ray repair cross complementary 7.

Author details

Qiang Xia[1*], Xiao-Ying Huang[2], Feng Xue[1], Jian-Jun Zhang[1], Bo Zhai[1], De-Chun Kong[3], Chao Wang[2], Zhao-Quan Huang[2] and Xi-Dai Long[1*]

*Address all correspondence to: xiaqiang@medmail.com.cn

*Address all correspondence to: xiaolonglong200166@yahoo.com.cn

1 Department of Liver Surgery, the Affiliated Renji Hospital, Shanghai Jiao Tong University School of Medicine, Shanghai, P.R. China

2 Department of Pathology, the Affiliated Hospital, Youjiang Medical College for Nationalities, Baise, P.R. China

3 Department of Physiology, the Basic Medicine, Shanghai Jiao Tong University School of Medicine, Shanghai, P.R. China

References

[1] Kensler TW, Roebuck BD, Wogan GN, et al. Aflatoxin: a 50-year odyssey of mechanistic and translational toxicology. Toxicological sciences : an official journal of the Society of Toxicology, 2011; 120(Suppl 1):(S28-48.

[2] Wang JS, Groopman JD. DNA damage by mycotoxins. Mutat Res, 1999; 424(1-2): 167-181.

[3] Groopman JD, Croy RG, Wogan GN. In vitro reactions of aflatoxin B1-adducted DNA. Proc Natl Acad Sci U S A, 1981; 78(9): 5445-5449.

[4] Long XD, Yao JG, Zeng Z, et al. DNA repair capacity-related to genetic polymorphisms of DNA repair genes and aflatoxin B1-related hepatocellular carcinoma among Chinese population. In: Kruman I (ed)). DNA Repair (Amst). Rijeka: InTech, 2011; 505-524.

[5] Jemal A, Bray F, Center MM, et al. Global cancer statistics. CA: a cancer journal for clinicians, 2011; 61(2): 69-90.

[6] Yu MC, Yuan JM, Lu SC. Alcohol, cofactors and the genetics of hepatocellular carcinoma. J Gastroenterol Hepatol, 2008; 23(Suppl 1): (S92-97.

[7]   Yeh FS, Mo CC, Yen RC. Risk factors for hepatocellular carcinoma in Guangxi, People's Republic of China. Natl Cancer Inst Monogr, 1985; 69:47-48.

[8]   Long XD, Yao JG, Huang YZ, et al. DNA repair gene XRCC7 polymorphisms (rs#7003908 and rs#10109984) and hepatocellular carcinoma related to AFB1 exposure among Guangxi population, China. Hepatology research : the official journal of the Japan Society of Hepatology, 2011; 41(11): 1085-1093.

[9]   Wilson DM, 3rd, Thompson LH. Life without DNA repair. Proc Natl Acad Sci U S A, 1997; 94(24): 12754-12757.

[10] Guengerich FP, Johnson WW, Shimada T, et al. Activation and detoxication of aflatoxin B1. Mutat Res, 1998; 402(1-2): 121-128.

[11] Parkin DM, Bray F, Ferlay J, et al. Global cancer statistics, 2002. CA Cancer J Clin, 2005; 55(2): 74-108.

[12] Long XD, Ma Y, Zhou YF, et al. XPD codon 312 and 751 polymorphisms, and AFB1 exposure, and hepatocellular carcinoma risk. BMC Cancer, 2009; 9(1):400.

[13] Wang F, Chang D, Hu FL, et al. DNA repair gene XPD polymorphisms and cancer risk: a meta-analysis based on 56 case-control studies. Cancer Epidemiol Biomarkers Prev, 2008; 17(3): 507-517.

[14] Sreeja L, Syamala VS, Syamala V, et al. Prognostic importance of DNA repair gene polymorphisms of XRCC1 Arg399Gln and XPD Lys751Gln in lung cancer patients from India. J Cancer Res Clin Oncol, 2008; 134(6): 645-652.

[15] Long XD, Ma Y, Qu de Y, et al. The polymorphism of XRCC3 codon 241 and AFB1-related hepatocellular carcinoma in Guangxi population, China. Ann Epidemiol, 2008; 18(7): 572-578.

[16] Yuan JM, Lu SC, Van Den Berg D, et al. Genetic polymorphisms in the methylenetetrahydrofolate reductase and thymidylate synthase genes and risk of hepatocellular carcinoma. Hepatology, 2007; 46(3): 749-758.

[17] Suarez-Martinez EB, Ruiz A, Matias J, et al. Early-onset sporadic basal-cell carcinoma: germline mutations in the TP53, PTCH, and XPD genes. P R Health Sci J, 2007; 26(4): 349-354.

[18] Shao J, Gu M, Xu Z, et al. Polymorphisms of the DNA gene XPD and risk of bladder cancer in a Southeastern Chinese population. Cancer Genet Cytogenet, 2007; 177(1): 30-36.

[19] Yin J, Vogel U, Ma Y, et al. Polymorphism of the DNA repair gene ERCC2 Lys751Gln and risk of lung cancer in a northeastern Chinese population. Cancer Genet Cytogenet, 2006; 169(1): 27-32.

[20] Stern MC, Conway K, Li Y, et al. DNA repair gene polymorphisms and probability of p53 mutation in bladder cancer. Mol Carcinog, 2006; 45(9): 715-719.

[21] Manuguerra M, Saletta F, Karagas MR, et al. XRCC3 and XPD/ERCC2 single nucleo-tide polymorphisms and the risk of cancer: a HuGE review. Am J Epidemiol, 2006; 164(4): 297-302.

[22] Long XD, Ma Y, Wei YP, et al. The polymorphisms of GSTM1, GSTT1, HYL1*2, and XRCC1, and aflatoxin B1-related hepatocellular carcinoma in Guangxi population, China. Hepatol Res, 2006; 36(1): 48-55.

[23] Chen G, Luo DZ, Liu L, et al. Hepatic local micro-environmental immune status in hepatocellular carcinoma and cirrhotic tissues. West Indian Med J, 2006; 55(6): 403-408.

[24] Carnaghan RB, Hartley RD, O'Kelly J. Toxicity and Fluorescence Properties of the Aflatoxins. Nature, 1963; 200: 1101.

[25] Long XD, Tang YH, Qu DY, et al. The toxicity and role of aflatoxin B1 and DNA re-pair (corresponding DNA repair enzymes) Youjiang Medical College for Nationali-ties Xue Bao, 2006; 28(2): 278-280.

[26] Wild CP, Turner PC. The toxicology of aflatoxins as a basis for public health deci-sions. Mutagenesis, 2002; 17(6): 471-481.

[27] Wild CP, Montesano R. A model of interaction: aflatoxins and hepatitis viruses in liv-er cancer aetiology and prevention. Cancer Lett, 2009; 286(1): 22-28.

[28] Ben Rejeb I, Arduini F, Arvinte A, et al. Development of a bio-electrochemical assay for AFB1 detection in olive oil. Biosens Bioelectron, 2009; 24(7): 1962-1968.

[29] Hsu CY, Chen YH, Chao PY, et al. Naturally occurring chlorophyll derivatives inhib-it aflatoxin B1-DNA adduct formation in hepatoma cells. Mutat Res, 2008; 657(2): 98-104.

[30] Scholl PF, McCoy L, Kensler TW, et al. Quantitative analysis and chronic dosimetry of the aflatoxin B1 plasma albumin adduct Lys-AFB1 in rats by isotope dilution mass spectrometry. Chem Res Toxicol, 2006; 19(1): 44-49.

[31] Habib SL, Said B, Awad AT, et al. Novel adenine adducts, N7-guanine-AFB1 ad-ducts, and p53 mutations in patients with schistosomiasis and aflatoxin exposure. Cancer Detect Prev, 2006; 30(6): 491-498.

[32] Giri I, Stone MP. Wobble dC.dA pairing 5' to the cationic guanine N7 8,9-dihydro-8-(N7-guanyl)-9-hydroxyaflatoxin B1 adduct: implications for nontargeted AFB1 muta-genesis. Biochemistry, 2003; 42(23): 7023-7034.

[33] Wang JS, Shen X, He X, et al. Protective alterations in phase 1 and 2 metabolism of aflatoxin B1 by oltipraz in residents of Qidong, People's Republic of China. Journal of the National Cancer Institute, 1999; 91(4): 347-354.

[34] Johnson WW, Guengerich FP. Reaction of aflatoxin B1 exo-8,9-epoxide with DNA: ki-netic analysis of covalent binding and DNA-induced hydrolysis. Proc Natl Acad Sci U S A, 1997; 94(12): 6121-6125.

[35] Chou MW, Chen W. Food restriction reduces aflatoxin B1 (AFB1)-DNA adduct formation, AFB1-glutathione conjugation, and DNA damage in AFB1-treated male F344 rats and B6C3F1 mice. The Journal of nutrition, 1997; 127(2): 210-217.

[36] Croy RG, Wogan GN. Temporal patterns of covalent DNA adducts in rat liver after single and multiple doses of aflatoxin B1. Cancer Res, 1981; 41(1): 197-203.

[37] Croy RG, Essigmann JM, Reinhold VN, et al. Identification of the principal aflatoxin B1-DNA adduct formed in vivo in rat liver. Proc Natl Acad Sci U S A, 1978; 75(4): 1745-1749.

[38] Martin CN, Garner RC. Aflatoxin B -oxide generated by chemical or enzymic oxidation of aflatoxin B1 causes guanine substitution in nucleic acids. Nature, 1977; 267(5614): 863-865.

[39] Lin JK, Miller JA, Miller EC. 2,3-Dihydro-2-(guan-7-yl)-3-hydroxy-aflatoxin B1, a major acid hydrolysis product of aflatoxin B1-DNA or -ribosomal RNA adducts formed in hepatic microsome-mediated reactions and in rat liver in vivo. Cancer Res, 1977; 37(12): 4430-4438.

[40] Long XD, Ma Y, Zhou YF, et al. XPD Codon 312 and 751 Polymorphisms, and AFB1 Exposure, and Hepatocellular Carcinoma Risk. BMC Cancer, 2009; 9(1): 400.

[41] Collins AR, Dusinska M, Gedik CM, et al. Oxidative damage to DNA: do we have a reliable biomarker? Environ Health Perspect, 1996; 104(Suppl 3): 465-469.

[42] Katafuchi A, Nohmi T. DNA polymerases involved in the incorporation of oxidized nucleotides into DNA: their efficiency and template base preference. Mutat Res, 2010; 703(1): 24-31.

[43] Kohda K, Tada M, Kasai H, et al. Formation of 8-hydroxyguanine residues in cellular DNA exposed to the carcinogen 4-nitroquinoline 1-oxide. Biochemical and biophysical research communications, 1986; 139(2): 626-632.

[44] Kasai H, Crain PF, Kuchino Y, et al. Formation of 8-hydroxyguanine moiety in cellular DNA by agents producing oxygen radicals and evidence for its repair. Carcinogenesis, 1986; 7(11): 1849-1851.

[45] Shen HM, Shi CY, Shen Y, et al. Detection of elevated reactive oxygen species level in cultured rat hepatocytes treated with aflatoxin B1. Free radical biology & medicine, 1996; 21(2): 139-146.

[46] Shen HM, Ong CN, Lee BL, et al. Aflatoxin B1-induced 8-hydroxydeoxyguanosine formation in rat hepatic DNA. Carcinogenesis, 1995; 16(2): 419-422.

[47] Shen HM, Ong CN, Shi CY. Involvement of reactive oxygen species in aflatoxin B1-induced cell injury in cultured rat hepatocytes. Toxicology, 1995; 99(1-2): 115-123.

[48] Wu HC, Wang Q, Wang LW, et al. Urinary 8-oxodeoxyguanosine, aflatoxin B1 exposure and hepatitis B virus infection and hepatocellular carcinoma in Taiwan. Carcinogenesis, 2007; 28(5): 995-999.

[49] Fortini P, Dogliotti E. Base damage and single-strand break repair: mechanisms and functional significance of short- and long-patch repair subpathways. DNA Repair (Amst), 2007; 6(4): 398-409.

[50] Caldecott KW. Mammalian single-strand break repair: mechanisms and links with chromatin. DNA Repair (Amst), 2007; 6(4): 443-453.

[51] Thompson LH, West MG. XRCC1 keeps DNA from getting stranded. Mutat Res, 2000; 459(1): 1-18.

[52] Fortini P, Ferretti C, Pascucci B, et al. DNA damage response by single-strand breaks in terminally differentiated muscle cells and the control of muscle integrity. Cell death and differentiation, 2012.

[53] Morgan WF, Corcoran J, Hartmann A, et al. DNA double-strand breaks, chromosomal rearrangements, and genomic instability. Mutat Res, 1998; 404(1-2): 125-128.

[54] Bressac B, Kew M, Wands J, et al. Selective G to T mutations of p53 gene in hepatocellular carcinoma from southern Africa. Nature, 1991; 350(6317): 429-431.

[55] Aguilar F, Hussain SP, Cerutti P. Aflatoxin B1 induces the transversion of G-->T in codon 249 of the p53 tumor suppressor gene in human hepatocytes. Proc Natl Acad Sci U S A, 1993; 90(18): 8586-8590.

[56] Coursaget P, Depril N, Chabaud M, et al. High prevalence of mutations at codon 249 of the p53 gene in hepatocellular carcinomas from Senegal. Br J Cancer, 1993; 67(6): 1395-1397.

[57] Liu YP, Lin Y, Ng ML. Immunochemical and genetic analysis of the p53 gene in liver preneoplastic nodules from aflatoxin-induced rats in one year. Ann Acad Med Singapore, 1996; 25(1): 31-36.

[58] Shen HM, Ong CN. Mutations of the p53 tumor suppressor gene and ras oncogenes in aflatoxin hepatocarcinogenesis. Mutat Res, 1996; 366(1): 23-44.

[59] Soini Y, Chia SC, Bennett WP, et al. An aflatoxin-associated mutational hotspot at codon 249 in the p53 tumor suppressor gene occurs in hepatocellular carcinomas from Mexico. Carcinogenesis, 1996; 17(5): 1007-1012.

[60] Hainaut P, Vahakangas K. p53 as a sensor of carcinogenic exposures: mechanisms of p53 protein induction and lessons from p53 gene mutations. Pathol Biol (Paris), 1997; 45(10): 833-844.

[61] Lunn RM, Zhang YJ, Wang LY, et al. p53 mutations, chronic hepatitis B virus infection, and aflatoxin exposure in hepatocellular carcinoma in Taiwan. Cancer Res, 1997; 57(16): 3471-3477.

[62] Mace K, Aguilar F, Wang JS, et al. Aflatoxin B1-induced DNA adduct formation and p53 mutations in CYP450-expressing human liver cell lines. Carcinogenesis, 1997; 18(7): 1291-1297.

[63] Deng ZL, Ma Y. Aflatoxin sufferer and p53 gene mutation in hepatocellular carcino-
      ma. World J Gastroenterol, 1998; 4(1): 28-29.

[64] Ghebranious N, Sell S. The mouse equivalent of the human p53ser249 mutation
      p53ser246 enhances aflatoxin hepatocarcinogenesis in hepatitis B surface antigen
      transgenic and p53 heterozygous null mice. Hepatology, 1998; 27(4): 967-973.

[65] Ghebranious N, Sell S. Hepatitis B injury, male gender, aflatoxin, and p53 expression
      each contribute to hepatocarcinogenesis in transgenic mice. Hepatology, 1998; 27(2):
      383-391.

[66] Nouspikel T. DNA repair in mammalian cells : Nucleotide excision repair: variations
      on versatility. Cellular and molecular life sciences : CMLS, 2009; 66(6): 994-1009.

[67] Rechkunova NI, Lavrik OI. Nucleotide excision repair in higher eukaryotes: mecha-
      nism of primary damage recognition in global genome repair. Sub-cellular biochem-
      istry, 2010; 50: 251-277.

[68] Robertson AB, Klungland A, Rognes T, et al. DNA repair in mammalian cells: Base
      excision repair: the long and short of it. Cellular and molecular life sciences : CMLS,
      2009; 66(6): 981-993.

[69] Donigan KA, Sweasy JB. Sequence context-specific mutagenesis and base excision re-
      pair. Molecular Carcinogenesis, 2009; 48(4): 362-368.

[70] Mills KD, Ferguson DO, Alt FW. The role of DNA breaks in genomic instability and
      tumorigenesis. Immunological reviews, 2003; 194: 77-95.

[71] Sonoda E, Hochegger H, Saberi A, et al. Differential usage of non-homologous end-
      joining and homologous recombination in double strand break repair. DNA Repair
      (Amst), 2006; 5(9-10): 1021-1029.

[72] Cahill D, Connor B, Carney JP. Mechanisms of eukaryotic DNA double strand break
      repair. Front Biosci, 2006; 11: 1958-1976.

[73] Iliakis G, Wu W, Wang M. DNA double strand break repair inhibition as a cause of
      heat radiosensitization: re-evaluation considering backup pathways of NHEJ. Int J
      Hyperthermia, 2008; 24(1): 17-29.

[74] van den Bosch M, Lohman PH, Pastink A. DNA double-strand break repair by ho-
      mologous recombination. Biol Chem, 2002; 383(6): 873-892.

[75] Kanaar R, Hoeijmakers JH, van Gent DC. Molecular mechanisms of DNA double
      strand break repair. Trends Cell Biol, 1998; 8(12): 483-489.

[76] Pastwa E, Blasiak J. Non-homologous DNA end joining. Acta Biochim Pol, 2003;
      50(4): 891-908.

[77] Shrivastav M, De Haro LP, Nickoloff JA. Regulation of DNA double-strand break re-
      pair pathway choice. Cell Res, 2008; 18(1): 134-147.

[78] Long XD, Ma Y, Deng ZL. GSTM1 and XRCC3 polymorphisms: Effects on levels of aflatoxin B1-DNA adducts. Chinese Journal of Cancer Research, 2009; 21(3): 177-184.

[79] Hsieh LL, Hsieh TT. Detection of aflatoxin B1-DNA adducts in human placenta and cord blood. Cancer Res, 1993; 53(6): 1278-1280.

[80] Lunn RM, Langlois RG, Hsieh LL, et al. XRCC1 polymorphisms: effects on aflatoxin B1-DNA adducts and glycophorin A variant frequency. Cancer Res, 1999; 59(11): 2557-2561.

[81] Long XD, Ma Y, Wei YP, et al. X-RAY REPAIR CROSS-COMPLEMENTING GROUP 1 (XRCC1) Arg 399 Gin POLYMORPHISM AND AFLATOXIN B1 (AFB1)-RELATED HEPATOCELLULAR CARCINOMA (HCC) IN GUANGXI POPULATION. 2005.

[82] Zhao L, Long XD, Yao JG, et al. Genetic polymorphism of XRCC3 codon 241 and Helicobacter pylori infection-related gastric antrum adenocarcinoma in Guangxi Population, China: A hospital-based case-control study. Cancer epidemiology, 2011; 35(6): 564-568.

[83] Long XD, Ma Y, Wei YP, et al. X-ray repair cross-complementing group 1 (XRCC1) Arg 399 Gln polymorphism and aflatoxin B1 (AFB1)-related hepatocellular carcinoma (HCC) in Guangxi population. Chinese Journal of Cancer Research, 2005; 17(1): 17-21.

[84] Long XD, Ma Y, Zhou YF, et al. Polymorphism in xeroderma pigmentosum complementation group C codon 939 and aflatoxin B1-related hepatocellular carcinoma in the Guangxi population. Hepatology, 2010; 52(4): 1301-1309.

[85] Long XD, Ma Y, Huang HD, et al. Polymorphism of XRCC1 and the frequency of mutation in codon 249 of the p53 gene in hepatocellular carcinoma among Guangxi population, China. Mol Carcinog, 2008; 47(4): 295-300.

[86] Long XD, Ma Y, Huang YZ, et al. Genetic polymorphisms in DNA repair genes XPC, XPD, and XRCC4, and susceptibility to Helicobacter pylori infection-related gastric antrum adenocarcinoma in Guangxi population, China. Mol Carcinog, 2010; 49(6): 611-618.

[87] Long XD, Ma Y, Wei YP, et al. [A study about the association of detoxication gene GSTM1 polymorphism and the susceptibility to aflatoxin B1-related hepatocellular carcinoma]. Zhonghua Gan Zang Bing Za Zhi, 2005; 13(9): 668-670.

[88] Long XD, Ma Y, Wei YP, et al. [Study on the detoxication gene gstM1-gstT1-null and susceptibility to aflatoxin B1 related hepatocellular carcinoma in Guangxi]. Zhonghua Liu Xing Bing Xue Za Zhi, 2005; 26(10): 777-781.

[89] X.M. W, Ma Y, Deng ZL, et al. The polymorphism at codon 939 of xeroderma pigmentosum C gene and hepatocellular carcinoma among Guangxi population. Zhonghua Xiaohua Zazhi, 2010; 30(11): 846-848.

[90] Long XD, Ma Y, Deng ZL, et al. [Association of the Thr241Met polymorphism of DNA repair gene XRCC3 with genetic susceptibility to AFB1-related hepatocellular

carcinoma in Guangxi population]. Zhonghua Yi Xue Yi Chuan Xue Za Zhi, 2008; 25(3): 268-271.

[91] Sugasawa K. XPC: its product and biological roles. Adv Exp Med Biol, 2008; 637: 47-56.

[92] Khan SG, Yamanegi K, Zheng ZM, et al. XPC branch-point sequence mutations disrupt U2 snRNP binding, resulting in abnormal pre-mRNA splicing in xeroderma pigmentosum patients. Hum Mutat, 2010; 31(2): 167-175.

[93] van der Spek PJ, Eker A, Rademakers S, et al. XPC and human homologs of RAD23: intracellular localization and relationship to other nucleotide excision repair complexes. Nucleic Acids Res, 1996; 24(13): 2551-2559.

[94] Wang XW, Vermeulen W, Coursen JD, et al. The XPB and XPD DNA helicases are components of the p53-mediated apoptosis pathway. Genes Dev, 1996; 10(10): 1219-1232.

[95] Yokoi M, Masutani C, Maekawa T, et al. The xeroderma pigmentosum group C protein complex XPC-HR23B plays an important role in the recruitment of transcription factor IIH to damaged DNA. J Biol Chem, 2000; 275(13): 9870-9875.

[96] Boulikas T. Xeroderma pigmentosum and molecular cloning of DNA repair genes. Anticancer Res, 1996; 16(2): 693-708.

[97] Araujo SJ, Wood RD. Protein complexes in nucleotide excision repair. Mutat Res, 1999; 435(1): 23-33.

[98] Bohr VA. Preferential DNA repair in active genes. Dan Med Bull, 1987; 34(6): 309-320.

[99] Qiu L, Wang Z, Shi X, et al. Associations between XPC polymorphisms and risk of cancers: A meta-analysis. Eur J Cancer, 2008; 44(15): 2241-2253.

[100] Sakano S, Matsumoto H, Yamamoto Y, et al. Association between DNA repair gene polymorphisms and p53 alterations in Japanese patients with muscle-invasive bladder cancer. Pathobiology, 2006; 73(6): 295-303.

[101] Li LM, Zeng XY, Ji L, et al. [Association of XPC and XPG polymorphisms with the risk of hepatocellular carcinoma]. Zhonghua gan zang bing za zhi = Zhonghua ganzangbing zazhi = Chinese journal of hepatology, 2010; 18(4): 271-275.

[102] Zhou RM, Li Y, Wang N, et al. [Correlation of XPC Ala499Val and Lys939Gln polymorphisms to risks of esophageal squamous cell carcinoma and gastric cardiac adenocarcinoma]. Ai Zheng, 2006; 25(9): 1113-1119.

[103] Ryk C, Kumar R, Sanyal S, et al. Influence of polymorphism in DNA repair and defence genes on p53 mutations in bladder tumours. Cancer Lett, 2006; 241(1): 142-149.

[104] Zhang D, Chen C, Fu X, et al. A meta-analysis of DNA repair gene XPC polymorphisms and cancer risk. J Hum Genet, 2008; 53(1): 18-33.

[105] Kazimirova A, Barancokova M, Dzupinkova Z, et al. Micronuclei and chromosomal aberrations, important markers of ageing: possible association with XPC and XPD polymorphisms. Mutat Res, 2009; 661(1-2): 35-40.

[106] Laczmanska I, Gil J, Karpinski P, et al. Polymorphism in nucleotide excision repair gene XPC correlates with bleomycin-induced chromosomal aberrations. Environ Mol Mutagen, 2007; 48(8): 666-671.

[107] De Ruyck K, Szaumkessel M, De Rudder I, et al. Polymorphisms in base-excision repair and nucleotide-excision repair genes in relation to lung cancer risk. Mutat Res, 2007; 631(2): 101-110.

[108] Mechanic LE, Millikan RC, Player J, et al. Polymorphisms in nucleotide excision repair genes, smoking and breast cancer in African Americans and whites: a population-based case-control study. Carcinogenesis, 2006; 27(7): 1377-1385.

[109] Shore RE, Zeleniuch-Jacquotte A, Currie D, et al. Polymorphisms in XPC and ERCC2 genes, smoking and breast cancer risk. Int J Cancer, 2008; 122(9): 2101-2105.

[110] Dong Z, Guo W, Zhou R, et al. Polymorphisms of the DNA repair gene XPA and XPC and its correlation with gastric cardiac adenocarcinoma in a high incidence population in North China. J Clin Gastroenterol, 2008; 42(8): 910-915.

[111] Zhang L, Zhang Z, Yan W. Single nucleotide polymorphisms for DNA repair genes in breast cancer patients. Clin Chim Acta, 2005; 359(1-2): 150-155.

[112] Hansen RD, Sorensen M, Tjonneland A, et al. XPA A23G, XPC Lys939Gln, XPD Lys751Gln and XPD Asp312Asn polymorphisms, interactions with smoking, alcohol and dietary factors, and risk of colorectal cancer. Mutat Res, 2007; 619(1-2): 68-80.

[113] Liang J, Gu A, Xia Y, et al. XPC gene polymorphisms and risk of idiopathic azoospermia or oligozoospermia in a Chinese population. Int J Androl, 2009; 32(3): 235-241.

[114] Francisco G, Menezes PR, Eluf-Neto J, et al. XPC polymorphisms play a role in tissue-specific carcinogenesis: a meta-analysis. Eur J Hum Genet, 2008; 16(6): 724-734.

[115] Cheo DL, Burns DK, Meira LB, et al. Mutational inactivation of the xeroderma pigmentosum group C gene confers predisposition to 2-acetylaminofluorene-induced liver and lung cancer and to spontaneous testicular cancer in Trp53-/- mice. Cancer Res, 1999; 59(4): 771-775.

[116] Fautrel A, Andrieux L, Musso O, et al. Overexpression of the two nucleotide excision repair genes ERCC1 and XPC in human hepatocellular carcinoma. J Hepatol, 2005; 43(2): 288-293.

[117] Cai XL, Gao YH, Yu ZW, et al. [A 1:1 matched case-control study on the interaction between HBV, HCV infection and DNA repair gene XPC Ala499Val, Lys939Gln for primary hepatocellular carcinoma]. Zhonghua Liu Xing Bing Xue Za Zhi, 2009; 30(9): 942-945.

[118] Henry SH, Bosch FX, Troxell TC, et al. Reducing Liver Cancer--Global Control of Aflatoxin. Science, 1999; 286(5449): 2453-2454.

[119] Aloyz R, Xu ZY, Bello V, et al. Regulation of cisplatin resistance and homologous recombinational repair by the TFIIH subunit XPD. Cancer Res, 2002; 62(19): 5457-5462.

[120] Bienstock RJ, Skorvaga M, Mandavilli BS, et al. Structural and functional characterization of the human DNA repair helicase XPD by comparative molecular modeling and site-directed mutagenesis of the bacterial repair protein UvrB. J Biol Chem, 2003; 278(7): 5309-5316.

[121] Keriel A, Stary A, Sarasin A, et al. XPD mutations prevent TFIIH-dependent transactivation by nuclear receptors and phosphorylation of RARalpha. Cell, 2002; 109(1): 125-135.

[122] Sturgis EM, Castillo EJ, Li L, et al. XPD/ERCC2 EXON 8 Polymorphisms: rarity and lack of significance in risk of squamous cell carcinoma of the head and neck. Oral Oncol, 2002; 38(5): 475-477.

[123] Benhamou S, Sarasin A. ERCC2/XPD gene polymorphisms and cancer risk. Mutagenesis, 2002; 17(6): 463-469.

[124] Qiao Y, Spitz MR, Shen H, et al. Modulation of repair of ultraviolet damage in the host-cell reactivation assay by polymorphic XPC and XPD/ERCC2 genotypes. Carcinogenesis, 2002; 23(2): 295-299.

[125] Kobayashi T, Uchiyama M, Fukuro S, et al. Mutations in the XPD gene in xeroderma pigmentosum group D cell strains: confirmation of genotype-phenotype correlation. Am J Med Genet, 2002; 110(3): 248-252.

[126] Rzeszowska-Wolny J, Polanska J, Pietrowska M, et al. Influence of polymorphisms in DNA repair genes XPD, XRCC1 and MGMT on DNA damage induced by gamma radiation and its repair in lymphocytes in vitro. Radiat Res, 2005; 164(2): 132-140.

[127] Spitz MR, Wu X, Wang Y, et al. Modulation of nucleotide excision repair capacity by XPD polymorphisms in lung cancer patients. Cancer Res, 2001; 61(4): 1354-1357.

[128] Stern MC, Johnson LR, Bell DA, et al. XPD codon 751 polymorphism, metabolism genes, smoking, and bladder cancer risk. Cancer Epidemiol Biomarkers Prev, 2002; 11(10 Pt 1): 1004-1011.

[129] Tang D, Cho S, Rundle A, et al. Polymorphisms in the DNA repair enzyme XPD are associated with increased levels of PAH-DNA adducts in a case-control study of breast cancer. Breast Cancer Res Treat, 2002; 75(2): 159-166.

[130] Terry MB, Gammon MD, Zhang FF, et al. Polymorphism in the DNA repair gene XPD, polycyclic aromatic hydrocarbon-DNA adducts, cigarette smoking, and breast cancer risk. Cancer epidemiology, biomarkers & prevention : a publication of the American Association for Cancer Research, cosponsored by the American Society of Preventive Oncology, 2004; 13(12): 2053-2058.

[131] Mechanic LE, Marrogi AJ, Welsh JA, et al. Polymorphisms in XPD and TP53 and mutation in human lung cancer. Carcinogenesis, 2005; 26(3): 597-604.

[132] Zhang H, He BC, He FC. [Impact of DNA repair gene hOGG1 Ser326Cys polymorphism on the risk of hepatocellular carcinoma]. Shijie Huaren xiaohua zazhi, 2006; 14(23): 2311-2314.

[133] Zhang H, Hao BT, He FC. [Impact of DNA repair gene hOGG1 Ser326Cys polymorphism on the risk of hepatocellular carcinoma]. 2006, 2006; 23(

[134] Zhang H, HAO B, HE F. Studies on Association between Susceptibility of Hepatocellular Carcinoma and Genetic Polymorphism Ser326Cys of DNA Repair Gene hOGG1. CHINESE JOURNAL OF CLINICAL ONCOLOGY, 2005; 32(15): 841-843.

[135] Ye X, Peng T, Li L. Study on polymorphisms in metabolic enzyme genes, DNA repair genes and individual susceptibility to hepatocellular carcinoma. Weisheng Yanjiu, 2006; 35(6): 805-807.

[136] Weiss JM, Goode EL, Ladiges WC, et al. Polymorphic variation in hOGG1 and risk of cancer: a review of the functional and epidemiologic literature. Molecular Carcinogenesis, 2005; 42(3): 127-141.

[137] Sakamoto T, Higaki Y, Hara M, et al. hOGG1 Ser326Cys polymorphism and risk of hepatocellular carcinoma among Japanese. Journal of epidemiology / Japan Epidemiological Association, 2006; 16(6): 233-239.

[138] Peng T, Shen HM, Liu ZM, et al. Oxidative DNA damage in peripheral leukocytes and its association with expression and polymorphisms of hOGG1: a study of adolescents in a high risk region for hepatocellular carcinoma in China. World J Gastroenterol, 2003; 9(10): 2186-2193.

[139] Nishimura S. Involvement of mammalian OGG1(MMH) in excision of the 8-hydroxyguanine residue in DNA. Free radical biology & medicine, 2002; 32(9): 813-821.

[140] Kohno T, Shinmura K, Tosaka M, et al. Genetic polymorphisms and alternative splicing of the hOGG1 gene, that is involved in the repair of 8-hydroxyguanine in damaged DNA. Oncogene, 1998; 16(25): 3219-3225.

[141] Cheng B, Jungst C, Lin J, et al. [Potential role of human DNA-repair enzymes hMTH1, hOGG1 and hMYHalpha in the hepatocarcinogenesis]. Journal of Huazhong University of Science and Technology Medical sciences, 2002; 22(3): 206-211, 215.

[142] Boiteux S, Radicella JP. The human OGG1 gene: structure, functions, and its implication in the process of carcinogenesis. Archives of biochemistry and biophysics, 2000; 377(1): 1-8.

[143] Aizhong W, CONG W, HE X, et al. [A hOGGl gene polymorphism and genetic susceptibility to colorectal cancer and hepatocellular carcinoma]. China J Gastroenterol, 2008; 17(10): 854-857.

[144] Yuan T, Wei J, Luo J, et al. Polymorphisms of Base-Excision Repair Genes hOGG1 326cys and XRCC1 280His Increase Hepatocellular Carcinoma Risk. Dig Dis Sci, 2012; 57(9): 2451-2457.

[145] Lakshmipathy U, Campbell C. Mitochondrial DNA ligase III function is independent of Xrcc1. Nucleic Acids Res, 2000; 28(20): 3880-3886.

[146] Marintchev A, Robertson A, Dimitriadis EK, et al. Domain specific interaction in the XRCC1-DNA polymerase beta complex. Nucleic Acids Res, 2000; 28(10): 2049-2059.

[147] Taylor RM, Moore DJ, Whitehouse J, et al. A cell cycle-specific requirement for the XRCC1 BRCT II domain during mammalian DNA strand break repair. Mol Cell Biol, 2000; 20(2): 735-740.

[148] Yamane K, Katayama E, Tsuruo T. The BRCT regions of tumor suppressor BRCA1 and of XRCC1 show DNA end binding activity with a multimerizing feature. Biochem Biophys Res Commun, 2000; 279(2): 678-684.

[149] Bhattacharyya N, Banerjee S. A novel role of XRCC1 in the functions of a DNA polymerase beta variant. Biochemistry, 2001; 40(30): 9005-9013.

[150] Dulic A, Bates PA, Zhang X, et al. BRCT domain interactions in the heterodimeric DNA repair protein XRCC1-DNA ligase III. Biochemistry, 2001; 40(20): 5906-5913.

[151] Vidal AE, Boiteux S, Hickson ID, et al. XRCC1 coordinates the initial and late stages of DNA abasic site repair through protein-protein interactions. EMBO J, 2001; 20(22): 6530-6539.

[152] Yacoub A, Park JS, Qiao L, et al. MAPK dependence of DNA damage repair: ionizing radiation and the induction of expression of the DNA repair genes XRCC1 and ERCC1 in DU145 human prostate carcinoma cells in a MEK1/2 dependent fashion. Int J Radiat Biol, 2001; 77(10): 1067-1078.

[153] Abdel-Rahman SZ, El-Zein RA. The 399Gln polymorphism in the DNA repair gene XRCC1 modulates the genotoxic response induced in human lymphocytes by the tobacco-specific nitrosamine NNK. Cancer Lett, 2000; 159(1): 63-71.

[154] Matullo G, Palli D, Peluso M, et al. XRCC1, XRCC3, XPD gene polymorphisms, smoking and (32)P-DNA adducts in a sample of healthy subjects. Carcinogenesis, 2001; 22(9): 1437-1445.

[155] Ratnasinghe D, Yao SX, Tangrea JA, et al. Polymorphisms of the DNA repair gene XRCC1 and lung cancer risk. Cancer Epidemiol Biomarkers Prev, 2001; 10(2): 119-123.

[156] Nelson HH, Kelsey KT, Mott LA, et al. The XRCC1 Arg399Gln polymorphism, sunburn, and non-melanoma skin cancer: evidence of gene-environment interaction. Cancer Res, 2002; 62(1): 152-155.

[157] Park JY, Lee SY, Jeon HS, et al. Polymorphism of the DNA repair gene XRCC1 and risk of primary lung cancer. Cancer Epidemiol Biomarkers Prev, 2002; 11(1): 23-27.

[158] Hsieh LL, Chien HT, Chen IH, et al. The XRCC1 399Gln polymorphism and the frequency of p53 mutations in Taiwanese oral squamous cell carcinomas. Cancer Epidemiol Biomarkers Prev, 2003; 12(5): 439-443.

[159] Shu XO, Cai Q, Gao YT, et al. A population-based case-control study of the Arg399Gln polymorphism in DNA repair gene XRCC1 and risk of breast cancer. Cancer Epidemiol Biomarkers Prev, 2003; 12(12): 1462-1467.

[160] Kelsey KT, Park S, Nelson HH, et al. A population-based case-control study of the XRCC1 Arg399Gln polymorphism and susceptibility to bladder cancer. Cancer Epidemiol Biomarkers Prev, 2004; 13(8): 1337-1341.

[161] Matsuo K, Hamajima N, Suzuki R, et al. Lack of association between DNA base excision repair gene XRCC1 Gln399Arg polymorphism and risk of malignant lymphoma in Japan. Cancer Genet Cytogenet, 2004; 149(1): 77-80.

[162] Yang JL, Han YN, Zheng SG. Influence of human XRCC1-399 single nucleotide polymorphism on primary hepatocytic carcinoma. Zhongliu, 2004; 24(4): 322-324.

[163] Jeon YT, Kim JW, Park NH, et al. DNA repair gene XRCC1 Arg399Gln polymorphism is associated with increased risk of uterine leiomyoma. Hum Reprod, 2005; 20(6): 1586-1589.

[164] Kirk GD, Turner PC, Gong Y, et al. Hepatocellular carcinoma and polymorphisms in carcinogen-metabolizing and DNA repair enzymes in a population with aflatoxin exposure and hepatitis B virus endemicity. Cancer epidemiology, biomarkers & prevention : a publication of the American Association for Cancer Research, cosponsored by the American Society of Preventive Oncology, 2005; 14(2): 373-379.

[165] Chung HH, Kim MK, Kim JW, et al. XRCC1 R399Q polymorphism is associated with response to platinum-based neoadjuvant chemotherapy in bulky cervical cancer. Gynecol Oncol, 2006; 103(3): 1031-1037.

[166] Kocabas NA, Karahalil B. XRCC1 Arg399Gln genetic polymorphism in a Turkish population. Int J Toxicol, 2006; 25(5): 419-422.

[167] Long XD, Ma Y, Wei YP, et al. The polymorphisms of GSTM1, GSTT1, HYL1*2, and XRCC1, and aflatoxin B1-related hepatocellular carcinoma in Guangxi population, China. Hepatology research : the official journal of the Japan Society of Hepatology, 2006; 36(1): 48-55.

[168] Bau DT, Hsieh YY, Wan L, et al. Polymorphism of XRCC1 codon arg 399 Gln is associated with higher susceptibility to endometriosis. Chin J Physiol, 2007; 50(6): 326-329.

[169] Deligezer U, Akisik EE, Dalay N. Lack of association of XRCC1 codon 399Gln polymorphism with chronic myelogenous leukemia. Anticancer Res, 2007; 27(4B): 2453-2456.

[170] Monaco R, Rosal R, Dolan MA, et al. Conformational effects of a common codon 399 polymorphism on the BRCT1 domain of the XRCC1 protein. Protein J, 2007; 26(8): 541-546.

[171] Sobti RC, Singh J, Kaur P, et al. XRCC1 codon 399 and ERCC2 codon 751 polymorphism, smoking, and drinking and risk of esophageal squamous cell carcinoma in a North Indian population. Cancer Genet Cytogenet, 2007; 175(2): 91-97.

[172] Geng J, Zhang YW, Huang GC, et al. XRCC1 genetic polymorphism Arg399Gln and gastric cancer risk: A meta-analysis. World J Gastroenterol, 2008; 14(43): 6733-6737.

[173] Ren y, Wang DS, Li Z, et al. Study on the Relationship between Gene XRCC1 Codon 399 Single Nucleotide Polymorphisms and Primary Hepatic Carcinoma in Han Nationality. Linchuang Gandang Bing Zazhi, 2008; 24(5): 361-364.

[174] Geng J, Zhang Q, Zhu C, et al. XRCC1 genetic polymorphism Arg399Gln and prostate cancer risk: a meta-analysis. Urology, 2009; 74(3): 648-653.

[175] Kiran M, Saxena R, Chawla YK, et al. Polymorphism of DNA repair gene XRCC1 and hepatitis-related hepatocellular carcinoma risk in Indian population. Mol Cell Biochem, 2009; 327(1-2): 7-13.

[176] Li Y, Long C, Lin G, et al. Effect of the XRCC1 codon 399 polymorphism on the repair of vinyl chloride metabolite-induced DNA damage. J Carcinog, 2009; 8(14.

[177] Saadat M, Ansari-Lari M. Polymorphism of XRCC1 (at codon 399) and susceptibility to breast cancer, a meta-analysis of the literatures. Breast Cancer Res Treat, 2009; 115(1): 137-144.

[178] Zeng X, Yu H, Qiu X, et al. [A case-control study of polymorphism of XRCC1 gene and the risk of hepatocellular carcinoma]. Chin J Dis Contr ol Pr ev, 2010; 14(8): 760-763.

[179] Qiu Y, Zhu S, Liu J, et al. [Study of susceptibility of chromosomal damage induced by vinyl chloride monomer associated with genetic polymorphism in APE1, XRCC1]. Wei Sheng Yan Jiu, 2007; 36(2): 132-136.

[180] Yu MW, Yang SY, Pan IJ, et al. Polymorphisms in XRCC1 and glutathione S-transferase genes and hepatitis B-related hepatocellular carcinoma. Journal of the National Cancer Institute, 2003; 95(19): 1485-1488.

[181] Pan HZ, Liang J, Yu Z, et al. Polymorphism of DNA repair gene XRCC1 and hepatocellular carcinoma risk in Chinese population. Asian Pacific journal of cancer prevention : APJCP, 2011; 12(11): 2947-2950.

[182] Liu F, Li B, Wei Y, et al. XRCC1 genetic polymorphism Arg399Gln and hepatocellular carcinoma risk: a meta-analysis. Liver international : official journal of the International Association for the Study of the Liver, 2011; 31(6): 802-809.

[183] Kiran M, Chawla YK, Jain M, et al. Haplotypes of microsomal epoxide hydrolase and x-ray cross-complementing group 1 genes in Indian hepatocellular carcinoma patients. DNA and cell biology, 2009; 28(11): 573-577.

[184] Han X, Xing Q, Li Y, et al. Study on the DNA Repair Gene XRCC1 and XRCC3 Polymorphism in Prediction and Prognosis of Hepatocellular Carcinoma Risk. Hepatogastroenterology, 2012; 59(119).

[185] Chen CC, Yang SY, Liu CJ, et al. Association of cytokine and DNA repair gene polymorphisms with hepatitis B-related hepatocellular carcinoma. Int J Epidemiol, 2005; 34(6): 1310-1318.

[186] Borentain P, Gerolami V, Ananian P, et al. DNA-repair and carcinogen-metabolising enzymes genetic polymorphisms as an independent risk factor for hepatocellular carcinoma in Caucasian liver-transplanted patients. European journal of cancer, 2007; 43(17): 2479-2486.

[187] Chen BP, Long XD, Fu GH. Meta-analysis of XRCC1 Codon 399 polymorphism and susceptibility to hepatocellular carcinoma. Shanghai Jiao Tong Xaxue Xuebao (Medical Version), 2011; 31(11): 1588-1591.

[188] Liu Y, Tarsounas M, O'Regan P, et al. Role of RAD51C and XRCC3 in genetic recombination and DNA repair. J Biol Chem, 2007; 282(3): 1973-1979.

[189] Xu ZY, Loignon M, Han FY, et al. Xrcc3 induces cisplatin resistance by stimulation of Rad51-related recombinational repair, S-phase checkpoint activation, and reduced apoptosis. J Pharmacol Exp Ther, 2005; 314(2): 495-505.

[190] Forget AL, Bennett BT, Knight KL. Xrcc3 is recruited to DNA double strand breaks early and independent of Rad51. J Cell Biochem, 2004; 93(3): 429-436.

[191] Bishop DK, Ear U, Bhattacharyya A, et al. Xrcc3 is required for assembly of Rad51 complexes in vivo. J Biol Chem, 1998; 273(34): 21482-21488.

[192] Brenneman MA, Weiss AE, Nickoloff JA, et al. XRCC3 is required for efficient repair of chromosome breaks by homologous recombination. Mutat Res, 2000; 459(2): 89-97.

[193] Wang J, Zhao Y, Jiang J, et al. Polymorphisms in DNA repair genes XRCC1, XRCC3 and XPD, and colorectal cancer risk: a case-control study in an Indian population. J Cancer Res Clin Oncol, 2010; 136(10): 1517-1525.

[194] Jiang Z, Li C, Xu Y, et al. A meta-analysis on XRCC1 and XRCC3 polymorphisms and colorectal cancer risk. Int J Colorectal Dis, 2010; 25(2): 169-180.

[195] Fang F, Wang J, Yao L, et al. Relationship between XRCC3 T241M polymorphism and gastric cancer risk: a meta-analysis. Med Oncol, 2010.

[196] Economopoulos KP, Sergentanis TN. XRCC3 Thr241Met polymorphism and breast cancer risk: a meta-analysis. Breast Cancer Res Treat, 2010; 121(2): 439-443.

[197] Zhang ZQ, Yang L, Zhang Y, et al. [Relationship between NQO1C(609T), RAD51(G135C), XRCC3(C241T) single nucleotide polymorphisms and acute lympho-blastic leukemia]. Zhongguo Shi Yan Xue Ye Xue Za Zhi, 2009; 17(3): 523-528.

[198] Sobczuk A, Romanowicz-Makowska H, Fiks T, et al. XRCC1 and XRCC3 DNA repair gene polymorphisms in breast cancer women from the Lodz region of Poland. Pol J Pathol, 2009; 60(2): 76-80.

[199] Krupa R, Synowiec E, Pawlowska E, et al. Polymorphism of the homologous recom-bination repair genes RAD51 and XRCC3 in breast cancer. Exp Mol Pathol, 2009; 87(1): 32-35.

[200] Andreassi MG, Foffa I, Manfredi S, et al. Genetic polymorphisms in XRCC1, OGG1, APE1 and XRCC3 DNA repair genes, ionizing radiation exposure and chromosomal DNA damage in interventional cardiologists. Mutat Res, 2009; 666(1-2): 57-63.

[201] Yen CY, Liu SY, Chen CH, et al. Combinational polymorphisms of four DNA repair genes XRCC1, XRCC2, XRCC3, and XRCC4 and their association with oral cancer in Taiwan. J Oral Pathol Med, 2008; 37(5): 271-277.

[202] Tekeli A, Isbir S, Ergen A, et al. APE1 and XRCC3 polymorphisms and myocardial infarction. In Vivo, 2008; 22(4): 477-479.

[203] Mateuca RA, Roelants M, Iarmarcovai G, et al. hOGG1(326), XRCC1(399) and XRCC3(241) polymorphisms influence micronucleus frequencies in human lympho-cytes in vivo. Mutagenesis, 2008; 23(1): 35-41.

[204] Liu L, Yang L, Zhang Y, et al. [Polymorphisms of RAD51(G135C) and XRCC3(C241T) genes and correlations thereof with prognosis and clinical outcomes of acute myeloid leukemia]. Zhonghua Yi Xue Za Zhi, 2008; 88(6): 378-382.

[205] Jacobsen NR, Raaschou-Nielsen O, Nexo B, et al. XRCC3 polymorphisms and risk of lung cancer. Cancer Lett, 2004; 213(1): 67-72.

[206] Sakata K, Someya M, Matsumoto Y, et al. Ability to repair DNA double-strand breaks related to cancer susceptibility and radiosensitivity. Radiation medicine, 2007; 25(9): 433-438.

[207] Fujimoto M, Matsumoto N, Tsujita T, et al. Characterization of the promoter region, first ten exons and nine intron-exon boundaries of the DNA-dependent protein kin-ase catalytic subunit gene, DNA-PKcs (XRCC7). DNA Res, 1997; 4(2): 151-154.

[208] Blunt T, Finnie NJ, Taccioli GE, et al. Defective DNA-dependent protein kinase activ-ity is linked to V(D)J recombination and DNA repair defects associated with the mur-ine scid mutation. Cell, 1995; 80(5): 813-823.

[209] van der Burg M, van Dongen JJ, van Gent DC. DNA-PKcs deficiency in human: long predicted, finally found. Current opinion in allergy and clinical immunology, 2009; 9(6): 503-509.

[210] Smider V, Chu G. The end-joining reaction in V(D)J recombination. Semin Immunol, 1997; 9(3): 189-197.

[211] Jolly CJ, Cook AJ, Manis JP. Fixing DNA breaks during class switch recombination. The Journal of experimental medicine, 2008; 205(3): 509-513.

[212] Sipley JD, Menninger JC, Hartley KO, et al. Gene for the catalytic subunit of the human DNA-activated protein kinase maps to the site of the XRCC7 gene on chromosome 8. Proc Natl Acad Sci U S A, 1995; 92(16): 7515-7519.

[213] Johnston PJ, MacPhail SH, Stamato TD, et al. Higher-order chromatin structure-dependent repair of DNA double-strand breaks: involvement of the V(D)J recombination double-strand break repair pathway. Radiat Res, 1998; 149(5): 455-462.

Chapter 2

DNA Damage, DNA Repair and Cancer

Carol Bernstein, Anil R. Prasad,
Valentine Nfonsam and Harris Bernstein

Additional information is available at the end of the chapter

1. Introduction

DNA damage appears to be a fundamental problem for life. In this chapter we review evidence indicating that DNA damages are a major primary cause of cancer. DNA damages give rise to mutations and epimutations that, by a process of natural selection, can cause progression to cancer. First, we describe the distinguishing characteristics of DNA damage, mutation and epimutation.

DNA damage is a change in the basic structure of DNA that is not itself replicated when the DNA is replicated. A DNA damage can be a chemical addition or disruption to a base of DNA (creating an abnormal nucleotide or nucleotide fragment) or a break in one or both chains of the DNA strands. When DNA carrying a damaged base is replicated, an incorrect base can often be inserted opposite the site of the damaged base in the complementary strand, and this can become a mutation in the next round of replication. Also DNA double-strand breaks may be repaired by an inaccurate repair process leading to mutations. In addition, a double strand break can cause rearrangements of the chromosome structure (possibly disrupting a gene, or causing a gene to come under abnormal regulatory control), and, if such a change can be passed to successive cell generations, it is also a form of mutation. Mutations, however, can be avoided if accurate DNA repair systems recognize DNA damages as abnormal structures, and repair the damages prior to replication. As illustrated in Figure 1, when DNA damages occur, DNA repair is a crucial protective process blocking entry of cells into carcinogenesis.

We note that DNA damages occur in both replicating, proliferative cells (e.g. those forming the internal lining of the colon or blood forming "hematopoietic" cells), and in differentiated, non-dividing cells (e.g. neurons in the brain or myocytes in muscle). Cancers occur primarily in proliferative tissues. If DNA damages in proliferating cells are not repaired due to inadequate expression of a DNA repair gene, this increases the risk of cancer. In contrast, when DNA damages occur in non-proliferating cells and are not repaired due to inadequate expression of a DNA repair gene, the damages can accumulate and cause premature aging. As examples, deficiencies in DNA repair genes *ERCC1* or *XPF* [1] or in *WRN* [2, 3] cause both increased risk of cancer as well as premature aging. In Figure 1, DNA repair is indicated as a crucial process impeding both cancer and premature aging.

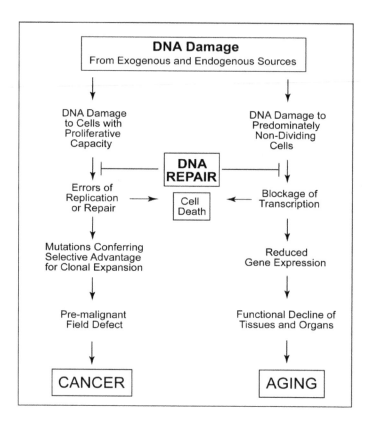

Figure 1. The roles of DNA damage and DNA repair in cancer and aging.

A mutation is a change in the DNA sequence in which normal base pairs are substituted, added, deleted or rearranged. The DNA containing a mutation still consists of a sequence of standard base pairs, and the altered DNA sequence can be copied when the DNA is replicated. A mutation can prevent a gene from carrying out its function, or it can cause a gene to be translated into a protein that functions abnormally. Mutations can activate oncogenes, inactivate tumor suppressor genes or cause genomic instability in replicating cells, and an assemblage of such mutations, together in the same cell, can lead to cancer. Cancers usually arise from an assemblage of mutations conferring a selective advantage that leads to clonal expansion (Figure 1). Colon cancers, for example, have an average of 15 "driver" mutations (mutations occurring repeatedly in different colon cancers) and about 75 "passenger" mutations (mutations occurring infrequently in colon cancers) [4, 5]. Colon cancers also were found to have an average of 9 duplications or deletions of chromosome segments [6] or, more recently, 17 focal amplifications, 28 recurrent deletions and up to 10 translocations [5]. Since mutations have normal DNA structure, they cannot be recognized or removed by DNA repair processes in living cells. Removal of a mutation only occurs if it is sufficiently deleterious to cause the death of the cell.

Another type of inheritable alteration, similar in some ways to a mutation, is an epigenetic change. An epigenetic change refers to a functionally relevant modification of the DNA, or of the histone proteins controlling the relaxation or tightened winding of the DNA within their nucleosome structures. Some epigenetic changes involve specific alterations of the DNA nucleotides. Examples of such changes include methylation of the DNA at particular sites (CpG islands) where the DNA starts to be transcribed into RNA. These changes may inhibit transcription. Other epigenetic changes involve modification of histones associated with particular regions of the DNA. These may inhibit or promote the ability of these regions to be transcribed into mRNA. Methylation of CpG islands or modification of histones can directly alter transcription of gene-encoded mRNAs but they can also occur in parts of the genome that code for microRNAs (miRNAs). MiRNAs are endogenous short non-protein coding RNAs (~22 nucleotides long) that post-transcriptionally regulate mRNA expression in a sequence specific manner. miRNAs either cause degradation of mRNAs or block their translation. Epigenetic modifications can play a role similar to mutation in carcinogenesis, and about 280 cancer prone epigenetic alterations are listed by Schnekenburger and Diederich [7]. Epigenetic alterations are usually copied onto the daughter chromosomes when the parental chromosome replicates.

Although epigenetic changes can be passed down from one cell generation to the next, they are not regarded as true mutations. Most epigenetic changes appear to be part of the differentiation program of the cell and are necessary to allow different types of cells to carry out different functions. In most cells of a human body, only about 5% of genes are active at any one time, often due to epigenetic modifications. However, abnormal unprogrammed epigenetic changes may also occur that alter the functioning of a cell and these changes are referred to as "epimutations." Programmed epigenetic changes can be reversed. During development, as daughter cells of a stem cell differentiate, some epigenetic changes are programmed for reversal. However, a double strand break in DNA (a type of DNA damage) can initiate unprogrammed epigenetic gene silencing both by causing methylation of a CpG island as well as by promoting silencing types of histone modifications [8]. Another form of epigenetic silencing may occur during DNA repair. The enzyme Parp1 (poly(ADP)-ribose polymerase) and its product poly(ADP)-ribose (PAR) accumulate at sites of DNA damage as part of a repair process [9]. This, in turn, directs recruitment and activation of the chromatin remodeling protein ALC1 that may cause nucleosome remodeling [10]. Nucleosome remodeling has been found to cause, for instance, epigenetic silencing of DNA repair gene *MLH1* [11]. Chemicals previously identified as DNA damaging agents, including benzene, hydroquinone, styrene, carbon tetrachloride and trichloroethylene, were shown to cause considerable hypomethylation of DNA, some through the activation of oxidative stress pathways [12]. Dietary agents also have been shown to affect DNA methylation or histone modification by numerous pathways [13]. Recent evidence indicates that epimutations occur in DNA repair genes that reduce their function. Epimutations in DNA repair genes allow DNA damages to accumulate, and are a cause of progression to cancer [14].

2. DNA damages are frequent, and DNA repair processes can be overwhelmed

Tens of thousands of DNA damages occur per day per cell, on average, in humans, due to reactive molecules produced by metabolism or by hydrolytic reactions in the warm aqueous cellular media. Some types of such endogenous damages, and their rates of occurrence, are shown in Table 1.

A considerable number of other types of endogenous DNA damages have been identified, many of which are mutagenic. These include propano-, etheno- and malondialdehyde-derived DNA adducts, base propenals, estrogen-DNA adducts, alkylated bases, deamination of each of cytosine, adenine and guanine (to form uracil, hypoxanthine and xanthine, respectively) and adducts formed with DNA by reactive carbonyl species [15].

While there are repair pathways that act on these DNA damages, the repair processes are not 100% efficient, and further damages occur even as current DNA damages are being repaired. Thus there is a steady state level of many DNA damages, reflecting the efficiencies of repair and the frequencies of occurrence. For instance, Helbock et al. [16] estimated the steady state level of oxidative adducts in rat liver as 24, 000 adducts per cell in young rats and 66, 000 adducts per cell in old rats. Nakamura and Swenberg [17] determined the number of AP sites (apurinc and apyrimidinic sites) in normal tissues of the rat (i.e. in lung, kidney, liver, testis, heart, colon and brain). The data indicated that the number of AP sites ranged from about 50, 000 per cell in liver, kidney and lung to about 200, 000 per cell in the brain. These steady state numbers of AP sites in genomic DNA were considered to represent the balance between formation and repair of AP sites.

DNA damages	Reported rate of occurrence	Ref.
Oxidative	86,000 per cell per day in rats	[18]
	10,000 per cell per day in humans 100,000 per cell per day in rats	[19]
	11,500 per cell per day for humans 74,000 per cell per day for rats	[16]
Specific oxidative damage products 8-hydroxyguanine, 8-hydroxydeoxyguanosine, 5-(hydroxymethyl) uracil	2,800 per cell per day in humans 34,800 per cell per day in mice	[20]
Depurinations	10,000 per cell during 20-hour generation period	[21]
	13,920 per cell per day (580/cell/hr)	[22]
	2,000 to 10,000 per cell per day	[23,24]
	9,000 per cell per day	[25]
Depyrimidinations	500 pyrimidines per cell during 20-hour generation period	[21]
	696 per cell per day (29/cell/hr)	[22]
Single-strand breaks	55,200 per cell per day (2,300/cell/hr)	[22]
Double-strand breaks	~10 per cell cycle in humans	[26]
	~50 per cell cycle in humans	[27]
O^6-methylguanine	3,120 per cell per day (130/cell/hr)	[22]
Cytosine deamination	192 per cell per day (8/cell/hr)	[22]

Table 1. DNA damages due to natural endogenous causes in mammalian cells

DNA repair pathways are usually able to keep up with the endogenous damages in replicating cells, in part by halting DNA replication at the site of damage until repair can occur [28, 29]. In contrast, non-replicating cells have a build-up of DNA damages, causing aging [30, 31].

However, some exogenous DNA damaging agents, such as those in tobacco smoke, discussed below, may overload the repair pathways, either with higher levels of the same type of DNA damages as those occurring endogenously or with novel types of damage that are repaired more slowly. In addition, if DNA repair pathways are deficient, due to inherited mutations or sporadic somatic epimutations in DNA repair genes in replicating somatic cells, unrepaired endogenous and exogenous damages will increase due to insufficient repair. Increased DNA damages would likely give rise to increased errors of replication past the damages (by trans-lesion synthesis) or increased error prone repair (e.g. by non-homologous end-joining), causing mutations. Increased mutations that activate oncogenes, inactivate tumor suppressor genes, cause genomic instability or give rise to other driver mutations in replicating cells would increase the risk of cancer.

3. Cancers are often caused by exogenous DNA damaging agents

Cancer incidence, in different areas of the world, varies considerably. Thus, the incidence of colon cancer among Black Native-Africans is less than 1 person out of 100, 000, while among male Black African-Americans it is 72.9 per 100, 000, and this difference is likely due to differences in diet [32, 33]. Rates of colon cancer incidence among populations migrating from lower-incidence to higher-incidence countries change rapidly, and within one generation can reach the rate in the higher-incidence country. This is observed, for instance, in migrants from Japan to Hawaii [34].

The most common cancers for men and women and their rates of incidence per 100, 000, averaged over the more developed areas and less developed areas of the world, are shown in Table 2 (from [35]). Overall, worldwide, cancer incidence in all organs combined is 300.1 per 100, 000 per year in more developed areas and 160.3 per 100, 000 per year in less developed areas [35]. The differences in cancer incidence between more developed areas of the world and less developed areas are likely due, in large part, to differences in exposure to exogenous carcinogenic factors. The lowest rates of cancers in a given organ (Table 2) may be due, at least in part, to endogenous DNA damages (as described in the previous section) that cause errors of replication (trans-lesion synthesis) or error prone repair (e.g. non-homologous end-joining), leading to carcinogenic mutations. The higher rates (Table 2) are likely largely attributable to exogenous factors, such as higher rates of tobacco use or diets higher in saturated fats that directly, or indirectly, increase the incidence of DNA damage.

It is interesting to note in Table 2 that, in cases where cancers occur in the same organs of men and women, men consistently have a higher rate of cancer than women. The basis for this is currently unknown.

	More developed areas		Less developed areas	
	Incidence	*Mortality*	*Incidence*	*Mortality*
Breast (women)	66.4	15.3	27.3	10.8
Prostate (men)	62.0	10.6	12.0	5.6
Lung (men)	47.4	39.4	27.8	24.6
Lung (women)	18.6	13.6	11.1	9.7
Colorectum (men)	37.6	15.1	12.1	6.9
Colorectum (women)	24.2	9.7	9.4	5.4
Esophagus (men)	6.5	5.3	11.8	10.1
Esophagus (women)	1.2	1.0	5.7	4.7
Stomach (men)	16.7	10.4	21.1	16.0
Stomach (women)	7.3	4.7	10.0	8.1
Liver (men)	8.1	7.2	18.9	17.4
Liver (women)	2.7	2.5	7.6	7.2
Bladder (men)	16.6	4.6	5.4	2.6
Bladder (women)	3.6	1.0	1.4	0.7
Cervix/Uterine (women)	12.9	2.4	5.9	1.7
Kidney (men)	11.8	4.1	2.5	1.3
Kidney (women)	5.8	1.7	1.4	0.8
Non-Hodgkin lymphoma (men)	10.3	3.6	4.2	3.0
Non-Hodgkin lymphoma (women)	7.0	2.2	2.8	1.9
Melanoma (men)	9.5	1.8	0.7	0.3
Melanoma (women)	8.6	1.1	0.6	0.3
Ovarian (women)	9.4	5.1	5.0	3.1

Table 2. Incidence and mortality rates for the most common cancers in age standardized rates per 100, 000 (excluding non-melanoma skin cancer) (Adapted from Jemal et al. [35]).

4. Exogenous DNA damaging agents in carcinogenesis

Carcinogenic exogenous factors have been identified as a major cause of many common cancers, including cancers of the lung, colorectum, esophagus, stomach, liver, cervix/uterus and melanoma. Often such exogenous factors have been shown to cause DNA damage, as described below.

5. Exogenous DNA damaging agents in lung cancer

In both developed and undeveloped countries, lung cancer is the most frequent cause of cancer mortality (Table 2, data for men and women combined). Lung cancer is large-ly caused by tobacco smoke, since risk estimates for lung cancer indicate that, in the United States, tobacco smoke is responsible for 90% of lung cancers. Also implicated in lung cancer (and somewhat overlapping with smoking) are occupational exposure to carcinogens (approximately 9 to 15%), radon (10%) and outdoor air pollution (perhaps 1 to 2%) [36].

Acrolein	122.4
Formaldehyde	60.5
Acrylonitrile	29.3
1,3-butadiene	105.0
Acetaldehyde	1448.0
Ethylene oxide	7.0
Isoprene	952.0
Benzo[a]pyrene	0.014

Table 3. Weight, in μg per cigarette, of several likely carcinogenic DNA damaging agents in tobacco smoke (from [37] Cunningham et al., 2011])

Tobacco smoke is a complex mixture of over 5, 300 identified chemicals, of which 150 are known to have specific toxicological properties (see partial summary by Cunningham [37]). A "Margin of Exposure" approach has recently been established to determine the most im-portant exogenous carcinogenic factors in tobacco smoke [37]. This quantitative-type of measurement is based on published dose response data for mutagenicity or carcinogenicity and the concentrations of these components in tobacco smoke (Table 3). Using the "Margin of Exposure" approach, Cunningham et al. [37] found the most important tumorigenic com-pounds in tobacco smoke to be, in order of importance, acrolein, formaldehyde, acryloni-trile, 1, 3-butadiene, acetaldehyde, ethylene oxide and isoprene.

Acrolein, the first agent in Table 3, is the structurally simplest α, β-unsaturated aldehyde (Figure 2). It can rapidly penetrate through the cell membrane and bind to the nucleophilic N^2-amine of deoxyguanine (dG) followed by cyclization of N1, to give the exocyclic DNA adduct α-hydroxy-1, N^2-propano-2'-deoxyguanine (α-HOPdG) (shown in Figure 2) and an-other product designated γ-HOPdG. The adducts formed by acrolein are a major type of DNA damage caused by tobacco smoke, and acrolein has been found to be mutagenic [38].

In tobacco smoke, acrolein has a concentration >8, 000 fold higher than benzo[a]pyrene (re-viewed in [38]), with 122.4 μg of acrolein per cigarette. Benzo[a]pyrene has long been thought to be an important carcinogen in tobacco smoke [39]. As reviewed by Alexandrov et al. [39], benzo[a]pyrene damages DNA by forming DNA adducts at the N^2 position of guanine (similar to where acrolein forms adducts). However, by the "Margin of Exposure" approach, based on published dose response data and its concentration in cigarette smoke of 0.014 μg per cigarette, benzo[a]pyrene is thought to be a much less important mutagen for lung tissue than acrolein and the other six highly likely carcinogens in tobacco smoke listed in Table 3 [37].

The other agents in Table 3 cause DNA damages in different ways. Formaldehyde, the second agent in Table 3, primarily causes DNA damage by introducing DNA-protein cross-links. These cross-links, in turn, cause mutagenic deletions or other small-scale chromosomal rearrangements [40] and may also cause mutations through single-nucleotide insertions [41]. Acrylonitrile, the third agent in Table 3, appears to cause DNA damage indirectly by increasing oxidative stress, leading to increased levels of 8'-hydroxyl-2-deoxyguanosine (8-OHdG) in DNA [42]. Oxidative stress also causes lipid peroxidation that generates malondialdehyde (MDA), and MDA forms DNA adducts with guanine, adenine and cytosine [43]. The fourth agent in Table 3, 1, 3-butadiene, causes genotoxicity both directly by forming a DNA adduct as well as indirectly by causing global loss of DNA methylation and histone methylation leading to epigenetic alterations [44]. The fifth agent in Table 3, acetaldehyde, reacts with 2'-deoxyguanosine in DNA to form DNA adducts [45]. The sixth agent in Table 3, ethylene oxide, forms mutagenic hydroxyethyl DNA adducts with adenine and guanine [46]. The seventh agent in Table 3, isoprene, is normally produced endogenously by humans, and is the main hydrocarbon of non-smoking human breath [47]. However, smoking one cigarette causes an increase in breath isoprene levels by an average of 70% [48]. Isoprene, after being metabolized to mono-epoxides, causes DNA damage measured as single and double strand breaks in DNA [49].

A large number of studies have been published in which the levels and characteristics of DNA adducts in the lung and bronchus of smokers and non-smokers have been compared, as reviewed by Phillips [50]. In most of these studies, significantly elevated levels of DNA adducts were detected in the peripheral lung, bronchial epithelium or bronchioalveolar lavage cells of the smokers, especially for total bulky DNA adducts. As further discussed by Phillips [50], mean levels of DNA adducts in ex-smokers (usually with at least a 1 year interval since smoking cessation) are found generally to be intermediate between the levels of smokers and life-long non-smokers. From these comparisons, the half-life of some DNA adducts in lung tissue are estimated to be ~1–2 years.

6. Exogenous DNA damaging agents in colorectal cancer

Up to 20% of current colorectal cancers in the United States may be due to tobacco smoke [51]. Presumably tobacco smoke causes colon cancer due to the DNA damaging agents described above for lung cancer. These agents may be taken up in the blood and carried to organs of the body.

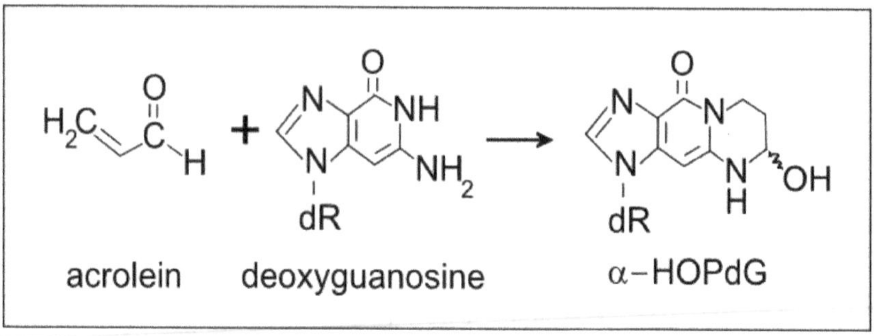

Figure 2. Reaction of acrolein with deoxyguanosine

The human colon is exposed to many compounds that are either of direct dietary origin or re-sult from digestive and/or microbial processes. Four different classes of colonic mutagenic compounds were analysed by de Kok and van Maanen [52] and evaluated for fecal mutagenici-ty. These included (1) pyrolysis compounds from food (heterocyclic aromatic amines and poly-cyclic aromatic hydrocarbons), (2) N-nitroso-compounds (from high meat diets, from drinking water with high nitrates or produced during ulcerative colitis), (3) fecapentaenes (produced by the colonic bacteria *Bacteriodes* in the presence of bile acids) and (4) bile acids (increased in the colon in response to a high fat diet and metabolized to genotoxic form by bacteria in the colon). Many of these diet-related mutagenic compounds were analysed by Pearson et al. [53] in terms of their presence in fecal water, and their effect on the cytotoxic or genotoxic activity of fecal water. Evidence in both of these studies was insufficient to evaluate the colorectal cancer risk as a result of specific exposures in quantitative terms.

However, substantial evidence implicates bile acids (the 4th possibility above) in colon caner. Bernstein et al. [54], summarized 12 studies indicating that the bile acids deoxycholic acid (DCA) and/or lithocholic acid (LCA) induce production of DNA damaging reactive oxygen species and/or reactive nitrogen species in colon cells of animal or human origin. They also tabulated 14 studies showing that DCA and LCA induce DNA damage in colon cells. In ad-dition to causing DNA damage, bile acids may also generate genomic instability by causing mitotic perturbations and reduced expression of spindle checkpoint proteins, giving rise to micro-nuclei, chromosome bridges and other structures that are precursors to aneuploidy [55]. Furthermore, at high physiological concentrations, bile acids cause frequent apoptosis, and those cells in the exposed populations with reduced apoptosis capability tend to survive and selectively proliferate [54, 56]. Cells with reduced ability to undergo apoptosis in response to DNA damage would tend to accumulate mutations when replication occurs past those damages, and such cells may give rise to colon cancers. In addition, 7 epidemiological studies between 1971 and 1990 (reviewed by Bernstein et al. [54]), found that fecal bile acid concentrations are increased in populations with a high incidence of colorectal cancer. A similar 2012 epidemiological study showed that concentrations of fecal LCA and DCA, respectively, were 4-fold and 5-fold higher in a population at 65-fold higher risk of colon cancer compared to a population at lower risk of colon cancer [32]. This evidence points to bile acids DCA and LCA as centrally important DNA-damaging carcinogens in colon cancer.

Dietary total fat intake and dietary saturated fat intake is significantly related to incidence of colon cancer [57]. Increasing total fat or saturated fat in human diets results in increases in DCA and LCA in the feces [58, 59], indicating increased contact of the colonic epithelium with DCA and LCA. Bernstein et al. [60] added the bile acid DCA to the standard diet of wild-type mice. This supplement raised the level of DCA in the feces of mice from the standard-diet fed mouse level of 0.3 mg DCA/g dry weight to 4.6 mg DCA/g dry weight, a level similar to that for humans on a high fat diet of 6.4 mg DCA/g dry weight. After 8 or 10 months on the DCA-supplemented diet, 56% of the mice developed invasive colon cancer. This directly indicates that DCA, a DNA damaging agent, at levels present in humans after a high fat diet, can cause colorectal cancer.

7. Exogenous DNA damaging agents implicated in other major cancers

It is beyond the scope of this chapter to detail the evidence implicating DNA damaging agents as etiologic agents in all of the significant cancers. Therefore, in Table 4 we indicate with a single reference the major DNA damaging agent in five additional prevalent cancers, in order to illustrate the generality of exogenous DNA damaging agents as causes of cancer. In particular, we point out, as reviewed by Handa et al. [61], *Helicobacter pylori* infection increases the production of reactive oxygen and reactive nitrogen species (RNS) in the human stomach, which, in turn, significantly increases DNA damage in the gastric epithelial cells. Thus, *H. pylori* infection acts as a DNA damaging agent. In the case of human papillomavirus (HPV) infection, Wei et al. [62] showed that cervical cells could resist RNS stress when not infected with HPV. However, cervical cells infected by HPV and exposed to RNS had higher levels of DNA double strand breaks as well as a higher mutation rate. This appeared to occur due to the ability of HPV to greatly reduce protein expression of the DNA damage repair/response gene *P53* when infected cells were stressed by RNS. Since reduced *P53* expression leads to greater RNS-induced DNA damage, HPV infection acts as a DNA damaging agent in the presence of RNS stress.

Cancer	Exogenous DNA damaging agent	Ref.
Esophagus	Bile acids	[63]
Stomach	*Helicobacter pylori* infection	[61]
Liver	*Aspergillus* metabolite aflatoxin B(1)	[64]
Cervix/Uterus	Human papillomavirus plus increased nitric oxide from tobacco smoke or other infection	[62]
Melanoma	UV light from solar radiation	[65]

Table 4. Selected cancers and relevant implicated exogenous DNA damaging agents

8. Deficient DNA repair due to a germ line mutation allows DNA damages to increase, leading to increased frequencies of mutation, epimutation and cancer

Expression of DNA repair genes may be reduced by inherited germ line mutations or genetic polymorphisms, or by epigenetic alterations or mutations in somatic cells, and these reductions may substantially increase the risk of cancer. Overall, about 30% of cancers are considered to be familial (largely due to inherited germ line mutations or genetic polymorphisms) and 70% are considered to be sporadic [66].

In 2 overlapping databases [67, 68] 167 and169 human genes (depending on the database) are listed that are directly employed in DNA repair or influence DNA repair processes. The lists were originally devised by Wood et al. [69, 70]. The genes are distributed in groups of DNA repair pathways and in related functions that affect DNA repair (Table 5). Bernstein et al. [71] illustrate many of the steps and order of action of the gene products involved for the first five DNA repair pathways listed in Table 5.

Individuals with an inherited impairment in DNA repair capability are often at considerably increased risk of cancer. If an individual has a germ line mutation in a DNA repair gene or a DNA damage response gene (that recognizes DNA damage and activates DNA repair), usually one abnormal copy of the gene is inherited from one of the parents and then the other copy is inactivated at some later point in life in a somatic cell. The inactivation may be due, for example, to point mutation, deletion, gene conversion, epigenetic silencing or other mechanisms [72]. The protein encoded by the gene will either not be expressed or be expressed in a mutated form. Consequently the DNA repair or DNA damage response function will be deficient or impaired, and damages will accumulate. Such DNA damages can cause errors during DNA replication or inaccurate repair, leading to mutations that can give rise to cancer.

Increased oxidative DNA damages also cause increased gene silencing by CpG island hypermethylation, a form of epimutation. These oxidative DNA damages induce formation and relocalization of a silencing complex that may result in cancer-specific aberrant DNA methylation and transcriptional silencing [73]. As pointed out above, the enzyme Parp1 (poly(ADP)-ribose polymerase) and its product poly(ADP)-ribose (PAR) accumulate at sites of DNA damage as part of a repair process [9], recruiting chromatin remodeling protein ALC1, causing nucleosome remodeling [10] that has been shown to direct epigenetic silencing of DNA repair gene *MLH1* [11]. If silencing of genes necessary for DNA repair occurs, the repair of further DNA damages will be deficient and more damages will accumulate. Such additional DNA damages will cause increased errors during DNA synthesis, leading to mutations that can give rise to cancer.

	Number of genes listed in the two databases
Homologous Recombinational Repair (HRR)	21,21
Non-homologous End Joining (NHEJ)	8,7
Nucleotide Excision Repair (NER)	30,29
Base Excision Repair (including PARP enzymes) (BER)	19,20
Mis-Match Repair (MMR)	11,10
Fanconi Anemia (FANC) [affects HRR (above) and translesion synthesis (TLS)]	10,16
Direct reversal of damage	3,3
DNA polymerases (act in various pathways)	17,15
Editing and processing nucleases (act in various pathways)	6,8
Ubiquitination and modification/Rad6 pathway including TLS	11,5
DNA damage response	12,14
Modulation of nucleotide pools	3,3
Chromatin structure	2,3
Defective in diseases and syndromes	4,5
DNA-topoisomerase crosslinks	2,1
Other genes	8,9

Table 5. DNA repair pathways and other processes affecting DNA repair [67, 68]

9. Inherited mutations in genes employed in DNA repair that give rise to syndromes characterized by increased risk of cancer.

Table 6 lists 36 genes for which an inherited mutation results in an increased risk of cancer. The proteins encoded by 35 of these genes are involved in DNA repair and in some cases also in other aspects of the DNA damage response such as cell cycle arrest and apoptosis. The polymerase coded for by the 36th gene, XPV (POLH), is involved in bypass (rather than repair) of DNA damage, called translesion synthesis. The genes listed in Table 6, when mutated in the germ line, give rise to a considerably increased lifetime risk of cancer, of up to 100% (e.g. p53 mutations [74]). Thus defects in DNA repair cause progression to cancer.

In addition to mutations in genes that may substantially raise lifetime cancer risk, there appear to be many weakly effective genetically inherited polymorphisms [single nucleotide polymorphisms (SNPs) and copy number variants (CNVs)]. By the HapMap Project, more than 3 million SNPs have been found, and by Genome Wide Association studies (GWAs), about 30 SNPs were found to increase risk of cancers. However the added risk of cancer by these SNPs is usually small, i.e. less than a factor of 2 increase [75]. A large twin study [66], involving 44, 788 pairs of twins, evaluated the risk of the same cancer before the age of 75 for monozygotic twins (identical genomes with the same polymorphisms) and dizygotic twins (having a 50% chance of the same polymorphisms). If one twin had colorectal, breast or prostate cancer, the monozygotic twin had an 11 to 18 percent chance of developing the same cancer while the dizygotic twin had only a 3 to 9% risk. The differences in monozygotic and dizygotic rates of paired cancer were not significant for the other 24 types of cancer evaluated in this study. Polymorphisms of the DNA repair gene ERCC1 will be discussed below in relation to targeted chemotherapy.

10. Epimutations may repress DNA repair gene expression, allowing DNA damages to increase, leading to increased frequency of further epimutation, mutation and cancer

While germ line (familial) mutations in DNA repair genes cause a high risk of cancer, somatic mutations in DNA repair genes are rarely found in sporadic (non-familial) cancers [4]. Much more often, DNA repair genes are found to have epigenetic alterations in cancers.

One example of the epigenetic down-regulation of a DNA repair gene in cancers comes from studies of the MMR protein MLH1. Truninger et al. [76] assessed 1, 048 unselected consecutive colon cancers. Of these, 103 were deficient in protein expression of MLH1, with 68 of these cancers being sporadic (the remaining MLH1 deficient cancers were due to germ line mutations). Of the 68 sporadic MLH1 protein-deficient colon cancers, 65 (96%) were found to be deficient due to epigenetic methylation of the CpG island of the MLH1 gene. Deficient protein expression of MLH1 may also have been caused, in the remaining 3 sporadic MLH1 protein-deficient cancers (which did not have germ line mutations), by over expression of the microRNA miR-155. When miR-155 was transfected into cells it caused reduced expression of MLH1 [77]. Overexpression of miR-155 was found in colon cancers in which protein expression of MLH1 was deficient and the MLH1 gene was neither mutated nor hypermethylated in its CpG island [77].

DNA repair gene(s)	Encoded protein	Repair pathway(s) affected	Ref.	Cancers with increased risk	Ref.
breast cancer 1 & 2	BRCA1, BRCA2	HRR of double strand breaks and daughter strand gaps	[85]	Breast, Ovarian	[86]
ataxia telangiectasia mutated	ATM	Different mutations in ATM reduce HRR, single strand annealing (SSA), NHEJ or homology directed double strand break rejoining (HDR)	[87]	Leukemia, Lymphoma, Breast	[87,88]
Nijmegen breakage syndrome	NBS	NHEJ	[89]	Lymphoid cancers	[89]
meiotic recombination 11	MRE11	HRR and NHEJ	[90]	Breast	[91]
Bloom's Syndrome (helicase)	BLM	HRR	[92]	Leukemia, Lymphoma, Colon, Breast, Skin, Auditory canal, Tongue, Esophagus, Stomach, Tonsil, Larynx, Lung, Uterus	[93]
Werner Syndrome (helicase)	WRN	HRR, NHEJ, long patch BER	[94]	Soft tissue sarcoma, Colorectal, Skin, Thyroid, Pancreatic	[95]
Rothman Thomson syndrome Rapadilino syndrome Baller Gerold syndrome	RECQ4	Helicase likely active in HRR	[96]	Basal cell carcinoma, Squamous cell carcinoma, Intraepidemial carcinoma	[97]
Fanconi's anemia gene FANC A,B,C,D1,D2,E,F,G,I,J,L,M,N	FANCA etc.	HRR and TLS	[98]	Leukemia, Liver tumors, Solid tumors many areas	[99]
xeroderma pigmentosa C, E [DNA damage binding protein 2 (DDB2)]	XPC XPE	Global genomic NER repairs damage in both transcribed and untranscribed DNA	[100, 101]	Skin cancer (melanoma and non-melanoma)	[100, 101]
xeroderma pigmentosa A, B, D, F, G	XPA XPB XPD XPF XPG	Transcription coupled NER repairs the transcribed strands of transcriptionally active genes	[102]	Skin cancer (melanoma and non-melanoma), Central nervous system cancers	[102]
xeroderma pigmentosa V (also called polymerase H)	XPV (POLH)	Translesion Synthesis (TLS)	[102]	Skin cancer (melanoma and non-melanoma)	[102]

DNA repair gene(s)	Encoded protein	Repair pathway(s) affected	Ref.	Cancers with increased risk	Ref.
mutS (E. coli) homolog 2 mutS (E. coli) homolog 6 mutL (E. coli) homolog 1 postmeiotic segregation increased 2 (S. cerevisiae)	MSH2 MSH6 MLH1 Pms2	MMR	[76]	Colorectal, endometrial. ovarian	[103]
mutY homolog (E. coli)	MUTYH	BER of A mispaired with 8OHdG, G, FapydG and C	[104]	Colon	[105]
ataxia telaniectsia and RAD3 related	ATR	DNA damage response likely affects HRR, not NHEJ	[106]	Oropharyngeal cancer	[107]
Li Fraumeni syndrome	P53	HRR, BER, NER and DNA Damage Response for those and for NHEJ and MMR	[108]	Sarcoma, Breast, Lung, Skin, Pancreas, Leukemia, Brain	[74]

Table 6. Inherited mutations in DNA repair genes that increase the risk of cancer

Another example of the epigenetic down-regulation of a DNA repair gene in cancer comes from studies of the direct reversal of methylated guanine bases by methyl guanine methyl transferase (MGMT). In the most common form of brain cancer, glioblastoma, the DNA repair gene *MGMT* is epigenetically methylated in 29% [78] to 66% [79] of tumors, thereby reducing protein expression of MGMT. However, for 28% of glioblastomas, the MGMT protein is deficient but the *MGMT* promoter is not methylated [79]. Zhang et al. [78] found, in the glioblastomas without methylated *MGMT* promoters, that the level of microRNA miR-181d is inversely correlated with protein expression of MGMT and that the direct target of miR-181d is the MGMT mRNA 3'UTR (the three prime untranslated region of MGMT mRNA), though they indicated that other miRNAs may also be involved in the reduction of protein expression of MGMT.

Almost all DNA repair deficiencies found, so far, in sporadic cancers, and in precancerous tissues surrounding cancers (field defects) are due to epigenetic changes. Examples of such epigenetic alterations in DNA repair genes in different types of cancer are shown in Table 7. A recent review [80] lists 41 reports (mostly not overlapping with those listed in Table 7) indicating methylation of 20 DNA repair genes in various cancers. In Table 7 data are also shown on DNA repair gene deficiencies for the field defects associated with colorectal, gastric, laryngeal and non-small cell lung cancer.

As summarized above, epimutations can result from oxidative DNA damages. Such damages cause formation and relocalization of a silencing complex that in turn causes increased gene silencing by CpG island hypermethylation [73]. Epigenetic nucleosome remodeling during DNA repair can also silence gene expression [11]. When CpG island methylation or nucleosome remodeling or other types of epigenetic alterations (e.g. micro RNAs or histone modifications) inhibit DNA repair genes, more damages will accumulate. Accumulated DNA damages cause increased errors during DNA synthesis and repair. Thus epigenetic deficiencies in DNA repair genes can have a cascading effect (a mutator phenotype), leading to genomic instability and accumulation of mutations and epimutations that can give rise to cancer.

Cancer	Epigenetic changes in cancer (mechanism)	% sporadic cancers with epimutations	Epigenetic changes in field defect (mechanism)	% field defects with epi-mutations	Ref.
Breast	BRCA1 (CGI*)	13% unselected 67% medullary 55% mucinous			[108]
	WRN (CGI)	17% unselected			[2]
Ovarian	BRCA1 (CGI)	31% of those with loss of heterozygosity			[108]
Colorectal	WRN (CGI)	38%			[2]
	MGMT (CGI)	46%	MGMT (CGI)	23%	[109]
	MGMT (CGI)	90%			[110]
	MLH1 (CGI)	65%			
	MLH1 (CGI)	10%			[76]
	MLH1 (CGI)	2%			
	MSH2 (CGI)	13%	MSH2	5%	[111]
	MGMT (CGI)	47%	MGMT	11%	
	ERCC1	100%	ERCC1	40%	
	PMS2	88%	PMS2	50%	[112]
XPF	XPF	55%	XPF	40%	
Gastric	MGMT (CGI)	88%	MGMT (CGI)	29%	[113]
	WRN (CGI)	25%			[2]
Esophageal squamous cell carcininoma	MLH1 (CGI)	49%			[114,115]
	MLH2 (CGI)	35%			
	MGMT (CGI)	41%			
Larynx	MGMT (CGI)	54%	MGMT (CGI)	38%	[116]

Cancer	Epigenetic changes in cancer (mechanism)	% sporadic cancers with epimutations	Epigenetic changes in field defect (mechanism)	% field defects with epi-mutations	Ref.
Non-small cell Lung	WRN (CGI)	38%			[2]
	MGMT (CGI)	70%	MGMT (CGI)	40%	[117]
Prostate	WRN (CGI)	20%			[2]
Thyroid	WRN (CGI)	13%			[2]
Non-Hodgkin lymphoma	WRN (CGI)	24%			[2]
Leukemias	WRN (CGI)	5-10%			[2]
Chondrosarcomas	WRN (CGI)	33%			[2]
Osteosarcomas	WRN (CGI)	11%			[2]
Brain glioblastoma	MGMT (CGI)	51%			[118]
	MGMT (miRNA)	28%			[78]
Liver hepatocellular carcinoma	P53 (non-CGI promoter site specific methylation)	100%			[119]
Papillary thyroid (tested 23 DNA repair genes for CGI)	MLH1 (CGI)	21%			[120]
	PCNA (CGI)	13%			
	OGG1 (CGI)	2%			

*CGI=CpG island methlyation

Table 7. Examples of epigenetic alterations (epimutations) of DNA repair genes in cancers and in field defects, with mechanisms indicated where known.

Deficiencies in DNA repair genes cause increased mutation rates. Mutations rates increase in MMR defective cells [81, 82] and in HRR defective cells [83]. Chromosomal rearrangements and aneuploidy also increase in HRR defective cells [84]. Thus, deficiency in DNA repair causes genomic instability and genomic instability is the likely main underlying cause of the genetic alterations leading to tumorigenesis. Deficient DNA repair permits the acquisition of a sufficient number of alterations in tumor suppressor genes and oncogenes to fuel carcinogenesis. Deficiencies in DNA repair appear to be central to the genomic and epigenomic instability characteristic of cancer.

Figure 3 illustrates the chain of consequences of exposure of cells to endogenous and exogenous DNA damaging agents that lead to cancer. The role of germ line defects in DNA repair genes in familial cancer are also indicated. The large role of DNA damage and consequent epigenetic DNA repair defects leading to sporadic cancer are emphasized. The roles of germ line mutation and directly induced somatic mutation in sporadic cancer are indicated as well.

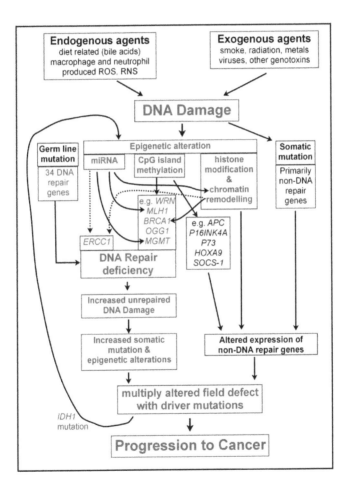

Figure 3. The roles of DNA damage, epigenetic deficiencies in DNA repair and mutation in progression to cancer.

11. Epigenetic alterations caused by micro RNAs

MicroRNAs (miRNAs) are endogenous non-coding RNAs, 19-25 nucleotides in length, that can have substantial effects on DNA repair. miRNAs can either directly or indirectly reduce expression of DNA repair or DNA damage response genes. As discussed above, over-expression of miR-155 causes reduced expression of DNA repair protein MLH1, and miR-155 is overexpressed in colon cancers [77] (curved arrow in Figure 3). Similarly, miR-181d is overexpressed in glioblastomas, causing reduced expression of DNA repair protein MGMT [78]. Although miRNAs can epigenetically regulate DNA repair gene expression, the expression levels of many miRNAs may themselves be subject to epigenetic regulation. One mechanism of epigenetic regulation of miRNA expression is hypomethylation of the promoter region of the DNA sequence that codes for the miRNA. Schnekenburger and Diederich [7] list miR-155 as one of a long list of mi-RNAs whose expression is increased by hypomythylation in colorectal cancers. In particular, hypomethylated miR-155 (the hypomethylation making it more active) targets genes *MLH1, MSH2* and *MSH6,* causing each of them to have reduced expression [7].

Wan et al. [121] referred to 6 further DNA repair genes that are directly targeted by miR-NAs. *ATM, RAD52, RAD23B, MSH2, BRCA1* and *P53*, are each specifically targeted by one or two of the 8 miRNAs miR-21, miR-24, miR-125b, miR-182, miR-210, miR-373, miR-421 and miR-504, with all but miR-210, miR-421 and miR-504 among those identified by Schnekenburger and Diederich [7] as overexpressed through epigenetic hypomethylation. Overexpression of any one of these miRNAs leads to reduced expression of its target DNA repair gene. Wan et al. [121] further listed 16 DNA damage response genes targeted by specific miRNAs. Wan et al. [121] indicated miR-15a, miR-16, miR-17, miR-20a, miR-21, miR-24, miR-29, miR-34a, miR-106a, miR-93, miR-124a, miR-125b, miR-192, miR-195, miR-215, miR-182, miR-373 as among those targeting DNA damage response genes. Of these, all but miR-124a were identified by Schnekenburger and Diederich [7], (and Malumbres [122] further identified miR-34a and miR-124a) as being among miRNAs whose expression is subject to epigenetic alteration in tumors. Other miRNAs whose expression is subject to epimutation in colorectal cancers (and their target DNA repair or DNA damage response genes) include miR-17 (*E2F1*), miR-34b/c (*P53*), miR-106a (*E2F1*), miR-200a and miR-200b (*MLH1, MSH2*) and miR-675 (*Rb*) [7].

12. Epigenetic alterations caused by chromosome remodeling and histone modification

Specific miRNAs can also indirectly (and strongly) reduce protein expression of DNA repair genes through their role in repression of proteins designated High Mobility Group A1 (HMGA1) and HMGA2 (the names come from the proteins' high electrophoretic mobility on acrylamide gels). HMGA1 and HMGA2 cause chromatin remodeling at specific sites in DNA and reduce expression at those sites. In particular, these proteins appear to control

DNA repair genes *BRCA1* and *ERCC1*. BRCA1 And ERCC1 proteins have key roles in DNA repair, particularly of double-strand breaks and interstrand crosslinks. *HMGA1* and *HMGA2* genes are usually active in embryogenesis, but normally have very low expression levels in adult tissues. Their expression levels in adult tissues are kept low by the actions of specific miRNAs. If expression of these miRNAs is reduced, then the repressive HMGA1 and HMGA2 proteins become highly expressed and, in particular, can reduce expression of *BRCA1* or *ERCC1* respectively.

As reviewed by Resar [123], all HMG proteins share an acidic carboxyl terminus and associate with chromatin. As an example, HMGA1A, in particular, has three AT-hook domains that allow it to bind to AT-rich regions and recruit an "enhanceosome" that may displace histones and cause chromosome remodeling and reduce gene expression. Baldassarre et al. [124] showed that HMGA1B protein binds to the promoter region of *BRCA1* and inhibits *BRCA1* promoter activity (indicated in Figure 3 as chromatin remodeling causing reduced *BRCA1*). In 12 surgically removed human breast carcinomas, there was an inverse correlation between HGMA1 protein and *BRCA1* mRNA levels. HGMA1 was almost undetectable in normal breast tissue, highly expressed in the tumor samples, and BRCA1 protein was strongly diminished in tumor samples. Baldassarre et al. [124] suggested that while only 11% of breast tumors had hypermethylation of the *BRCA1* gene, 82% of aggressive breast cancer specimens have low BRCA1 protein, and most of these could be due to chromatin remodeling by high levels of HMGA1 protein.

Similarly, HMGA2 binds to an *ERCC1* promoter site and represses *ERCC1* promoter activity [125]. The miRNAs miR-23a, miR-26a and miR-30a inhibit *HMG2A* protein expression [126] though it has not been reported whether these miRNAs are under epigenetic control. In Figure 3, one of two dotted lines is used to indicate possible repression of *ERCC1* by epigenetically induced chromatin remodeling.

Resar [123] and Baldassarre et al. [124] summarized reports indicating that *HGMA1* is widely overexpressed in aggressive malignancies including cancers of the thyroid, head and neck, colon, lung, breast, pancreas, hematopoetic system, cervix, uterine corpus, prostate and central nervous system. Palmieri et al. [127] showed that *HGMA1* and *HMGA2* are targeted (and thus strongly reduced in expression) by miR-15, miR-16, miR-26a, miR-196a2 and Let-7a. The promoter regions associated with miR-16, miR-196a2 and Let-7a miRNAs are epimutated by hypomethylation [7, 122] while Sampath et al. [128] showed, in addition, that the coding regions for miR-15 and miR-16 were epigenetically silenced due to histone deacetylase activity. Palmieri et al. [127] further showed that these 5 miRNAs are drastically reduced in a panel of 41 pituitary adenomas, accompanied by increases in *HMGA1* and *HMGA2* specific mRNAs. In a more recent study on pituitary adenomas by D'Angelo et al. [129], reduced expression of 18 miRNAs was found, with 5 of them targeting *HMGA1* or *HMGA2*. In this recent study, among the 18 miRNAs with reduced expression, the reduced expression of miR-26b, miR-34b, miR-432 and miR-592 was known to be due to epigenetic alteration [7, 122]. Thus, epigenetic miRNA silencing, causing strong expression of *HMGA1* and *HMGA2*, occurs in many types of cancer and this may be related to reductions found in expression of DNA repair genes *BRCA1, BRCA2* and *ERCC1*.

Suzuki et al. [130], using genome wide profiling, found 174 primary transcription units for miRNAs, called "pri-miRNAs" (large precursor RNAs which may encode multiple miR-NAs), of which they identified 37 as potential targets for epigenetic silencing. Of these 37 pri-miRNAs, 22 were encoded by DNA sequences with CpG islands (all of which were hypermethylated in colorectal cancer cells) while the other pri-miRNAs were subject to regulation by epigenetic "activating marks" without evidence of deregulated methylation.

Activating marks are alterations on histones that cause transcriptional activation of the genes associated with those altered histones (reviewed by Tchou-Wong et al. [131]). In particular, the nucleosome, the fundamental subunit of chromatin, is composed of 146 bp of DNA wrapped around an octamer of four core histone proteins (H3, H4, H2A, and H2B). Posttranslational modifications (i.e., acetylation, methylation, phosphorylation, and ubiquitination) of the N- and C-terminal tails of the four core histones play an important role in regulating chromatin biology. These specific histone modifications, and their combinations, are translated, through protein interactions, into distinct effects on nuclear processes, such as activation or inhibition of transcription. In eukaryotes, methylation of lysine 4 in histone H3 (H3K4), which interacts with the promoter region of genes, is linked to transcriptional activation. There is a strong positive correlation between trimethylation of H3K4, transcription rates, active polymerase II occupancy and histone acetylation. Thus trimethylation of H3K4 is an activating mark.

In addition to pri-miRNAs being regulated by activating marks, some miRNAs appear to be directly regulated by these histone modifications. As summarized by Sampath et al. [128], histone deacetylases catalyze the removal of acetyl groups on specific lysines around gene promoters to trigger demethylation of otherwise methylated lysine 4 on histones (H3K4me2/3) and this causes loss of these activating marks, promoting chromatin compaction, and leading to epigenetic silencing. Sampath et al. [128] showed that such histone deacetylase activity mediates the epigenetic silencing of miRNAs miR-15a, miR-16, and miR-29b. As indicated above, miR-15, miR-16 specifically target HGMA1 and HMGA2. If miR-15 and miR-16 lose their activating marks, they have reduced expression, causing HGMA1 and HGMA2 to be transcriptionally activated, thus reducing expression of DNA repair genes BRCA1 and ERCC1.

In Figure 3, histone modification and chromatin remodeling are indicated as epigenetically altering the expression of many genes in progression to cancer, and specifically causing reduced BRCA1 and possibly (as indicated by one dotted line) reduced expression of ERCC1. In addition, a second dotted line is used to indicate possible repression of ERCC1 by an miRNA. Klase et al. [132] showed that a particular virally coded miRNA down regulates ERCC1 protein expression at the p-body level (a p-body is a cytoplasmic granule "processing body" that interacts with miRNAs to repress translation or trigger degradation of target mRNAs). A survey of human miRNA homology regions to ERCC1 mRNA indicates at least 21 human coded miRNAs that could act to decrease ERCC1 mRNA translation (shown in Microcosm Targets [133]). ERCC1 protein expression, assessed by immunohistochemical staining, is deficient due to an epigenetic mechanism in colon cancers [110], and this could be due to action of one or more miRNAs, acting directly on ERCC1 mRNA.

13. Driver mutations and pathways to cancer progression

Recent research indicates a mechanism by which an early driver mutation may cause subsequent epigenetic alterations or mutations in pathways leading to cancer. Wang et al. [134] point out that isocitrate dehydrogenase genes *IDH1* and *IDH2* are the most frequently mutated metabolic genes in human cancer. A gene frequently mutated in cancer is considered to be a driver mutation [4] so that mutations in *IDH1* and *IDH2* would be driver mutations. Wang et al. [134] further point out that *IDH1* and *IDH2* mutant cells produce an excess metabolic intermediate, 2-hydroxyglutarate, which binds to catalytic sites in key enzymes that are important in altering histone and DNA promoter methylation. Thus, mutations in *IDH1* and *IDH2* generate a DNA CpG island methylator phenotype that causes promoter hypermethylation and concomitant silencing of tumor suppressor genes such as the DNA repair genes *MLH1*, *MGMT* and *BRCA1*. As shown in Figure 3, a driver mutation in *IDH1* can cause a feedback loop leading to increased DNA repair deficiency, further mutations and epimutations, and consequent accelerated tumor progression.

A study, involving 51 patients with brain gliomas who had two or more biopsies over time, showed that mutation in the *IDH1* gene occurred prior to the occurrence of a *p53* mutation or a 1p/19q loss of heterozygosity, indicating that *IDH1* mutation is an early driver mutation [135]. Work by Turcan et al. [136] showed that *IDH1* mutation alone is sufficient to establish the brain glioma CpG island methylator phenotype. Carillo et al. [137] showed that when an *IDH1* mutation was present in glioblastoma tumors, 64% of these were hypermethylated in the promoter regions of *MGMT*.

Other initial driver mutations can cause progression to glioblastoma as well. As pointed out above, increased levels of miR-181d also cause reduced expression of MGMT protein in glioblastoma. Nelson et al. [138] indicate that a single type of miRNA may target hundreds of different mRNAs, causing alterations in multiple pathways. Patients with a glioblastoma that does not harbor an *IDH1* mutation have an overall fairly short survival time, while patients with both mutated *IDH1* and methylated *MGMT* have a subtype of glioblastoma with a much longer survival time (implying a different pathway of cancer progression) [137].

An *IDH1* mutation that gives rise to a CpG island methylator phenotype that causes promoter hypermethylation and concomitant silencing of *MGMT* also causes promoter silencing of other genes as well. In addition to silencing of genes, the CpG island methylator phenotype can cause methylation of the promoter regions of long interspersed nuclear element-1 (LINE-1) DNA sequences. Ohka et al. [139] point out that LINE-1 is a class of retroposons that are the most successful integrated mobile elements in the human genome, and account for about 18% of human DNA. Ohka et al. [139] found that LINE-1 methylation is directly proportional to *MGMT* promoter methylation in gliomas and suggested that LINE-1 methylation could be used as a proxy to indicate the CpG island methylator phenotype status in glioblastomas. This phenotype, likely associated with methylation of the *MGMT* promoter, in turn, indicates whether treatment with the DNA alkylating agent temozolomide will be beneficial in treatment of a patient with a glioblastoma, since MGMT removes the alkyl groups added to guanine by temozolomide.

14. Field defects

Field defects have been described in many types of gastrointestinal cancers [140]. A field defect arises when an epimutation or mutation occurs in a stem cell that causes that stem cell to give rise to a number of daughter stem cells that can out-compete neighboring stem cells. These initial mutated cells form a patch of somewhat more rapidly growing cells (an initial field defect). That patch then enlarges at the expense of neighboring cells, followed by, at some point, an additional mutation or epimutation arising in one of the field defect stem cells so that this new stem cell with two advantageous mutations can generate daughter stem cells that can out-compete the surrounding field defect of cells that have just one advantageous mutation. As illustrated in Figure 4, this process of expanding sub-patches within earlier patches will occur multiple times until a particular constellation of mutations results in a cancer (represented by the small dark patch in Figure 4. It should also be noted that a cancer, once formed, continues to evolve and continues to produce sub clones. A renal cancer, sampled in 9 areas, had 40 ubiquitous mutations, 59 mutations shared by some, but not all regions, and 29 "private" mutations only present in one region [141].

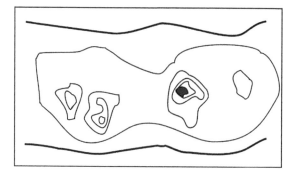

Figure 4. Schematic of a field defect in progression to cancer

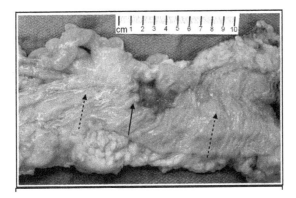

Figure 5. Colon resection including a colon cancer. Dashed arrows indicate grossly unremarkable colonic mucosa. Ulcerated hemorrhagic mass represents a moderately differentiated invasive adenocarcinoma. Solid arrow indicates the heaped up edge of the malignant ulcer

Figure 5 shows an opened resected segment of a human colon that has a colon cancer. As illustrated by Bernstein et al. [142], there are about 100 colonic microscopic epithelial crypts per sq mm in the colonic epithelium. The resection shown in Figure 5 has an area of about 6.5 cm by 23 cm, or 150 sq cm, or 15, 000 sq mm. Thus this area has about 1.5 million crypts. There are 10-20 stem cells at the base of each colonic crypt [143, 144]. Therefore there are likely about 15 million stem cells in the grossly unremarkable colonic mucosal epithelium shown in Figure 5. Evidence reported by Facista et al. [112], and listed in Table 7, indicates that in many such resections, most of the stem cells in such an area up to 10 cm distant (in each direction) from a colon cancer (such as in the grossly unremarkable area shown in Figure 5), and the majority of their differentiated daughter cells, are epigenetically deficient for protein expression of the DNA repair genes *ERCC1, PMS2* and/or *XPF*, although the epithelium is histologically normal.

The stem cells most distant from the cancer, deficient for *ERCC1, PMS2* and/or *XPF*, can be considered to constitute an outer ring, and be deficient as well in the inner rings, of a field defect schematically illustrated in Figure 4. The outer ring in Figure 4 includes, within its circumscribed area, on the order of 15 million stem cells, presumably arising from an initial progenitor stem cell deficient in DNA repair (due to epigenetic silencing). As a result of this repair deficit, the initial stem cell was genetically unstable, giving rise to an increased frequency of mutations in its decendents. One daughter stem cell among its decendents had a mutation that, by chance, provided a replicative advantage. This descendent then underwent clonal expansion because of its replicative advantage. Among the further decendents of the clone, new mutations arose frequently, since these descendents had a mutator phenotype [145], due to the repair deficiency passed down epigenetically from the original repair-defective stem cell. Among these new mutations, some would provide further replicative advantages, giving rise to a succession of more aggressively growing sub clones (inner rings), and eventually a cancer.

15. Exogenous carcinogenic agents cause reduced expression of DNA repair genes

Many known carcinogenic agents cause reduced expression of DNA repair genes or directly inhibit the actions of DNA repair proteins. Table 8 lists examples of carcinogens that have such effects. Due to space limitations, many other such carcinogens are not listed. These findings further link DNA damage to cancer.

Carcinogens	Inhibit DNA Repair Gene	Mechanism Shown	Ref.
Arsenic compounds	PARP		[146,147]
	XRCC1		
	Ligase 3		[146,148]
	Ligase 4		
	DNA POLB, XRCC4		[146]
	DNA PKCS, TOPO2B		
	OGG1, ERCC1, XPF		[149]
	XPB, XPC, XPE		[150]
	P53	Inhibition of P53 serine 15 phosphorylation	[151]
Cadmium compounds	MSH2, ERCC1, XRCC1	Promoter methylation	[152]
	OGG1		
	MSH2, MSH6 proteins	Cd²⁺ binds to proteins	[153]
	OGG1 protein	Oxidation of Ogg1	[154]
	DNA-PK, XPD		[155]
	XPC		[156]
Bile acids			
deoxycholate	MUTYH, OGG1	mRNA reduced	[157]
	BRCA1		[158]
lithocholate	DNA POLB		[159]
Lipid peroxidation compounds			
4-hydroxy-2-nonenal (4-HNE)	Nuc. Excision Repair	NER protein adducts	[160]
Malondialdehyde	Nuc. Excision Repair	NER protein adducts	[161]

Carcinogens	Inhibit DNA Repair Gene	Mechanism Shown	Ref.
Oxidative stress	MisMatch Repair	Oxidative damage to MMR proteins	[162]
	ERCC1 protein	Oxidative attack	[163]
	OGG1 protein	Degraded by calpain	[164]
Gamma irrad.	OGG1, XRCC1	mRNA reduced	[165]
Benzo(a)pyrene	BRCA1	miR-638 increased	[166]
Methylcholanthrene/ diethylnitrosamine	BRCA1, ERCC1, XRCC1, MLH1		[167]
Styrene	XRCC1, OGG1, XPC	mRNA reduced	[168]
Aristolochic acid	P53, PARP1, OGG1, ERCC1, MGMT	mRNA reduced	[169]
Antimony	XPE	mRNA reduced	[170]
Nickel	MGMT	Promoter methylation	[171]

Table 8. Examples of carcinogenic agents that cause reduced expression of DNA repair genes

16. Polyphenols can epigenetically increase expression of DNA repair genes

Some polyphenols affect expression of many genes, including DNA repair genes, through epigenetic alterations, as reviewed by Link et al. [172]. Examples of DNA repair genes expression increased by epigenetic alteration are listed in Table 9.

Phytochemical	Plant source	Mechanism	Targeted DNA Repair Genes	Ref.
Epigalocatechin-3-gallate	Green tea	Reversal of CpG island methylation	MGMT, MLH1	[173]
Dihydrocoumarin	Yellow sweet clover	p53 acetylation	P53	[174]
Genistein	Soy	Reversal of CpG island methylation	MGMT	[173]
Genistein	Soy	Histone acetylation	P53	[175]

Table 9. Examples of phytochemicals that increase expression of DNA repair genes by an epigenetic mechanism

17. Possible protection against cancer by phytochemicals that increase DNA repair by unknown mechanisms

A recent review article by Collins et al. [176] summarizes some examples of micronutrients that affect DNA repair gene expression, though by unknown mechanisms. Table 10 lists such phytochemicals, without defined mechanisms, that increase DNA repair gene expression, along with commonly known foods that are high in those phytochemicals [177, 178, 179].

Phytochemical (test system)	Examples of foods high in nutrient	Increased DNA Repair Gene Expression	Ref.
Ellagic acid (mice)	Raspberries, pomeganate	*XPA, ERCC5, DNA Ligase 3*	[180]
Silymarin (cells *in vitro*)	Artichoke, milk thistle	*MGMT*	[181]
Curcumin (cells *in vitro*)	Turmeric	*MGMT*	
Chlorogenic acid (cells *in vitro*)	Blueberries, coffee, sunflower seeds, artichoke	*PARP*	
Caffeic acid (cells *in vitro*)	coffee, cranberry, carrot	*PMS2*	
m-coumaric acid (cells *in vitro*)	olives (and metabolite of caffeic acid)	*PARP, PMS2*	[182]
3-(m-hydroxyphenyl) propionic acid (cells *in vitro*)	(major metabolite of caffeic acid and degradation product of proanthocyanidins in chocholate)	*PARP, PMS2*	

Table 10. Examples of phytochemicals that increase expression of DNA repair genes by unknown mechanisms

Bernstein et al. [182] evaluated antioxidants based on their ability to increase DNA repair proteins PARP-1 and Pms2 *in vitro*. They tested 19 anti-oxidant compounds and of these 19 compounds only chlorogenic acid and its metabolic products: chlorogenic acid, caffeic acid, *m*-coumaric acid and 3-(*m*-hydroxyphenyl) propionic acid, increased expression of the two tested DNA repair genes in HCT-116 cells (Table 10).

Chlorogenic acid (CGA) (high in blueberries, coffee, sunflower seeds, artichoke) [177, 183, 184] was then tested as a preventive agent in the recently devised diet-related mouse model of colon cancer [60]. As described above in the section Exogenous DNA damaging agents in colorectal cancer, deoxycholic acid (DCA), a DNA damaging agent, at levels present after a high fat diet, can cause colorectal cancer. When DCA is added to the diet of wild-type mice to raise the level of DCA in the mouse feces to the level in feces of humans on a high fat diet, by 10 months of feeding 94% of the mice develop tumors in their colons with 56% developing colonic adenocarcinomas [60]. This mouse model develops tumors solely in the colon, phenotypically similar to development of colon cancer in humans. When CGA, equivalent to 3 cups of coffee a day for humans, was added to the DCA supplemented diet it was dramatically protective against development of colon cancer, reducing incidence of colon cancer significantly from 56% to 18% [60].

18. Targeting of chemotherapeutic agents to cancers deficient in DNA repair

As discussed above, DNA repair deficiency often arises early in progression to cancer and can give rise to genomic instability, a general feature of cancers. If cancer cells are deficient in DNA repair they are likely to be more vulnerable than normal cells to inactivation by DNA damaging agents. This vulnerability of cancer cells can be exploited to the benefit of the patient. Some of the most clinically effective chemotherapeutic agents currently used in cancer treatment are DNA damaging agents, and their therapeutic effectiveness appears to often depend on deficient DNA repair in cancer cells.

In the next four sections we discuss repair deficiencies in cancer cells that can be effectively targeted by DNA damaging chemotherapeutic agents. In addition, deficiency in a DNA repair pathway that arises during tumor development may make cancer cells more reliant on a remaining reduced set of DNA repair pathways for survival. Recent studies indicate that drugs that inhibit one of these alternative pathways in such cancers cells can be useful in cancer therapy. Targeting cancer cells having a repair deficiency with specific DNA damaging agents, or with agents that inhibit alternative repair pathways, offers a new promising approach for treating a variety of cancers.

19. Targeting cancers deficient in BRCA1

The BRCA1 (breast cancer 1 early onset) protein is employed in an important DNA repair pathway, homologous recombinational repair (HRR). This pathway removes a variety of types of DNA damages, and is the only pathway that can accurately remove double-strand damages such as double-strand breaks and inter-strand cross-links. BRCA1 also has other functions related to preservation of genome integrity (reviewed by Yun and Hiom [185]). Individuals with a germ-line inherited defect in the *BRCA1* gene are at increased risk of breast, ovarian and other cancers. In addition to inherited germ-line defects in *BRCA1*, deficiencies in expression of this gene may arise in somatic cells either by mutation or by epimutation during progression to sporadic (non-germline) cancer.

Patients with a variety of types of cancer are treated effectively with chemotherapeutic agents that cause double-strand breaks (e.g. the topoisomerase inhibitor etoposide), or cause inter-strand cross-links (e.g. the platinum compound cisplatin). These damages can cause cancer cells to undergo apoptosis (a form of cell death). However, patients treated with these agents often prove to be intrinsically resistant, or develop resistance during treatment. Quinn et al. [186] demonstrated that BRCA1 expression is necessary for such resistance. This finding suggests that BRCA1-mediated DNA repair can protect cancer cells from therapeutic

DNA damaging drugs. Thus, although high expression of BRCA1 may be initially beneficial to the individual by reducing the risk of developing cancer, it also may be detrimental once cancer has developed by counteracting the therapeutic effect of DNA-damaging agents targeted to the cancer cells.

Patients with non-small cell lung cancer (NSLC) are often treated with DNA cross-linking platinum therapeutic compounds such as cisplatin, carboplatin or oxaliplatin. NSCLC is the leading cause of cancer deaths worldwide, and almost 70% of patients with NSCLC have locally advanced or metastatic disease at diagnosis. Improved survival after platinum-containing chemotherapy in metastatic NSCLC correlates with low BRCA1 expression in the primary tumor [187, 188]. This finding indicates that low BRCA1-mediated DNA repair is detrimental to the cancer upon treatment, and thus beneficial to the patient. BRCA1 likely protects cancer cells by participating in a pathway that removes the potentially lethal DNA cross-links introduced by the platinum drugs. Since low BRCA1 expression in the tumor appears to be beneficial to the patient, Taron et al. [187] and Papadek et al. [188] concluded that BRCA1 expression is potentially an important tool for use in cancer management and should be assessed for predicting chemosensitivity and tailoring chemotherapy in lung cancer.

Over 90% of ovarian cancers appear to arise sporadically in somatic cells and are associated with BRCA1 dysfunction. Weberpals et al. [189] showed for patients having sporadic ovarian cancer treated with platinum drugs, the median survival was longer for patients with lower expression of BRCA1 vs. higher BRCA1 expression (46 vs. 33 months).

20. Targeting cancers deficient in ERCC1

ERCC1 (Excision Repair Cross-Complementaion group 1) is a key protein needed to remove platinum adducts and repair inter- and intra-strand cross-links [190]. ERCC1 dimerizes with XPF (xeroderma pigmentosum complementation group F) protein to form a complex that can excise damaged DNA. Over-expression of ERCC1 is associated with cellular resistance to platinum compounds, whereas ERCC1 down-regulation sensitizes cells to cisplatin [191, 192].

Cisplatin has made a major impact in the chemotherapeutic treatment of testicular cancer. Over 90% of patients with newly diagnosed testicular germ cell cancer, and 70 to 80% of patients with metastatic testicular cancer, can be cured using cisplatin based combination chemotherapy [193]. Hypersensitivity of testicular cancer to cisplatin appears to be due to low levels of the three NER proteins ERCC1, XPF and XPA [194].

Simon et al. [195] evaluated ERCC1 mRNA expression in lung tumors as a predictor of survival of NSCLC patients. They found that patients with relatively low ERCC1 mRNA expression had poor overall survival. This finding suggests that low ERCC1-mediated DNA repair allows DNA damages to persist and give rise to carcinogenic mutations. However, they also noted that those NSCLC tumors with relatively low ERCC1 expression responded better to platinum based therapy. Lord et al. [196] found that low *ERCC1* mRNA expression in the primary tumor correlates with prolonged survival after cisplatin plus gemitabine chemotherapy in NSCLC. Median overall survival with low ERCC1 expression tumors was 61.6 weeks compared to 20.4 weeks for patients with high expression tumors.

Zhou et al. [197] reported that a particular genetic polymorphism that alters ERCC1 mRNA level predicts overall survival in advanced NSCLC patients treated with platinum based chemotherapy. Olaussen et al. [198] found that patients with completely resected NSCLC tumors that were ERCC1-negative benefited from adjuvant cisplatin-based chemotherapy, whereas patients with ERCC1-positive tumors did not benefit. They suggested that determination of ERCC1 expression in NSCLC cells before chemotherapy can make a contribution as an independent predictor of the effect of adjuvant chemotherapy. Papadaki et al. [188] found that *ERCC1* mRNA level in the primary tumor of patients with metastatic NSCLC could predict the effectiveness of cisplatin based chemotherapy. Low *ERCC1* mRNA level was significantly associated with higher response rate, longer median progression-free survival and median overall survival. Leng et al. [199] found that patients with ERCC1 negative expression had a longer progression free survival and overall survival than ERCC1 positive patients after receiving platinum based adjuvant therapy. Thus *ERCC1* mRNA level, like *BRCA1* mRNA level (discussed above), in the primary tumor at the time of diagnosis could be used to predict platinum sensitivity of NSCLC.

ERCC1 expression also appears to have predictive significance for ovarian cancer. Dabholkar et al. [200] found in ovarian tumor tissues that *ERCC1* mRNA expression levels were higher in patients who were resistant to platinum based therapy than in those patients who responded to such therapy. Kang et al. [201] observed that a particular polymorphism of the ERCC1 gene sequence was associated with clinical outcome of platinum based chemotherapy in patients with ovarian cancer. Weberpals et al. [189] also showed for ovarian cancer patients that higher ERCC1 mRNA level, alone, or especially in combination with higher *BRCA1* mRNA level in the tumor, predicted shorter overall patient survival after platinum therapy.

ERCC1 protein expression is often reduced within colon cancers and in a field defect surrounding these cancers [112]. For metastatic colorectal cancer patients receiving combination oxaliplatin and fluorouracil chemotherapy, lower *ERCC1* mRNA expression in the tumor predicts longer survival [202]. Viguier et al. [203] found that a particular ERCC1 genetic polymorphism predicts a better tumor response to oxaliplatin/5-fluorouracil combination chemotherapy in patients with metastatic colorectal cancer.

Low *ERCC1* mRNA levels also predict better response and survival for gastric cancer patients [204] and bladder cancer patients [205] receiving cisplatin-based chemotherapy.

Thus numerous studies involving cancer of the testis, lung, ovary, colon, stomach and bladder indicated that platinum based chemotherapy can enhance patient outcome when targeted specifically to tumors with low ERCC1 expression. Such tumors have diminished ability to repair the DNA damages, particularly the cross-links, induced in the tumors by the platinum compound.

21. Targeting cancers deficient in MGMT

Alkylating agents, including chloroethylnitrosoureas, procarbazine and temozolomide, are commonly used to treat malignant brain tumors. These agents cause DNA damage by adding alkyl groups to DNA. Such damages may then be repaired or, if unrepaired, trigger cell death. As an example, temozolomide methylates DNA at several sites generating mainly N^7-methylguanine and N^3-methyladenine adducts, which constitute nearly 90% of the total methylation events. However these adducts are efficiently removed and accurately replaced by the base excision repair pathway, and thus have low cytotoxic potential. About 5 to 10% of the methylation events caused by temozolomide produce O^6-methylguanine which is cytotoxic, and this adduct accounts for the beneficial therapeutic effect of temozolomide and other alkylating agents on malignant brain tumors.

O^6-methylguanine methyltransferase (MGMT) is a DNA repair enzyme that rapidly reverses alkylation (including methylation) at the O^6 position of guanine, thus neutralizing the cytotoxic effects of chemotherapeutic alkylating agents such as temozolomide. High MGMT activity in tumor tissue is associated with resistance to alkylating agents. MGMT activity is controlled by a promoter sequence, and methylation of the CpG island in the promoter silences the gene in cancer cells, so that these cells no longer produce MGMT. In addition, as described above, an increased level of miR-181d can also decrease MGMT expression and help the ability of temozolomide to give a beneficial therapeutic effect [78].

Esteller et al. [206] showed that methylation of the *MGMT* promoter increases the responsiveness of the gliomas (brain tumors) to chemotherapeutic alkylating agents, leading to regression of the tumors and prolonged overall and disease free survival. Paz et al. [207] showed that hypermethylation of CpG islands within the promoter sequence of the *MGMT* gene predicts a better clinical response to temozolomide in primary gliomas. They considered that their results might open up possibilities for more customized treatments of human brain tumors. Hegi et al. [208] demonstrated a significantly improved clinical outcome in patients with malignant glioma who had a methylated *MGMT* promoter and were treated with temozolomide. The 18-month survival rate was 62% among patients with a methylated *MGMT* promoter compared with only 8% in the absence of promoter methylation. Hegi et al. [209] reviewed further evidence that *MGMT* promoter methylation is associated with improved progression-free and overall survival in malignant glioma patients treated with alkylating agents. They also discussed strategies to overcome MGMT-mediated chemoresistance that are currently under investigation. Upon reviewing the relevant evidence, Weller et al. [210] concluded that *MGMT* promoter methylation is the key mechanism of *MGMT* gene silencing, and could be used as a biomarker for predicting a favorable outcome in patients with malignant glioma who are exposed to alkylating chemotherapy. They considered that this biomarker is on the verge of entering clinical decision-making.

22. Targeting cancers with a repair deficiency using a PARP inhibitor; synthetic lethality

If a tumor is deficient in an essential protein component of a DNA repair pathway, the cancer cells would likely be more reliant on remaining DNA repair pathways for survival. Drugs that inhibit one of these alternative pathways, in principle, might prove to be useful in cancer therapy by selectively killing the cancer cells. An example of such an approach is the use of poly(ADP-ribose) polymerase [PARP] inhibitors against tumors that are deficient in BRCA1 or BRCA2 [211]. This approach has provided proof-of-concept for an anticancer strategy termed "synthetic lethality." By this strategy the inhibition of a particular repair pathway in cancer cells that are already deficient in another repair pathway preferentially induces greater toxicity in repair deficient cancer cells than in normal non-cancer cells. Current research guided by this strategy is directed at finding new agents that inactivate protein components of major repair pathways, and thus could be targeted against cancers that are already deficient in another repair pathway [212].

A germ-line mutation in one BRCA1 or BRCA2 allele substantially increases the risk of developing several cancers, including breast, ovarian, and prostate cancer. Diploid cells heterozygous for either a BRCA1 or a BRCA2 mutant allele may lose expression of the remaining wild-type allele, resulting in deficient homologous recombinational repair. This loss causes an increase in unrepaired DSBs that can lead to mutations (through compensatory inaccurate repair) and chromosomal aberrations that drive carcinogenesis. Inactivation of the wild-type allele in the cell lineage leading to the tumor is thought to be an obligate step in this carcinogenesis pathway, a step that does not occur in the normal non-cancer tissues of the patient.

The deficiency in homologous recombinational repair is thus specific to the tumor, and can be exploited by employing PARP inhibitors. Ordinarily, single-strand breaks (SSBs), as distinct from DSBs, are repaired by the base excision repair pathway, in which the enzyme PARP1 plays a key role. The inhibition of PARP1 leads to the accumulation of DNA SSBs. Unrepaired SSBs can give rise to DSBs at replication forks during DNA replication. Thus PARP inhibition in tumor cells with deficient homologous recombinational repair (because of the absence of BRCA1 or BRCA2) generates unrepaired SSBs that are likely to cause an overwhelming accumulation of DSBs leading to tumor cell death. In contrast, the normal tissues of a patient consists of cells that are heterozygous for a BRCA1 or BRCA2 mutant allele and therefore retain homologous recombinational repair function, and have a sensitivity to PARP inhibitors similar to that of wild-type cells. Thus PARP inhibition induces selective tumor cell killing while sparing normal cells.

Fong et al. [213] conducted a preliminary clinical evaluation of the oral PARP inhibitor olaparib. They observed that 63% of patients carrying BRCA1 or BRCA2 mutations who had ovarian, breast or prostate cancer had a clinical benefit from treatment with olaparib with few adverse side effects. This is an example of the concept of "synthetic lethality" which occurs when there is a potent lethal synergy between two otherwise non-lethal events. The two events in this case are (1) a specific PARP inhibitor blocks repair of SSBs causing an increase in SSBs leading to an increase in DSBs; and (2) a tumor restricted genetic loss of function or homologous recombinational repair that is ordinarily needed to accurately repair these DSBs.

A subsequent trial of olaparib in BRCA mutation-associated breast cancer demonstrated objective positive response rates of 41%, again with limited toxicity [214]. About 10% of women with ovarian cancer carry a *BRCA1* or *BRCA2* mutant allele. Audeh et al. [215] showed that the oral PARP inhibitor olaparib has antitumor activity in women carriers of *BRCA1* or *BRCA2* alleles who have ovarian cancer. The objective positive response rate was 33%.

23. Overview of the role of DNA damage and repair in carcinogenesis

In this section we present a brief overview of the relationship of DNA damage and repair to carcinogenesis, and the implications of this relationship for strategies of prevention and therapy, emphasizing the evidence reviewed above. Carcinogenesis is generally viewed as a Darwinian process that occurs in a somatic cell lineage by mutation or epimutation and natural selection. Natural selection operates on the basis of the adaptive benefit to individual cells in the lineage of more rapid cell division or higher resistance to cell death (apoptosis) than occurs in neighboring cells. Most of the random mutations and epimutations that arise during progression to cancer are likely to be disadvantageous or neutral from the prospective of the emerging cancerous cells, and only those that promote more rapid overall growth are advantageous. The cell lineage that ultimately becomes a cancer probably passes through a series of evolutionary pre-cancerous stages involving sequential rounds of mutation/epimutation and selection [216]. The initial stage is probably a lineage of cells with a small selective advantage that forms an early field within a tissue. Within this defective field successive mutation and selection events occur which finally give rise to an invasive and then metastatic cell lineage. During this process the cell lineage acquires the hallmarks of cancer (summarized by Hanahan and Weinberg [217]). These include: sustaining proliferative signaling, evading growth suppressors, resisting cell death, enabling replicative immortality, inducing angiogenesis, reprogramming energy metabolism, and evading immune destruction.

Mutations arise from unrepaired DNA damages, either by translesion synthesis during DNA replication or by inaccurate repair of DNA damages, as in the inaccurate process of non-homologous end joining of double-strand breaks. Mutations may also arise by spontaneous replication errors without the intervention of DNA damage, but this source of mutation is likely less frequent than mutations caused by DNA damage. The primary cause(s) of epimutations (such as CpG island methylations) are not well understood, but evidence suggests that epimutations arise during the repair processes that remove DNA damages. The sources of DNA damage underlying carcinogenesis can be extrinsic or intrinsic. Epidemiologic evidence suggests that a large proportion of the DNA damages contributing to cancer arise from extrinsic stressful conditions, including such factors as smoking, high fat diet, certain infections and UV light exposure. The possible contribution from intrinsic causes, such as free radical production during normal metabolism, have not been assessed. A pervasive characteristic of human tumors is genomic instability [217]. A likely major source of this instability is loss of DNA repair capability. Germ line mutations in DNA repair genes generally lead to syndromes characterized by a greatly increased risk of cancer. The majority of cancers arise sporadically, i.e. are not primarily due to germ line mutations. A frequent characteristic of sporadic cancers is loss of expression of one or more DNA repair proteins through epigenetic silencing. The several different DNA repair pathways that occur in mammalian cells each specialize in removing different types of damage, but they are also partially overlapping. Thus reduction of a particular repair pathway may have different carcinogenic consequences from loss of another repair pathway [218]. However, the deleterious effect of loss of one pathway may be partially ameliorated by another functioning pathway.

This general view of the role of DNA damage and repair in carcinogenesis has implications for the prevention and treatment of cancer. Cancer incidence could be substantially reduced by a general avoidance of the known sources of DNA damage such as smoking. In addition to avoiding DNA damage, it should also be beneficial to increase DNA repair, or at least to avoid extrinsic factors that decrease repair. The factors affecting repair capability are less well studied than those causing DNA damage, but several are known, and a significant benefit may be derived from considering such factors as well.

The finding that DNA repair deficiency is a common feature of cancers, and is perhaps the underlying cause of the genetic instability of cancers, has implications for therapy. If a cancer is composed of cells deficient in DNA repair, it is, in principle, vulnerable to agents that cause DNA damage. Thus a chemotherapeutic DNA damaging agent can be targeted to cancers that lack the capability to repair the particular type of DNA damage caused by the agent. This can lead to a level of DNA damages that overwhelms the defenses of the cancer cells and causes their death. Non-cancerous cells with normal repair would not be targeted. Thus the toxicity of such DNA damaging agents to the treated patient would be limited. A dramatic example of such targeted therapy is the high cure rate of testicular cancer due to a defect in the ability of the cancer cells to repair DNA inter-strand cross-links, and the use of cross-linking platinum compounds to kill such cells.

Another strategy, which is currently the basis for numerous ongoing clinical trials, involves synthetic lethality. By this strategy cancers that are deficient in one DNA repair pathway can be made more vulnerable to DNA damage by treatment with agents that inhibit an additional repair pathway. Promising clinical results, so far, have been obtained in the treatment of patients with breast and ovarian cancer due to an inherited genetic defect in the homologous recombinational repair pathway. Such cancers are deficient in the ability to repair double-strand breaks. Treatment of these cancers with an agent that interferes with another pathway that ordinarily repairs single-strand breaks allows such breaks to accumulate and to be converted to double-strand breaks during DNA replication. The increase in double-strand breaks appears to overwhelm the cancer cells, while sparing normal cells, thus providing positive clinical benefit to the patient without much toxicity.

Author details

Carol Bernstein[1*], Anil R. Prasad[2], Valentine Nfonsam[3] and Harris Bernstein[4]

*Address all correspondence to: bernstein324@yahoo.com

1 Research Service Line, Southern Arizona Veterans Affairs Health Care System, Tucson, AZ, USA

2 Department of Pathology, University of Arizona, Tucson, AZ, USA

3 Department of Surgery, University of Arizona, Tucson, AZ, USA

4 Department of Cellular and Molecular Medicine, University of Arizona, Tucson, AZ, USA

References

[1] Gregg SQ, Robinson AR, Niedernhofer LJ. Physiological consequences of defects in ERCC1-XPF DNA repair endonuclease. DNA Repair (Amst) 2011;10(7) 781-791.

[2] Agrelo R, Cheng WH, Setien F, Ropero S, Espada J, Fraga MF, Herranz M, Paz MF, Sanchez-Cespedes M, Artiga J, Guerrero D, Castells A, von Kobbe C, Bohr VA, Estel-ler M. Epigenetic inactivation of the premature aging Werner syndrome gene in human cancer. Proc Natl Acad Sci U S A 2006;103(23) 8822-8827.

[3] Dreesen O, Stewart CL. Accelerated aging syndromes, are they relevant to normal human aging? Aging (Albany NY) 2011;3(9) 889-895.

[4] Wood LD, Parsons DW, Jones S, Lin J, Sjöblom T, Leary RJ, Shen D, Boca SM, Barber T, Ptak J, Silliman N, Szabo S, Dezso Z, Ustyanksky V, Nikolskaya T, Nikolsky Y, Karchin R, Wilson PA, Kaminker JS, Zhang Z, Croshaw R, Willis J, Dawson D, Shipit-sin M, Willson JK, Sukumar S, Polyak K, Park BH, Pethiyagoda CL, Pant PV, Ballin-ger DG, Sparks AB, Hartigan J, Smith DR, Suh E, Papadopoulos N, Buckhaults P, Markowitz SD, Parmigiani G, Kinzler KW, Velculescu VE, Vogelstein B. The genomic landscapes of human breast and colorectal cancers. Science 2007;318(5853) 1108-1113.

[5] Cancer Genome Atlas Network. Comprehensive molecular characterization of human colon and rectal cancer. Nature 2012;487(7407) 330-337. doi: 10.1038/nature11252.

[6] Leary RJ, Lin JC, Cummins J, Boca S, Wood LD, Parsons DW, Jones S, Sjöblom T, Park BH, Parsons R, Willis J, Dawson D, Willson JK, Nikolskaya T, Nikolsky Y, Ko-pelovich L, Papadopoulos N, Pennacchio LA, Wang TL, Markowitz SD, Parmigiani G, Kinzler KW, Vogelstein B, Velculescu VE. Integrated analysis of homozygous de-letions, focal amplifications, and sequence alterations in breast and colorectal can-cers. Proc Natl Acad Sci USA 2008;105(42) 16224-16229.

[7] Schnekenburger M, Diederich M. Epigenetics Offer New Horizons for Colorectal Cancer Prevention. Curr Colorectal Cancer Rep 2012;8(1) 66-81.

[8] O'Hagan HM, Mohammad HP, Baylin SB. Double strand breaks can initiate gene silencing and SIRT1-dependent onset of DNA methylation in an exogenous promoter CpG island. PLoS Genet 2008;4(8) e1000155.

[9] Malanga M, Althaus FR. The role of poly(ADP-ribose) in the DNA damage signaling network. Biochem Cell Biol 2005;83(3) 354-64.

[10] Gottschalk AJ, Timinszky G, Kong SE, Jin J, Cai Y, Swanson SK, Washburn MP, Florens L, Ladurner AG, Conaway JW, Conaway RC. Poly(ADP-ribosyl)ation directs recruitment and activation of an ATP-dependent chromatin remodeler. Proc Natl Acad Sci U S A 2009;106(33) 13770-13774.

[11] Lin JC, Jeong S, Liang G, Takai D, Fatemi M, Tsai YC, Egger G, Gal-Yam EN, Jones PA. Role of nucleosomal occupancy in the epigenetic silencing of the MLH1 CpG island. Cancer Cell 2007;12(5) 432-444.

[12] Tabish AM, Poels K, Hoet P, Godderis L. Epigenetic Factors in Cancer Risk: Effect of Chemical Carcinogens on Global DNA Methylation Pattern in Human TK6 Cells. PLoS One 2012;7(4) e34674. Epub 2012 Apr 11.

[13] Stefanska B, Karlic H, Varga F, Fabianowska-Majewska K, Haslberger AG. Epigenetic mechanisms in anti-cancer actions of bioactive food components-the implications in cancer prevention. Br J Pharmacol 2012 Apr 27. doi: 10.1111/j. 1476-5381.2012.02002.x. [Epub ahead of print]

[14] Jacinto FV, Esteller M. Mutator pathways unleashed by epigenetic silencing in human cancer. Mutagenesis 2007;22(4) 247-253.

[15] De Bont R, van Larebeke N. Endogenous DNA damage in humans: a review of quantitative data. Mutagenesis 2004;19(3) 169-185. http://mutage.oxfordjournals.org/content/19/3/169.full.pdf+html

[16] Helbock HJ, Beckman KB, Shigenaga MK, Walter PB, Woodall AA, Yeo HC, Ames BN. DNA oxidation matters: the HPLC-electrochemical detection assay of 8-oxo-deoxyguanosine and 8-oxo-guanine. Proc Natl Acad Sci U S A. 1998;95(1) 288-293.

[17] Nakamura J, Swenberg JA. Endogenous apurinic/apyrimidinic sites in genomic DNA of mammalian tissues. Cancer Res 1999;59(11) 2522-2526.

[18] Fraga CG, Shigenaga MK, Park JW, Degan P, Ames BN. Oxidative damage to DNA during aging: 8-hydroxy-2'-deoxyguanosine in rat organ DNA and urine. Proc Natl Acad Sci U S A 1990;87(12) 4533-4537.

[19] Ames BN, Shigenaga MK, Hagen TM. Oxidants, antioxidants, and the degenerative diseases of aging. Proc Natl Acad Sci U S A 1993;90(17) 7915-7922.

[20] Foksinski M, Rozalski R, Guz J, Ruszkowska B, Sztukowska P, Piwowarski M, Klungland A, Olinski R. Urinary excretion of DNA repair products correlates with

metabolic rates as well as with maximum life spans of different mammalian species. Free Radic Biol Med 2004;37(9) 1449-1454.

[21] Lindahl T. DNA repair enzymes acting on spontaneous lesions in DNA. In: Nichols WW and Murphy DG (eds.) DNA Repair Processes. Symposia Specialists, Miami, 1977. p225-240.

[22] Tice, R.R., and Setlow, R.B. DNA repair and replication in aging organisms and cells. In: Finch EE and Schneider EL (eds.) Handbook of the Biology of Aging. Van Nostrand Reinhold, New York. 1985. p173-224.

[23] Lindahl T, Nyberg B. Rate of depurination of native deoxyribonucleic acid. Biochemistry 1972;11(19) 3610-3618.

[24] Lindahl T. Instability and decay of the primary structure of DNA. Nature 1993;362(6422) 709-715.

[25] Nakamura J, Walker VE, Upton PB, Chiang SY, Kow YW, Swenberg JA. Highly sensitive apurinic/apyrimidinic site assay can detect spontaneous and chemically induced depurination under physiological conditions. Cancer Res 1998;58(2) 222-225.

[26] Haber JE. DNA recombination: the replication connection. Trends Biochem Sci 1999;24(7) 271-275.

[27] Vilenchik MM, Knudson AG. Endogenous DNA double-strand breaks: production, fidelity of repair, and induction of cancer. Proc Natl Acad Sci U S A 2003;100(22) 12871-12876. http://www.ncbi.nlm.nih.gov/pmc/articles/PMC240711/?tool=pubmed

[28] Nam EA, Cortez D. ATR signalling: more than meeting at the fork. Biochem J 2011;436(3) 527-536.

[29] Maher RL, Branagan AM, Morrical SW. Coordination of DNA replication and recombination activities in the maintenance of genome stability. J Cell Biochem 2011;112(10) 2672-82. doi: 10.1002/jcb.23211.

[30] Holmes GE, Bernstein C, Bernstein H. Oxidative and other DNA damages as the basis of aging: a review. Mutat Res 1992;275(3-6) 305-315.

[31] Hoeijmakers JH. DNA damage, aging, and cancer. N Engl J Med 2009;361(15) 1475-1485.

[32] Ou J, DeLany JP, Zhang M, Sharma S, O'Keefe SJ. Association between low colonic short-chain fatty acids and high bile acids in high colon cancer risk populations. Nutr Cancer 2012;64(1) 34-40.

[33] O'Keefe SJ, Kidd M, Espitalier-Noel G, Owira P. Rarity of colon cancer in Africans is associated with low animal product consumption, not fiber. Am J Gastroenterol 1999;94(5) 1373-1380.

[34] Maskarinec G, Noh JJ. The effect of migration on cancer incidence among Japanese in Hawaii. Ethn Dis 2004 Summer;14(3) 431-439.

[35] Jemal A, Bray F, Center MM, Ferlay J, Ward E, Forman D. Global cancer statistics. CA Cancer J Clin 2011;61(2) 69-90.

[36] Alberg AJ, Ford JG, Samet JM; American College of Chest Physicians. Epidemiology of lung cancer: ACCP evidence-based clinical practice guidelines (2nd edition). Chest 2007;132(3 Suppl) 29S-55S.

[37] Cunningham FH, Fiebelkorn S, Johnson M, Meredith C. A novel application of the Margin of Exposure approach: segregation of tobacco smoke toxicants. Food Chem Toxicol 2011;49(11) 2921-2933.

[38] Liu XY, Zhu MX, Xie JP. Mutagenicity of acrolein and acrolein-induced DNA adducts. Toxicol Mech Methods 2010;20(1) 36-44.

[39] Alexandrov K, Rojas M, Satarug S. The critical DNA damage by benzo(a)pyrene in lung tissues of smokers and approaches to preventing its formation. Toxicol Lett 2010;198(1) 63-68.

[40] Speit G, Merk O. Evaluation of mutagenic effects of formaldehyde in vitro: detection of crosslinks and mutations in mouse lymphoma cells. Mutagenesis 2002;17(3) 183-187.

[41] Grogan D, Jinks-Robertson S. Formaldehyde-induced mutagenesis in Saccharomyces cerevisiae: molecular properties and the roles of repair and bypass systems. Mutat Res 2012;731(1-2) 92-98.

[42] Pu X, Kamendulis LM, Klaunig JE. Acrylonitrile-induced oxidative stress and oxidative DNA damage in male Sprague-Dawley rats. Toxicol Sci 2009;111(1) 64-71.

[43] Marnett LJ. Oxy radicals, lipid peroxidation and DNA damage. Toxicology 2002;181-182 219-222.

[44] Koturbash I, Scherhag A, Sorrentino J, Sexton K, Bodnar W, Swenberg JA, Beland FA, Pardo-Manuel Devillena F, Rusyn I, Pogribny IP. Epigenetic mechanisms of mouse interstrain variability in genotoxicity of the environmental toxicant 1, 3-butadiene. Toxicol Sci. 2011;122(2) 448-456.

[45] Garcia CC, Angeli JP, Freitas FP, Gomes OF, de Oliveira TF, Loureiro AP, Di Mascio P, Medeiros MH. (13C2)-Acetaldehyde promotes unequivocal formation of 1, N2-propano-2'-deoxyguanosine in human cells. J Am Chem Soc 2011;133(24) 9140-9143.

[46] Tompkins EM, McLuckie KI, Jones DJ, Farmer PB, Brown K. Mutagenicity of DNA adducts derived from ethylene oxide exposure in the pSP189 shuttle vector replicated in human Ad293 cells. Mutat Res 2009;678(2) 129-137.

[47] Gelmont D, Stein RA, Mead JF. Isoprene-the main hydrocarbon in human breath. Biochem Biophys Res Commun 1981;99(4) 1456-1460.

[48] Senthilmohan ST, McEwan MJ, Wilson PF, Milligan DB, Freeman CG. Real time analysis of breath volatiles using SIFT-MS in cigarette smoking. Redox Rep 2001;6(3) 185-187.

[49] Fabiani R, Rosignoli P, De Bartolomeo A, Fuccelli R, Morozzi G. DNA-damaging ability of isoprene and isoprene mono-epoxide (EPOX I) in human cells evaluated with the comet assay. Mutat Res 2007;629(1) 7-13.

[50] Phillips DH. Smoking-related DNA and protein adducts in human tissues. Carcinogenesis 2002;23(12) 1979-2004.

[51] Giovannucci E, Martínez ME. Tobacco, colorectal cancer, and adenomas: a review of the evidence. J Natl Cancer Inst 1996;88(23) 1717-1730.

[52] de Kok TM, van Maanen JM. Evaluation of fecal mutagenicity and colorectal cancer risk. Mutat Res 2000; 463(1) 53-101.

[53] Pearson JR, Gill CI, Rowland IR. Diet, fecal water, and colon cancer--development of a biomarker. Nutr Rev 2009;67(9) 509-526.

[54] Bernstein H, Bernstein C, Payne CM, Dvorak K. Bile acids as endogenous etiologic agents in gastrointestinal cancer. World J Gastroenterol 2009;15(27) 3329-3340.

[55] Payne CM, Crowley-Skillicorn C, Bernstein C, Holubec H, Moyer MP, Bernstein H. Hydrophobic bile acid-induced micronuclei formation, mitotic perturbations, and decreases in spindle checkpoint proteins: relevance to genomic instability in colon carcinogenesis. Nutr Cancer 2010;62(6) 825-840.

[56] Bernstein H, Bernstein C, Payne CM, Dvorakova K, Garewal H. Bile acids as carcinogens in human gastrointestinal cancers. Mutat Res 2005;589(1) 47-65.

[57] Hursting SD, Thornquist M, Henderson MM. Types of dietary fat and the incidence of cancer at five sites. Prev Med 1990;19(3) 242-253.

[58] Reddy BS, Hanson D, Mangat S, Mathews L, Sbaschnig M, Sharma C, Simi B. Effect of high-fat, high-beef diet and of mode of cooking of beef in the diet on fecal bacterial enzymes and fecal bile acids and neutral sterols. J Nutr 1980;110(9) 1880-1887.

[59] Stadler J, Stern HS, Yeung KS, McGuire V, Furrer R, Marcon N, Bruce WR. Effect of high fat consumption on cell proliferation activity of colorectal mucosa and on soluble faecal bile acids. Gut 1988;29(10) 1326-1331.

[60] Bernstein C, Holubec H, Bhattacharyya AK, Nguyen H, Payne CM, Zaitlin B, Bernstein H. Carcinogenicity of deoxycholate, a secondary bile acid. Arch Toxicol 2011;85(8) 863-871

[61] Handa O, Naito Y, Yoshikawa T. Redox biology and gastric carcinogenesis: the role of Helicobacter pylori. Redox Rep 2011;16(1) 1-7.

[62] Wei L, Gravitt PE, Song H, Maldonado AM, Ozbun MA. Nitric oxide induces early viral transcription coincident with increased DNA damage and mutation rates in human papillomavirus-infected cells. Cancer Res 2009;69(11) 4878-4884.

[63] Goldman A, Shahidullah M, Goldman D, Khailova L, Watts G, Delamere N, Dvorak K. A novel mechanism of acid and bile acid-induced DNA damage involving Na+/H + exchanger: implication for Barrett's oesophagus. Gut 2010;59(12) 1606-1616.

[64] Smela ME, Hamm ML, Henderson PT, Harris CM, Harris TM, Essigmann JM. The aflatoxin B(1) formamidopyrimidine adduct plays a major role in causing the types of mutations observed in human hepatocellular carcinoma. Proc Natl Acad Sci U S A 2002;99(10) 6655-6660.

[65] Kanavy HE, Gerstenblith MR. Ultraviolet radiation and melanoma. Semin Cutan Med Surg 2011;30(4) 222-228

[66] Lichtenstein P, Holm NV, Verkasalo PK, Iliadou A, Kaprio J, Koskenvuo M, Pukkala E, Skytthe A, Hemminki K. Environmental and heritable factors in the causation of cancer--analyses of cohorts of twins from Sweden, Denmark, and Finland. N Engl J Med 2000;343(2) 78-85.

[67] https://dnapittcrew.upmc.com/db/orthologs.php

[68] http://sciencepark.mdanderson.org/labs/wood/dna_repair_genes.html

[69] Wood RD, Mitchell M, Sgouros J, Lindahl T. Human DNA repair genes. Science 2001;291(5507) 1284-1289.

[70] Wood RD, Mitchell M, Lindahl T. Human DNA repair genes, 2005. Mutat Res 2005;577(1-2) 275-283.

[71] Bernstein C, Bernstein H, Payne CM, Garewal H. DNA repair/pro-apoptotic dual-role proteins in five major DNA repair pathways: fail-safe protection against carcino-genesis. Mutat Res 2002;511(2) 145-178.

[72] Foulkes WD. Inherited susceptibility to common cancers. N Engl J Med 2008;359(20) 2143-2153.

[73] O'Hagan HM, Wang W, Sen S, Destefano Shields C, Lee SS, Zhang YW, Clements EG, Cai Y, Van Neste L, Easwaran H, Casero RA, Sears CL, Baylin SB. Oxidative damage targets complexes containing DNA methyltransferases, SIRT1, and poly-comb members to promoter CpG Islands. Cancer Cell. 2011;20(5) 606-619.

[74] Malkin D. Li-fraumeni syndrome. Genes Cancer 2011;2(4) 475-484.

[75] Roukos DH. Genome-wide association studies: how predictable is a person's cancer risk? Expert Rev Anticancer Ther 2009;9(4) 389-392.

[76] Truninger K, Menigatti M, Luz J, Russell A, Haider R, Gebbers JO, Bannwart F, Yurtsever H, Neuweiler J, Riehle HM, Cattaruzza MS, Heinimann K, Schär P, Jiricny J, Marra G. Immunohistochemical analysis reveals high frequency of PMS2 defects in colorectal cancer. Gastroenterology 2005;128(5) 1160-1171.

[77] Valeri N, Gasparini P, Fabbri M, Braconi C, Veronese A, Lovat F, Adair B, Vannini I, Fanini F, Bottoni A, Costinean S, Sandhu SK, Nuovo GJ, Alder H, Gafa R, Calore F,

Ferracin M, Lanza G, Volinia S, Negrini M, McIlhatton MA, Amadori D, Fishel R, Croce CM. Modulation of mismatch repair and genomic stability by miR-155. Proc Natl Acad Sci U S A 2010;107(15) 6982-6987.

[78]   Zhang W, Zhang J, Hoadley K, Kushwaha D, Ramakrishnan V, Li S, Kang C, You Y, Jiang C, Song SW, Jiang T, Chen CC. miR-181d: a predictive glioblastoma biomarker that downregulates MGMT expression. Neuro Oncol 2012;14(6) 712-719.

[79]   Spiegl-Kreinecker S, Pirker C, Filipits M, Lötsch D, Buchroithner J, Pichler J, Silye R, Weis S, Micksche M, Fischer J, Berger W.O6-Methylguanine DNA methyltransferase protein expression in tumor cells predicts outcome of temozolomide therapy in glioblastoma patients. Neuro Oncol 2010 Jan;12(1) 28-36.

[80]   Lahtz C, Pfeifer GP. Epigenetic changes of DNA repair genes in cancer. J Mol Cell Biol 2011;3(1) 51-58.

[81]   Narayanan L, Fritzell JA, Baker SM, Liskay RM, Glazer PM. Elevated levels of mutation in multiple tissues of mice deficient in the DNA mismatch repair gene Pms2. Proc Natl Acad Sci U S A 1997;94(7) 3122-3127.

[82]   Hegan DC, Narayanan L, Jirik FR, Edelmann W, Liskay RM, Glazer PM. Differing patterns of genetic instability in mice deficient in the mismatch repair genes Pms2, Mlh1, Msh2, Msh3 and Msh6. Carcinogenesis 2006;27(12) 2402-2408.

[83]   Tutt AN, van Oostrom CT, Ross GM, van Steeg H, Ashworth A. Disruption of Brca2 increases the spontaneous mutation rate in vivo: synergism with ionizing radiation. EMBO Rep 2002;3(3) 255-260.

[84]   German J. Bloom's syndrome. I. Genetical and clinical observations in the first twenty-seven patients. Am J Hum Genet 1969;21(2) 196-227.

[85]   Nagaraju G, Scully R. Minding the gap: the underground functions of BRCA1 and BRCA2 at stalled replication forks. DNA Repair (Amst) 2007;6(7) 1018-1031.

[86]   Lancaster JM, Powell CB, Kauff ND, Cass I, Chen LM, Lu KH, Mutch DG, Berchuck A, Karlan BY, Herzog TJ; Society of Gynecologic Oncologists Education Committee. Society of Gynecologic Oncologists Education Committee statement on risk assessment for inherited gynecologic cancer predispositions. Gynecol Oncol 2007;107(2) 159-162.

[87]   Keimling M, Volcic M, Csernok A, Wieland B, Dörk T, Wiesmüller L. Functional characterization connects individual patient mutations in ataxia telangiectasia mutated (ATM) with dysfunction of specific DNA double-strand break-repair signaling pathways. FASEB J 2011;25(11) 3849-3860.

[88]   Thompson LH, Schild D. Recombinational DNA repair and human disease. Mutat Res 2002;509(1-2) 49-78.

[89]   Chrzanowska KH, Gregorek H, Dembowska-Bagińska B, Kalina MA, Digweed M. Nijmegen breakage syndrome (NBS). Orphanet J Rare Dis 2012;7 13.

[90] Rapp A, Greulich KO. After double-strand break induction by UV-A, homologous recombination and nonhomologous end joining cooperate at the same DSB if both systems are available. J Cell Sci 2004;117(Pt 21) 4935-4945.

[91] Bartkova J, Tommiska J, Oplustilova L, Aaltonen K, Tamminen A, Heikkinen T, Mistrik M, Aittomäki K, Blomqvist C, Heikkilä P, Lukas J, Nevanlinna H, Bartek J. Aberrations of the MRE11-RAD50-NBS1 DNA damage sensor complex in human breast cancer: MRE11 as a candidate familial cancer-predisposing gene. Mol Oncol 2008;2(4) 296-316.

[92] Nimonkar AV, Ozsoy AZ, Genschel J, Modrich P, Kowalczykowski SC. Human exonuclease 1 and BLM helicase interact to resect DNA and initiate DNA repair. Proc Natl Acad Sci U S A 2008;105(44) 16906-16911.

[93] German J. Bloom's syndrome. XX. The first 100 cancers. Cancer Genet Cytogenet 1997;93(1) 100-106.

[94] Bohr VA. Deficient DNA repair in the human progeroid disorder, Werner syndrome. Mutat Res 2005;577(1-2) 252-259.

[95] Monnat RJ Jr. Human RECQ helicases: roles in DNA metabolism, mutagenesis and cancer biology. Semin Cancer Biol 2010;20(5) 329-339.

[96] Singh DK, Ahn B, Bohr VA. Roles of RECQ helicases in recombination based DNA repair, genomic stability and aging. Biogerontology 2009;10(3) 235-252.

[97] Anbari KK, Ierardi-Curto LA, Silber JS, Asada N, Spinner N, Zackai EH, Belasco J, Morrissette JD, Dormans JP. Two primary osteosarcomas in a patient with Rothmund-Thomson syndrome. Clin Orthop Relat Res 2000;(378) 213-223.

[98] Thompson LH, Hinz JM. Cellular and molecular consequences of defective Fanconi anemia proteins in replication-coupled DNA repair: mechanistic insights. Mutat Res 2009;668(1-2) 54-72.

[99] Alter BP. Cancer in Fanconi anemia, 1927-2001. Cancer 2003;97(2) 425-440.

[100] Lehmann AR, McGibbon D, Stefanini M. Xeroderma pigmentosum. Orphanet J Rare Dis 2011;6 70.

[101] Oh KS, Imoto K, Emmert S, Tamura D, DiGiovanna JJ, Kraemer KH. Nucleotide excision repair proteins rapidly accumulate but fail to persist in human XP-E (DDB2 mutant) cells. Photochem Photobiol 2011;87(3) 729-733.

[102] Manchanda R, Menon U, Michaelson-Cohen R, Beller U, Jacobs I. Hereditary nonpolyposis colorectal cancer or Lynch syndrome: the gynaecological perspective. Curr Opin Obstet Gynecol 2009;21(1) 31-38.

[103] David SS, O'Shea VL, Kundu S. Base-excision repair of oxidative DNA damage. Nature 2007;447(7147) 941-950.

[104] Cleary SP, Cotterchio M, Jenkins MA, Kim H, Bristow R, Green R, Haile R, Hopper JL, LeMarchand L, Lindor N, Parfrey P, Potter J, Younghusband B, Gallinger S.

Germline MutY human homologue mutations and colorectal cancer: a multisite case-control study. Gastroenterology 2009;136(4) 1251-1260.

[105] López-Contreras AJ, Fernandez-Capetillo O. The ATR barrier to replication-born DNA damage. DNA Repair (Amst) 2010;9(12) 1249-1255.

[106] Tanaka A, Weinel S, Nagy N, O'Driscoll M, Lai-Cheong JE, Kulp-Shorten CL, Knable A, Carpenter G, Fisher SA, Hiragun M, Yanase Y, Hide M, Callen J, McGrath JA. Germline mutation in ATR in autosomal- dominant oropharyngeal cancer syndrome. Am J Hum Genet 2012;90(3) 511-517.

[107] Viktorsson K, De Petris L, Lewensohn R. The role of p53 in treatment responses of lung cancer. Biochem Biophys Res Commun 2005;331(3) 868-880.

[108] Esteller M, Silva JM, Dominguez G, Bonilla F, Matias-Guiu X, Lerma E, Bussaglia E, Prat J, Harkes IC, Repasky EA, Gabrielson E, Schutte M, Baylin SB, Herman JG. Promoter hypermethylation and BRCA1 inactivation in sporadic breast and ovarian tumors. J Natl Cancer Inst 2000;92(7) 564-569.

[109] Shen L, Kondo Y, Rosner GL, Xiao L, Hernandez NS, Vilaythong J, Houlihan PS, Krouse RS, Prasad AR, Einspahr JG, Buckmeier J, Alberts DS, Hamilton SR, Issa JP. MGMT promoter methylation and field defect in sporadic colorectal cancer. J Natl Cancer Inst 2005;97(18) 1330-1338.

[110] Psofaki V, Kalogera C, Tzambouras N, Stephanou D, Tsianos E, Seferiadis K, Kolios G. Promoter methylation status of hMLH1, MGMT, and CDKN2A/p16 in colorectal adenomas. World J Gastroenterol 2010;16(28) 3553-3560.

[111] Lee KH, Lee JS, Nam JH, Choi C, Lee MC, Park CS, Juhng SW, Lee JH. Promoter methylation status of hMLH1, hMSH2, and MGMT genes in colorectal cancer associated with adenoma-carcinoma sequence. Langenbecks Arch Surg 2011;396(7) 1017-1026.

[112] Facista A, Nguyen H, Lewis C, Prasad AR, Ramsey L, Zaitlin B, Nfonsam V, Krouse RS, Bernstein H, Payne CM, Stern S, Oatman N, Banerjee B, Bernstein C. Deficient expression of DNA repair enzymes in early progression to sporadic colon cancer. Genome Integr 2012;3(1): 3.

[113] Zou XP, Zhang B, Zhang XQ, Chen M, Cao J, Liu WJ. Promoter hypermethylation of multiple genes in early gastric adenocarcinoma and precancerous lesions. Hum Pathol 2009;40(11) 1534-1542.

[114] Ling ZQ, Li P, Ge MH, Hu FJ, Fang XH, Dong ZM, Mao WM. Aberrant methylation of different DNA repair genes demonstrates distinct prognostic value for esophageal cancer. Dig Dis Sci 2011;56(10) 2992-3004.

[115] Lee KH, Lee JS, Nam JH, Choi C, Lee MC, Park CS, Juhng SW, Lee JH. Promoter methylation status of hMLH1, hMSH2, and MGMT genes in colorectal cancer associated with adenoma-carcinoma sequence. Langenbecks Arch Surg 2011;396(7) 1017-1026.

[116] Paluszczak J, Misiak P, Wierzbicka M, Woźniak A, Baer-Dubowska W. Frequent hy-permethylation of DAPK, RARbeta, MGMT, RASSF1A and FHIT in laryngeal squa-mous cell carcinomas and adjacent normal mucosa. Oral Oncol 2011;47(2) 104-107.

[117] Guo M, House MG, Hooker C, Han Y, Heath E, Gabrielson E, Yang SC, Baylin SB, Herman JG, Brock MV. Promoter hypermethylation of resected bronchial margins: a field defect of changes? Clin Cancer Res 2004;10(15) 5131-5136.

[118] Skiriute D, Vaitkiene P, Saferis V, Asmoniene V, Skauminas K, Deltuva VP, Tama-sauskas A. MGMT, GATA6, CD81, DR4, and CASP8 gene promoter methylation in glioblastoma. BMC Cancer 2012;12(1) 218.

[119] Pogribny IP, James SJ. Reduction of p53 gene expression in human primary hepato-cellular carcinoma is associated with promoter region methylation without coding region mutation. Cancer Lett 2002;176(2) 169-74.

[120] Guan H, Ji M, Hou P, Liu Z, Wang C, Shan Z, Teng W, Xing M. Hypermethylation of the DNA mismatch repair gene hMLH1 and its association with lymph node meta-stasis and T1799A BRAF mutation in patients with papillary thyroid cancer. Cancer 2008;113(2) 247-255.

[121] Wan G, Mathur R, Hu X, Zhang X, Lu X. miRNA response to DNA damage. Trends Biochem Sci 2011;36(9) 478-484.

[122] Malumbres M. miRNAs and cancer: An epigenetics view. Mol Aspects Med 2012 Jul 4 [Epub ahead of print]

[123] Resar LM. The high mobility group A1 gene: transforming inflammatory signals into cancer? Cancer Res 2010;70(2) 436-439.

[124] Baldassarre G, Battista S, Belletti B, Thakur S, Pentimalli F, Trapasso F, Fedele M, Pierantoni G, Croce CM, Fusco A. Negative regulation of BRCA1 gene expression by HMGA1 proteins accounts for the reduced BRCA1 protein levels in sporadic breast carcinoma. Mol Cell Biol 2003;23(7) 2225-2238.

[125] Borrmann L, Schwanbeck R, Heyduk T, Seebeck B, Rogalla P, Bullerdiek J, Wisniew-ski JR. High mobility group A2 protein and its derivatives bind a specific region of the promoter of DNA repair gene ERCC1 and modulate its activity. Nucleic Acids Res 2003;31(23) 6841-6851.

[126] Lee S, Jung JW, Park SB, Roh K, Lee SY, Kim JH, Kang SK, Kang KS. Histone deacety-lase regulates high mobility group A2-targeting microRNAs in human cord blood-derived multipotent stem cell aging. Cell Mol Life Sci 2011;68(2) 325-336.

[127] Palmieri D, D'Angelo D, Valentino T, De Martino I, Ferraro A, Wierinckx A, Fedele M, Trouillas J, Fusco A. Downregulation of HMGA-targeting microRNAs has a criti-cal role in human pituitary tumorigenesis. Oncogene 2011 Dec 5. doi: 10.1038/onc. 2011.557.

[128] Sampath D, Liu C, Vasan K, Sulda M, Puduvalli VK, Wierda WG, Keating MJ. Histone deacetylases mediate the silencing of miR-15a, miR-16, and miR-29b in chronic lymphocytic leukemia. Blood 2012;119(5) 1162-1172.

[129] D'Angelo D, Palmieri D, Mussnich P, Roche M, Wierinckx A, Raverot G, Fedele M, Croce CM, Trouillas J, Fusco A. Altered MicroRNA Expression Profile in Human Pituitary GH Adenomas: Down-Regulation of miRNA Targeting HMGA1, HMGA2, and E2F1. J Clin Endocrinol Metab 2012;97(7) E1128-1138.

[130] Suzuki H, Takatsuka S, Akashi H, Yamamoto E, Nojima M, Maruyama R, Kai M, Yamano HO, Sasaki Y, Tokino T, Shinomura Y, Imai K, Toyota M. Genome-wide profiling of chromatin signatures reveals epigenetic regulation of MicroRNA genes in colorectal cancer. Cancer Res 2011;71(17) 5646-5658.

[131] Tchou-Wong KM, Kiok K, Tang Z, Kluz T, Arita A, Smith PR, Brown S, Costa M. Effects of nickel treatment on H3K4 trimethylation and gene expression. PLoS One 2011;6(3) e17728.

[132] Klase Z, Winograd R, Davis J, Carpio L, Hildreth R, Heydarian M, Fu S, McCaffrey T, Meiri E, Ayash-Rashkovsky M, Gilad S, Bentwich Z, Kashanchi F: HIV-1 TAR miRNA protects against apoptosis by altering cellular gene expression. Retrovirology 2009;6 18.

[133] http://www.ebi.ac.uk/enright-srv/microcosm/htdocs/targets/v5/

[134] Wang P, Dong Q, Zhang C, Kuan PF, Liu Y, Jeck WR, Andersen JB, Jiang W, Savich GL, Tan TX, Auman JT, Hoskins JM, Misher AD, Moser CD, Yourstone SM, Kim JW, Cibulskis K, Getz G, Hunt HV, Thorgeirsson SS, Roberts LR, Ye D, Guan KL, Xiong Y, Qin LX, Chiang DY. Mutations in isocitrate dehydrogenase 1 and 2 occur frequently in intrahepatic cholangiocarcinomas and share hypermethylation targets with glioblastomas. Oncogene 2012 Jul 23. doi: 10.1038/onc.2012.315. [Epub ahead of print]

[135] Watanabe T, Nobusawa S, Kleihues P, Ohgaki H. IDH1 mutations are early events in the development of astrocytomas and oligodendrogliomas. Am J Pathol 2009;174(4) 1149-1153.

[136] Turcan S, Rohle D, Goenka A, Walsh LA, Fang F, Yilmaz E, Campos C, Fabius AW, Lu C, Ward PS, Thompson CB, Kaufman A, Guryanova O, Levine R, Heguy A, Viale A, Morris LG, Huse JT, Mellinghoff IK, Chan TA. IDH1 mutation is sufficient to establish the glioma hypermethylator phenotype. Nature 2012;483(7390) 479-483.

[137] Carrillo JA, Lai A, Nghiemphu PL, Kim HJ, Phillips HS, Kharbanda S, Moftakhar P, Lalaezari S, Yong W, Ellingson BM, Cloughesy TF, Pope WB. Relationship between Tumor Enhancement, Edema, IDH1 Mutational Status, MGMT Promoter Methylation, and Survival in Glioblastoma. AJNR Am J Neuroradiol 2012;33(7) 1349-1355.

[138] Nelson PT, Kiriakidou M, Mourelatos Z, Tan GS, Jennings MH, Xie K, Wang WX. High-throughput experimental studies to identify miRNA targets directly, with special focus on the mammalian brain. Brain Res 2010;1338 122-130.

[139] Ohka F, Natsume A, Motomura K, Kishida Y, Kondo Y, Abe T, Nakasu Y, Namba H, Wakai K, Fukui T, Momota H, Iwami K, Kinjo S, Ito M, Fujii M, Wakabayashi T. The global DNA methylation surrogate LINE-1 methylation is correlated with MGMT promoter methylation and is a better prognostic factor for glioma. PLoS One 2011;6(8) e23332.

[140] Bernstein C, Bernstein H, Payne CM, Dvorak K, Garewal H. Field defects in progression to gastrointestinal tract cancers. Cancer Lett 2008;260(1-2) 1-10.

[141] Gerlinger M, Rowan AJ, Horswell S, Larkin J, Endesfelder D, Gronroos E, Martinez P, Matthews N, Stewart A, Tarpey P, Varela I, Phillimore B, Begum S, McDonald NQ, Butler A, Jones D, Raine K, Latimer C, Santos CR, Nohadani M, Eklund AC, Spencer-Dene B, Clark G, Pickering L, Stamp G, Gore M, Szallasi Z, Downward J, Futreal PA, Swanton C. N Engl J Med 2012;366(10) 883-892.

[142] Bernstein C, Facista A, Nguyen H, Zaitlin B, Hassounah N, Loustaunau C, Payne CM, Banerjee B, Goldschmid S, Tsikitis VL, Krouse R, Bernstein H. Cancer and age related colonic crypt deficiencies in cytochrome c oxidase I. World J Gastrointest Oncol 2010;2(12) 429-442.

[143] Nicolas P, Kim KM, Shibata D, Tavaré S: The stem cell population of the human colon crypt: analysis via methylation patterns. PLoS Comput Biol 2007;3 e28.

[144] Willis ND, Przyborski SA, Hutchison CJ, Wilson RG: Colonic and colorectal cancer stem cells: progress in the search for putative biomarkers. J Anat 2008;213 59-65.

[145] Loeb LA. Human cancers express mutator phenotypes: origin, consequences and targeting. Nat Rev Cancer 2011;11(6) 450-457.

[146] Roy M, Sinha D, Mukherjee S, Biswas J. Curcumin prevents DNA damage and enhances the repair potential in a chronically arsenic-exposed human population in West Bengal, India. Eur J Cancer Prev 2011;20(2) 123-131.

[147] Qin XJ, Liu W, Li YN, Sun X, Hai CX, Hudson LG, Liu KJ. Poly(ADP-ribose) polymerase-1 inhibition by arsenite promotes the survival of cells with unrepaired DNA lesions induced by UV exposure. Toxicol Sci 2012;127(1) 120-129.

[148] Ebert F, Weiss A, Bültemeyer M, Hamann I, Hartwig A, Schwerdtle T. Arsenicals affect base excision repair by several mechanisms. Mutat Res 2011;715(1-2) 32-41.

[149] Andrew AS, Karagas MR, Hamilton JW. Decreased DNA repair gene expression among individuals exposed to arsenic in United States drinking water. Int J Cancer 2003;104(3) 263-268.

[150] Nollen M, Ebert F, Moser J, Mullenders LH, Hartwig A, Schwerdtle T. Impact of arsenic on nucleotide excision repair: XPC function, protein level, and gene expression. Mol Nutr Food Res 2009;53(5) 572-582.

[151] Shen S, Lee J, Weinfeld M, Le XC. Attenuation of DNA damage-induced p53 expression by arsenic: a possible mechanism for arsenic co-carcinogenesis. Mol Carcinog 2008;47(7) 508-518.

[152] Zhou ZH, Lei YX, Wang CX. Analysis of aberrant methylation in DNA repair genes during malignant transformation of human bronchial epithelial cells induced by cadmium. Toxicol Sci 2012;125(2) 412-417.

[153] Wieland M, Levin MK, Hingorani KS, Biro FN, Hingorani MM. Mechanism of cadmium-mediated inhibition of Msh2-Msh6 function in DNA mismatch repair. Biochemistry 2009;48(40) 9492-9502.

[154] Bravard A, Vacher M, Gouget B, Coutant A, de Boisferon FH, Marsin S, Chevillard S, Radicella JP. Redox regulation of human OGG1 activity in response to cellular oxidative stress. Mol Cell Biol 2006;26(20) 7430-7436.

[155] Viau M, Gastaldo J, Bencokova Z, Joubert A, Foray N. Cadmium inhibits non-homologous end-joining and over-activates the MRE11-dependent repair pathway. Mutat Res 2008;654(1) 13-21.

[156] Schwerdtle T, Ebert F, Thuy C, Richter C, Mullenders LH, Hartwig A. Genotoxicity of soluble and particulate cadmium compounds: impact on oxidative DNA damage and nucleotide excision repair. Chem Res Toxicol 2010;23(2) 432-442.

[157] Burnat G, Majka J, Konturek PC. Bile acids are multifunctional modulators of the Barrett's carcinogenesis. J Physiol Pharmacol 2010;61(2) 185-192.

[158] Romagnolo DF, Chirnomas RB, Ku J, Jeffy BD, Payne CM, Holubec H, Ramsey L, Bernstein H, Bernstein C, Kunke K, Bhattacharyya A, Warneke J, Garewal H. Deoxycholate, an endogenous tumor promoter and DNA damaging agent, modulates BRCA-1 expression in apoptosis-sensitive epithelial cells: loss of BRCA-1 expression in colonic adenocarcinomas. Nutr Cancer 2003;46(1) 82-92.

[159] Ogawa A, Murate T, Suzuki M, Nimura Y, Yoshida S. Lithocholic acid, a putative tumor promoter, inhibits mammalian DNA polymerase beta. Jpn J Cancer Res 1998;89(11) 1154-1159.

[160] Feng Z, Hu W, Tang MS. Trans-4-hydroxy-2-nonenal inhibits nucleotide excision repair in human cells: a possible mechanism for lipid peroxidation-induced carcinogenesis. Proc Natl Acad Sci U S A 2004;101(23) 8598-8602.

[161] Feng Z, Hu W, Marnett LJ, Tang MS. Malondialdehyde, a major endogenous lipid peroxidation product, sensitizes human cells to UV- and BPDE-induced killing and mutagenesis through inhibition of nucleotide excision repair. Mutat Res 2006;601(1-2) 125-136.

[162] Chang CL, Marra G, Chauhan DP, Ha HT, Chang DK, Ricciardiello L, Randolph A, Carethers JM, Boland CR. Oxidative stress inactivates the human DNA mismatch repair system. Am J Physiol Cell Physiol 2002;283(1) C148-154.

[163] Langie SA, Knaapen AM, Houben JM, van Kempen FC, de Hoon JP, Gottschalk RW, Godschalk RW, van Schooten FJ. The role of glutathione in the regulation of nucleotide excision repair during oxidative stress. Toxicol Lett 20075;168(3) 302-309.

[164] Hill JW, Hu JJ, Evans MK. OGG1 is degraded by calpain following oxidative stress and cisplatin exposure. DNA Repair (Amst) 2008;7(4) 648-654.

[165] Sudprasert W, Navasumrit P, Ruchirawat M. Effects of low-dose gamma radiation on DNA damage, chromosomal aberration and expression of repair genes in human blood cells. Int J Hyg Environ Health. 2006;209(6) 503-511.

[166] Li D, Wang Q, Liu C, Duan H, Zeng X, Zhang B, Li X, Zhao J, Tang S, Li Z, Xing X, Yang P, Chen L, Zeng J, Zhu X, Zhang S, Zhang Z, Ma L, He Z, Wang E, Xiao Y, Zheng Y, Chen W. Aberrant expression of miR-638 contributes to benzo(a)pyrene-induced human cell transformation. Toxicol Sci 2012;125(2) 382-391.

[167] Liu WB, Ao L, Cui ZH, Zhou ZY, Zhou YH, Yuan XY, Xiang YL, Cao J, Liu JY. Molecular analysis of DNA repair gene methylation and protein expression during chemical-induced rat lung carcinogenesis. Biochem Biophys Res Commun 2011;408(4) 595-601.

[168] Hanova M, Stetina R, Vodickova L, Vaclavikova R, Hlavac P, Smerhovsky Z, Naccarati A, Polakova V, Soucek P, Kuricova M, Manini P, Kumar R, Hemminki K, Vodicka P. Modulation of DNA repair capacity and mRNA expression levels of XRCC1, hOGG1 and XPC genes in styrene-exposed workers. Toxicol Appl Pharmacol 2010;248(3) 194-200.

[169] Chen YY, Chung JG, Wu HC, Bau DT, Wu KY, Kao ST, Hsiang CY, Ho TY, Chiang SY. Aristolochic acid suppresses DNA repair and triggers oxidative DNA damage in human kidney proximal tubular cells. Oncol Rep 2010;24(1) 141-153.

[170] Grosskopf C, Schwerdtle T, Mullenders LH, Hartwig A. Antimony impairs nucleotide excision repair: XPA and XPE as potential molecular targets. Chem Res Toxicol 2010;23(7) 1175-1183.

[171] Ji W, Yang L, Yu L, Yuan J, Hu D, Zhang W, Yang J, Pang Y, Li W, Lu J, Fu J, Chen J, Lin Z, Chen W, Zhuang Z. Epigenetic silencing of O6-methylguanine DNA methyltransferase gene in NiS-transformed cells. Carcinogenesis 2008;29(6) 1267-1275.

[172] Link A, Balaguer F, Goel A. Cancer chemoprevention by dietary polyphenols: promising role for epigenetics. Biochem Pharmacol 2010;80(12) 1771-1792.

[173] Fang M, Chen D, Yang CS. Dietary polyphenols may affect DNA methylation. J Nutr 2007;137(1 Suppl) 223S-228S.

[174] Olaharski AJ, Rine J, Marshall BL, Babiarz J, Zhang L, Verdin E, Smith MT. The flavoring agent dihydrocoumarin reverses epigenetic silencing and inhibits sirtuin deacetylases. PLoS Genet 2005;1(6) e77.

[175] Kikuno N, Shiina H, Urakami S, Kawamoto K, Hirata H, Tanaka Y, Majid S, Igawa M, Dahiya R. Genistein mediated histone acetylation and demethylation activates tumor suppressor genes in prostate cancer cells. Int J Cancer 2008;123(3) 552-560.

[176] Collins AR, Azqueta A, Langie SA. Effects of micronutrients on DNA repair. Eur J Nutr 2012;51(3) 261-279.

[177] http://www.phenol-explorer.eu

[178] http://www.ajcn.org/content/93/6/1220.full.pdf+html

[179] http://www.hmdb.ca/search/search?query=caffeic+acid

[180] Aiyer HS, Vadhanam MV, Stoyanova R, Caprio GD, Clapper ML, Gupta RC. Dietary berries and ellagic acid prevent oxidative DNA damage and modulate expression of DNA repair genes. Int J Mol Sci 2008;9(3) 327-341.

[181] Niture SK, Velu CS, Smith QR, Bhat GJ, Srivenugopal KS. Increased expression of the MGMT repair protein mediated by cysteine prodrugs and chemopreventative natural products in human lymphocytes and tumor cell lines. Carcinogenesis 2007;28(2) 378-389.

[182] Bernstein H, Crowley-Skillicorn C, Bernstein C, Payne CM, Dvorak K, Garewal H. Dietary Compounds that Enhance DNA Repair and their Relevance to Cancer and Aging. In: BR Landseer (ed.) New Research on DNA Repair. New York: Nova Science Publishers, Inc.; 2007. p99-113.

[183] Clifford MN. Chlorogenic acids and other cinnamates --- nature, occurrence and dietary burden. J Sci Food Agric 1999;79(3) 362-372.

[184] Mattila P, Kumpulainen J. Determination of free and total phenolic acids in plant-derived foods by HPLC with diode-array detection. J Agric Food Chem 2002;50(13) 3660-3667.

[185] Yun MH, Hiom K. Understanding the functions of BRCA1 in the DNA-damage response. Biochem Soc Trans 2009;37(Pt 3) 597-604.

[186] Quinn JE, Kennedy RD, Mullan PB, Gilmore PM, Carty M, Johnston PG, Harkin DP. BRCA1 functions as a differential modulator of chemotherapy-induced apoptosis. Cancer Res 2003;63(19) 6221-6228.

[187] Taron M, Rosell R, Felip E, Mendez P, Souglakos J, Ronco MS, Queralt C, Majo J, Sanchez JM, Sanchez JJ, Maesire J. BRCA1 mRNA expression levels as an indicator of chemoresistance in lung cancer. Human Molecular Genetics 2004;13(20) 2443-2449.

[188] Papadaki C, Sfakianaki M, Ioannidis G, Lagoudaki E, Trypaki M, Tryfonidis K, Mavroudis D, Stathopoulos E, Georgoulias V, Souglakos J. ERCC1 and BRCA1 mRNA expression levels in the primary tumor could predict the effectiveness of the second-line cisplatin-based chemotherapy in pretreated patents with metastatic non-small cell lung cancer. J Thorac Oncol 2012;7(4) 663-671.

[189] Weberpals J, Garbuio K, O'Brien A, Clark-Knowles K, Doucette S, Antoniouk O, Goss G, Dimitroulakos J. The DNA repair proteins BRCA1 and ERCC1 as predictive markers in sporadic ovarian cancer. Int J Cancer 2009;124(4) 806-815.

[190] Reed E. Platinum-DNA adduct, nucleotide excision repair and platinum based anti-cancer chemotherapy. Cancer Treat Rev 1998;24(5) 331-344.

[191] Kelland L. The resurgence of platinum-based cancer chemotherapy. Nat Rev Cancer 2007;7(8) 573-584.

[192] Martin LP, Hamilton TC, Schilder RJ. Platinum Resistance: The role of DNA repair pathways. Clin Cancer Res 2008;14(5) 1291-1295.

[193] Bosl GJ, Motzer RJ. Testicular germ-cell cancer. N Engl J Med 1997;337(4) 242-253.

[194] Welsh C, Day R, McGurk C, Masters JRW, Wood RD, Koberle B. Reduced levels of XPA, ERCC1 and XPF DNA repair proteins in testis tumor cell lines. Int J Cancer 2004;110(3) 352-361.

[195] Simon GR, Sharma S, Cantor A, Smith P, Bepler G. ERCC1 expression is a predictor of survival in resected patients with non-small cell lung cancer. Chest 2005;127(3) 978-983.

[196] Lord RVN, Brabender J, Gandara D, Alberola V, Camps C, Domine M, Cardenal F, Sanchez JM, Gumerlock PH, Taron M, Sanchez JJ, Danenberg KD, Danenberg PV, Rosell R. Low ERCC1 expression correlates with prolonged survival after cisplatin plus gemcitabine chemotherapy in non-small cell lung cancer. Clin Cancer Res 2002;8(7) 2286-2291.

[197] Zhou W, Gurubhagavatula S, Liu G, Park S, Neuberg DS, Wain JC, Lynch TJ, Su L, Christiani DC. Excision repair cross-complementation group 1 polymorphism predicts overall survival in advanced non-small cell lung cancer patients treated with platinum-based chemotherapy. Clin Cancer Res 2004;10(15) 4939-4943.

[198] Olaussen KA, Dunant A, Fouret P, Brambilla E, Andre F, Haddad V, Taranchon E, Filipits M, Pirker R, Popper HH, Stahel R, Sabatier L, Pignon J-P, Tursz T, Le Chava-lier T, Soria, J-C. DNA repair by ERCC1 in non-small-cell lung cancer and cisplatin-based adjuvant chemotherapy. N Engl J Med 2006;355(10) 983-991.

[199] Leng XF, Chen MW, Xian L, Dai L, Ma GY, Li MH. Combined analysis of mRNA expression of ERCC1, BAG-1, BRCA1, RRM1 and TUBB3 to predict prognosis in patients with non-small cell lung cancer who received adjuvant chemotherapy. J Exp Clin Cancer Res 2012;23 31:25.

[200] Dabholkar M, Bostick-Bruton F, Weber C, Bohr VA, Egwuagu C, Reed E. ERCC1 and ERCC2 expression in malignant tissues from ovarian cancer patients. J Natl Cancer Inst 1992;84(19) 1512-1517.

[201] Kang S, Ju W, Kim JW, Park N-H, Song Y-S, Kim SC, Park S-Y, Kang S-B, Lee H-P. Association between excision repair cross-complementation group 1 polymorphism

and clinical outcome of platinum based chemotherapy in patients with epithelial ovarian cancer. Exp Mol Med 2006;8(3) 320-324.

[202] Shirota Y, Stoehlmacher J, Brabender J, Xiong Y-P, Uetake H, Danenberg KD, Groshen S, Tsao-Wei DD, Danenberg PV, Lenz H-J. ERCC1 and thymidylate synthase mRNA levels predict survival for colorectal cancer patients receiving combination oxaliplatin and fluorouracil chemotherapy. J Clin Oncol 2001;19(23) 4298-4304.

[203] Viguier J, Boige V, Miquel C, Pocard M, Giraudeau B, Sabourin J-C, Ducreux M, Sarasin A, Praz F. ERCC1 codon 118 polymorphism is a predictive factor for the tumor response to oxaliplatin/5-fluorouracil combination chemotherapy in patients with advanced colorectal cancer. Clin Cancer Res 2005;11(17) 6212-6217.

[204] Metzger R, Leichman CG, Danenberg KD, Danenberg PV, Lenz H-J, Hayashi K, Groshen S, Salonga D, Cohen H, Laine L, Crookes P, Silberman H, Barando J, Konda B, Leichman L. ERCC1 mRNA levels complement thymidylate synthase mRNA levels in predicting response and survival for gastric cancer patients receiving combination cisplatin and fluorouracil chemotherapy. J Clin Oncol 1998;16(1) 309-316.

[205] Bellmunt J, Paz-Ares L, Cuello M, Cecere FL, Albiol S, Guillem V, Gallardo E, Carles J, Mendez P, de la Cruz JJ, Taron M, Rosell R, Baselga J; Spanish Oncology Genitourinary Group. Gene expression of ERCC1 as a novel prognostic marker in advanced bladder cancer patients receiving cisplatin-based chemotherapy. Ann Oncol 2007;18(3) 522-528.

[206] Esteller M, Garcia-Foncillas J, Andion E, Goodman SN, Hidalgo OF, Vanaclocha V, Baylin SB, Herman JG. Inactivation of the DNA-repair gene MGMT and the clinical response of gliomas to alkylating agents. N Engl J Med 2000;343(19) 1350-1354.

[207] Paz MF, Yaya-Tur R, Rojas-Marcos I, Reynes G, Pollan M, Aguirre-Cruz L, Garcia-Lopez JL, Piquer J, Safont M-J, Balana C, Sanchez-Cespedes M, Garcia-Villanueva M, Arribas L, Esteller M. CpG island hypermethylation of the DNA repair enzyme methyltransferase predicts response to temozolomide in primary gliomas. Clin Cancer Res 2004;10(15) 4933-4938.

[208] Hegi ME, Diserens A-C, Godard S, Dietrich P-Y, Regli L, Ostermann S, Otten P, Van Melle G, de Tribolet N, Stupp R. Clinical trial substantiates the predictive value of O-6-methylguanine-DNA methyltransferase promoter methylation in glioblastoma patients treated with temozolomide. Clin Cancer Res 2004;10(6) 1871-1874.

[209] Hegi ME, Liu L, Herman JG, Stupp R, Wick W, Weller M, Mehta MP, Gilbert MR. Correlation of O^6-methylguanine methyltransferase (MGMT) promoter methylation with clinical outcomes in glioblastoma and clinical strategies to modulate MGMT activity. J Clin Oncol 2008;26(25) 4189-4199.

[210] Weller M, Stupp R, Reifenberger G, Brandes AA, van den Bent MJ, Wick W, Hegi ME. MGMT promoter methylation in malignant gliomas: ready for personalized medicine? Nat Rev Neurol 2010;6(1) 39-51.

[211] Leung M, Rosen D, Fields S, Cesano A, Budman DR. Poly(ADP-ribose) polymerase-1 inhibition: preclinical and clinical development of synthetic lethality. Mol Med 2011;17(7-8) 854-862. doi: 10.2119/molmed.2010.00240.

[212] Basu B, Yap TA, Molife LR, de Bono JS. Targeting the DNA damage response in oncology: past, present and future perspectives. Curr Opin Oncol 2012;24(3) 316-324.

[213] Fong PC, Boss DS, Yap TA, Tutt A, Wu P, Roelvink, MM, Mortimer P, Swaisland H, Lau A, O'Connor MJ, Ashworth A, Carmichael J, Kaye SB, Schellens JHM, de Bono JS. Inhibition of poly(ADP-ribose) polymerase in tumors from BRCA mutation carriers. N Engl J Med 2009;361(2) 123-134.

[214] Tutt A, Robson M, Garber JE, Domchek SM, Audeh MW, Weitzel JN, Friedlander M, Arun B, Loman N, Schmutzler RK, Wardley A, Mitchell G, Earl H, Wickens M, Carmichael J. Oral poly(ADP-ribose) polymerase inhibitor olaparib in patients with BRCA1 or BRCA2 mutations and advanced breast cancer: a proof-of-concept trial. Lancet 2010;376(9737) 235-244.

[215] Audeh MW, Carmichael J, Penson RT, Friedlander M, Powell B, Bell-McGuinn KM, Scott C, Weitzel JN, Oaknin A, Loman N, Lu K, Schmutzler RK, Matulonis U, Wickens M, Tutt A. Oral poly(ADP-ribose) polymerase inhibitor olaparib in patients with BRCA1 or BRCA2 mutations and recurrent ovarian cancer: a proof-of-concept trial. Lancet 2010;376(9737) 245-251.

[216] Stratton MR. Exploring the genomes of cancer cells: progress and promise. Science 2011;331(6024) 1553-1558.

[217] Hanahan D, Weinberg RA. Hallmarks of cancer: the next generation. Cell 2011;144(5) 646-674.

[218] Lord CJ, Ashworth A. The DNA damage response and cancer therapy. Nature 2012;481(7382) 287-294.

Chapter 3

DNA Repair and Resistance to Cancer Therapy

António S. Rodrigues, Bruno Costa Gomes,
Célia Martins, Marta Gromicho, Nuno G. Oliveira,
Patrícia S. Guerreiro and José Rueff

Additional information is available at the end of the chapter

1. Introduction

Humans are constantly exposed to diverse chemical and physical agents that have the potential to damage DNA, such as reactive oxygen species (ROS), ionizing radiation (IR), UV light, and various environmental, dietary or pollutant chemical agents. The integrity and survival of a cell is critically dependent on genome stability, and cells possess multiple pathways to repair these DNA lesions. These pathways are diverse and target different types of lesions.

The critical role played by DNA repair in the maintenance of genome stability is highlighted by the fact that many enzymes involved have been conserved through evolution [1-4]. Very rarely germ line mutations occur in several of the DNA repair genes and are the cause of cancer predisposing syndromes, such as *Xeroderma pigmentosum* (XP), [5], Fanconi anemia (FA) and ataxia telangiectasia (AT) and are associated with inherent chromosome instability [2]. One of the most well-known examples of a defect in DNA repair leading to cancer is the association of germ-line *BRCA1/2* mutations with breast, ovarian and peritoneal malignancies [6]. These rare human DNA repair syndromes have been invaluable in providing mechanistic explanations for the involvement of DNA repair system in cancer. They have also been instrumental in the translation of these findings to the clinic.

On the other hand, recent studies have shown that defective DNA damage repair is present in virtually all sporadic tumours [7]. Mutations in DNA repair genes could be either responsible for the occurrence of tumours or could arise due to random accumulation of mutations during cycling of cancer cells. The presence of incorrect DNA repair in tumour cells predisposes them to accumulate even more genetic alterations. For example, colorectal and endometrial cancers with defective DNA mismatch repair (MMR) due to mutations in the *MLH1* and *MSH2* genes exhibit increased rates of acquisition of single nucleotide changes and small insertions/deletions [8]. Thus, the presence of a "mutator phenotype" [9] could increase the evolutionary acquisition of alterations that ultimately could lead to enhanced drug resistance.

A further reminder on the importance of DNA repair is the observation that mutations in specific genes can lead not to an increase in cancer but to accelerated aging syndromes [7]. An example of this is Cockayne's syndrome (CS), which causes severe progeroid syndromes [10]. Mutations in the genes that encode two proteins in a nucleotide excision repair (NER) sub-pathway called transcription coupled repair (TCR) cause global premature cell death through apoptosis. In this case apoptosis ensures that DNA mutations are not transmitted to daughter cells, albeit at the expense of cell viability, and highlights the importance of maintaining DNA integrity.

One major problem in cancer therapy is the fact that of the 7.6 million cancer deaths that occur every year worldwide (2008 data; http://www.who.int/cancer/en/), many are due to failure of cancer therapy associated with acquired and intrinsic resistance mechanisms. These mechanisms of resistance can be classified in different ways, but the most characterized are altered cellular drug transport, increased survival or decreased cell death, altered DNA repair, and alterations in drug targets [11, 12]. Over the last years the importance of DNA repair pathways in resistance to chemotherapy has been increasingly recognized, but translation to the clinic is still scarce. Since many classical cancer therapies target DNA, the influence of DNA repair systems in response to DNA damage which primarily result from chemotherapy and radiotherapy is critical to cell survival. The use of inhibitors of DNA repair or DNA damage signalling pathways provides an interesting opportunity to target the genetic differences that exist between normal and tumour tissue [13, 14].

The rationale underlying the use of DNA damaging agents in therapeutic strategies is to kill cancer cells while sparing normal tissues, due to increased cell cycling of cancer cells. Unfortunately highly cycling normal cells (e.g. bone marrow, hair follicles and gastrointestinal epithelia) are also targeted by DNA damaging therapeutic agents, giving rise to the secondary effects normally seen after cancer therapy (e.g. diarrhoea, mouth ulcers, hair loss, anaemia and susceptibility to infections). Nevertheless, DNA-damaging chemotherapeutic agents are effective and prolong survival of cancer patients [15]. Chemotherapeutic agents commonly used in cancer treatment produce a plethora of lesions that can be targets for cellular responses. For example, DNA double strand breaks (DSBs), single-strand breaks (SSBs), and oxidized bases are induced by ionizing radiation (IR), anthracyclines, platinum compounds and taxanes. Anthracyclines are topoisomerase II inhibitors and DNA intercalating agents, which when used can lead to DSBs. Platinum compounds are bifunctional alkylating agents that induce predominantly intra- and interstrand crosslinks (ICLs) and taxanes are mitotic inhibitors. All these lesions induce cellular responses that cover a multitude of pathways, including DNA repair pathways, DNA tolerance mechanisms, coordination networks that link repair and cell cycle progression, as well as apoptotic and other cell death pathways when DNA damage is irreparable [16-19].

The DNA repair pathways that respond to these lesions include: direct repair of alkyl adducts by O6-alkylguanine DNA alkyltransferase (MGMT); repair of base damage and SSBs by base excision repair (BER); repair of bulky DNA adducts by nucleotide excision repair (NER); repair of cross-links by DNA interstrand cross-link repair and repair of mismatches and insertion/deletion loops by DNA mismatch repair (MMR); repair of DSBs by homologous recombination (HR) and non-homologous end joining (NHEJ). Detailed description of the biochemical pathways of DNA repair is beyond the scope of this chapter as several reviews on the subject have been published [1, 17, 20-23].

The observation that a variety of tumours frequently present deregulated expression of DNA repair genes (e.g. *MGMT, PARP1*) rapidly lead to the notion that DNA repair pathways could be targeted in cancer treatment and lead to personalized therapy [24, 25]. Tumours with specific DNA repair defects could be completely dependent on back-up DNA repair pathways for their survival. This dependence could be exploited therapeutically to induce cell death and apoptosis in tumour cells [26, 27]. The genetic state in which simultaneous inactivation of 2 genes (or pathways) is lethal, while loss of one or the other alone is viable is called synthetic lethality (also known as conditional genetics). The rationale for inducing synthetic lethality in cancer is that certain cancer cells lack one pathway to repair their DNA (e.g. HR) but have alternative pathways (base excision or single-strand repair) that allow them to survive. Inhibition of these alternative pathways would then impair DNA repair and induce cell death [26, 27]. Therefore it predicts that genotoxic agents leading to a particular type of DNA damage will kill cancer cells with genetic deficits in repair of that type of damage. Recently, this specific anticancer strategy has been the focus of intense investigations [28, 29].

In the case of the hereditary *BRCA1/2*-deficient breast and ovarian cancer syndromes, mentioned earlier, this strategy has been translated into the clinic, in the form of PARP inhibitors. These *BRCA1/2* tumours are defective in the repair of DSBs by HR. When a replication fork in one of these tumours encounters a DNA SSB, it converts that into a DSB, but the presence of a DSB prevents progression of the replication apparatus. Since *BRCA1/2* are both required for DSB repair, the tumour cells with those mutated genes will depend on repair of SSBs to prevent DSBs from occurring. The DNA repair protein PARP1 is required for repair of SSBs, and small molecular inhibitors of PARP1 will prevent repair of SSBs, more specifically in cells that are deficient in *BRCA1/2*. Since normal cells have the ability to repair the DSBs generated at the replication fork, because they have at least one normal allele of *BRCA1/2*, the use of PARP inhibitors has the potential of targeting only tumour cells. This proof of concept proven clinically, where the PARP1 inhibitor olaparib improves the progression-free survival of familial breast cancer [30]. Following this lead several small molecule DNA repair inhibitors are being developed worldwide.

However, not all *BRCA1/2* defective tumours respond equally well to this type of therapy. Thus, in the past years evidence has accumulated that drug resistance is also linked to alterations in these pathways [31-33]. Thus, tumour cells may also acquire resistance by invoking biochemical mechanisms that reduce drug action or by acquiring additional alterations in

DNA damage response pathways [34]. Therefore, the focus has also been directed on DNA repair pathways that could be responsible for cancer drug resistance.

Resistance to chemotherapy limits the effectiveness of anti-cancer drug treatment. Tumours may be intrinsically drug-resistant or develop resistance to chemotherapy during treatment. Acquired resistance is a particular problem, as tumours not only become resistant to the drugs originally used to treat them, but may also become cross-resistant to other drugs with different mechanisms of action. Resistance to chemotherapy is believed to cause treatment failure in over 90% of patients with metastatic cancer [35]. Thus, drug resistance is clearly a major clinical problem.

The attempt to develop more targeted therapeutics has been a major objective in cancer research in last years, and more and more molecular targets are being identified (e.g. tyrosine kinase inhibitors, monoclonal antibodies targeting membrane receptor kinases). Some of these targeted therapies are in clinical use, while others are being evaluated in clinical trials to validate their efficacy. More recently, the quest for targeted therapies has also focused on DNA repair pathways. Unfortunately, resistance to these therapies is also likely to appear, as has occurred with other targeted therapies, such as the tyrosine kinase inhibitors of the fusion *BCR-ABL1* gene, responsible for most cases of chronic myeloid leukaemia (e.g. imatinib, dasatinib, nilotinib). The application of DNA repair inhibitors in the clinic has also shown to be fraught with difficulty, since they also target DNA repair pathways in normal cells. The early clinical trial with MGMT inhibitors in combination with temozolomide (TMZ) was stopped early because the combined treatments harmed bone marrow as well as cancer tissue, whereas the clinical success of PARP inhibitors transpired since PARP is not critical to cell survival. Hence, unlike past visions of a "magic bullet" towards cancer, future research on cancer therapy should more reasonably envisage cancer therapy as a "never ending story", in which novel targeted therapeutics are constantly being overcome by the evolutionary processes present in cancerous cells [36].

2. Targeting DNA repair pathways

As mentioned, DNA repair pathways include the direct reversal of lesions, essentially de-alkylation of alkylated bases by *MGMT*, NER, BER, MMR and the double strand break repair by HR and NHEJ. Alterations in all these pathways have been observed in drug resistant tumour cells; however, the clinical significance of the alterations is not completely understood. Numerous genes involved in each of these pathways have been shown to be up- or down-regulated in diverse types of tumours and constitute a potential source of biomarkers to evaluate drug resistance to cancer chemotherapeutics [25, 32, 33].

3. MGMT and drug resistance

Alkylating agents are widely used to treat cancers, and one of the major DNA lesions formed occurs essentially by the alkylation of DNA at the O^6-position of guanine, which subsequently can generate DNA breaks and cell death. TMZ, streptozotocin, procarbazine and dacarbazine are examples of cancer chemotherapeutics that methylate DNA [37].

Direct repair of alkylated guanine residues proceeds through the removal of the alkyl moiety by MGMT. MGMT is a conserved protein from prokaryotes through eukaryotes. The MGMT protein removes the alkyl group from O^6-alkylguanine by direct transfer to a cysteine residue in its active site to which the alkyl group becomes covalently attached, resulting in the inactivation of the protein. The MGMT protein is subsequently ubiquitinated and degraded by the proteasome [38, 39]. The O^6-alkylguanine adduct accounts for about 10% of total alkylations, but displays a strong mutagenic and cytotoxic potential, because O^6-alkylguanines exhibit distorted base pairing characteristics in pairing with thymine, thereby, resulting in G:C to A:T transitions upon DNA replication [40]. Hence the unique DNA repair mechanism which depends on the suicidal degradation of the MGMT protein.

Tumour expression of *MGMT* varies and correlates with therapeutic response to alkylating agents. Numerous studies have found a strong correlation between MGMT activity and drug resistance in primary tumours and established human tumour cell lines [16, 41, 42]. High levels of expression have also been noted in melanoma [43], pancreatic carcinoma [16] besides glioblastomas [44]. Resistance to alkylating agents such as TMZ has been linked to over-expression of *MGMT* [43]. Therefore MGMT levels are being studied as biomarkers of intrinsic chemosensitivity to alkylating agents, such as TMZ or BCNU (carmustine).

Conversely, reduced MGMT activity in cultured tumour cells and human tumours is often the result of epigenetic silencing by promoter methylation of CpG islands, which leads to the formation of inactive chromatin that limits transcription, and therefore higher chemosensitivity to alkylation. Hegi *et al.* reported that of 206 patients with glioblastoma that were treated with TMZ and radiotherapy, those with a methylated *MGMT* promoter (45%) had a significantly better survival [45]. Hence, *MGMT* promoter methylation status is emerging as a prognostic factor for tumour therapy and is currently being assessed for selecting glioblastoma chemosensitivity towards TMZ [46-48]. The mechanisms underlying increased *MGMT* promoter methylation are complex and not completely known, although it is one of the most studied DNA repair genes [38]. In normal cells *MGMT* promoter methylation is uncommon, but occurs frequently in tumours. Approximately 25% of tumours of many different types, including non-small-cell carcinoma of the lung, lymphoma, head and neck cancers, and up to 40% of glioma and colorectal tumours were found to present CpG island promoter methylation [49].

Since high *MGMT* expression results in drug resistance to alkylating agents, one strategy to overcome resistance and improve efficacy is to use pseudo substrates of MGMT, such as O^6-benzylguanine (O^6-BG) or O^6-(4-bromothenyl) guanine (O^6-BTG or lomeguatrib or PaTrin-2) which inactivate the enzyme and enhance cell death [50]. O^6-BG is a specific, potent, and nontoxic inhibitor and leads to sensitization of cancer cells to cisplatin, chloroethylating and methylating agents [51, 52]. Clinical trials are underway to test combinations of O^6-BG with carmustine or TMZ for the treatment of glioma, anaplastic glioma, lymphoma, myeloma, colon cancer, melanoma and sarcoma, among others [53]. O^6-BTG presents higher bioavailability than O^6-BG, but also presents higher haematological toxicity when co-administered with

TMZ compared to TMZ alone. Therefore full use of this inhibitor may be more distant [54, 55]. Haematological toxicity was also observed with O^6-BTG co-administered with dacarbazine in patients with advanced melanoma and other solid tumours [56]. The combination of O^6-BTG and TMZ was also evaluated in a phase I clinical trial for advanced solid tumours [57], and in a pilot study for refractory acute leukaemia [58]. A phase I clinical trial was also conducted associating O^6-BTG with Irinotecan for colorectal cancer [59]. A phase II clinical trial of O^6-BTG plus TMZ for stage IV metastatic colorectal cancer is already completed. The trial was considered completed after the recruitment of 19 patients due to the absence of responses and also because evidences from other studies suggest that the O^6-BTG dosing regimen was inappropriate [55]. These studies showed a consistent depletion of MGMT and provided non-toxic doses of O^6-BG or O^6-BTG to be used in further studies. The haematological toxicity observed with the combination of MGMT inhibitors and chemotherapeutic agents might be attributed to an effective depletion of MGMT in off-target cells [60]. Additionally, the administration of a sub-optimal dose of the MGMT inhibitor, a therapeutic dosing schedule that allows the recovery of the MGMT activity or the choice of an inadequate treatment for the type of cancer could explain the lack of effects in clinical trials. In view of this, tumour-targeted delivery of MGMT inhibitors by the development of specific formulations or local administration [61] could be adopted to improve the therapeutic efficacy of the chemotherapeutic drugs and to translate into the clinic the results obtained in preclinical studies. Nonetheless, it is not clear if clinical application of MGMT inhibitors is a viable therapy in all settings.

4. Targeting MMR in cancer drug resistance

MMR is involved in the detection and repair of base-base mispairs during DNA replication, small insertion/deletion mutations at repetitive microsatellite regions and also in the regulation of homologous recombination [62]. MMR proteins are also involved in the repair of DNA damage caused by ROS and alkylating agents. MMR proteins interact with components of other repair pathways, including NER, BER, and HR, thus signalling with other pathways in response to DNA damage.

The MMR system consists of various proteins. MSH2 heterodimerizes with MSH6 or MSH3 to form MutSα or MutSβ, respectively, both of which are ATPases that play a critical role in mismatch recognition and initiation of repair. This induces a conformational change in MutS, resulting in a clamp that translocates on DNA in a ATP dependent manner, recruits the MutL complex, which in humans is a heterodimer consisting of the MLH1 and PMS2 proteins, and displaces DNA polymerase and PCNA, thereafter recruiting an exonuclease (EXO1) that degrades the newly synthesized DNA strand [63]. Other MMR genes (*MLH1*, *MLH3*, *PMS1*, and *PMS2*) are involved in MMR. MLH1 also heterodimerizes with PMS2, PMS1, or MLH3 to form MutLα, MutLβ, or MutLγ, respectively [63]. Polymerase δ (pol δ) then polymerizes the DNA stretch and DNA Ligase I performs ligation.

MMR deficiency leads to a wide range of tumour types. Germline deficiency in MMR accounts for the Lynch syndrome (hereditary non-polyposis colorectal cancer -HNPCC), in which a large increase in frequency of insertion and deletion mutations in simple repeat (microsatellite) sequences, a phenomenon known as microsatellite instability (MSI), is observed [64]. DNA mismatch repair deficiency in sporadic tumours is seen in colonic, gastric, endometrial, and other solid tumours. MSI is also associated with a wide variety of non-HNPCC and non-colonic tumours, including endometrial, ovarian, gastric, cervical, breast, skin, lung, prostate, and bladder tumours as well as glioma, leukaemia, and lymphoma [65].

Defects in MMR are also associated with resistance to certain chemotherapeutic agents [66]. Resistance to alkylating agents such as TMZ and procarbazine occurs with inactivation of MMR in tumour cells [63]. MMR-deficient cells are relatively resistant to methylating agents (up to 100 fold), whereas cells with a functioning MMR system enter either G2 arrest or apoptosis, depending on the severity of the DNA damage [67]. Down regulation of proteins of the MMR pathway is associated with resistance to clinically important drugs including platinum-containing compounds, anthracyclines, alkylating agents, antimetabolites and epipodophyllotoxins [68].

For example, MSH2 protein deficiency by enhancing MSH2 degradation leads to substantial reduction in DNA mismatch repair and increased resistance to thiopurines. Somatic deletions of genes regulating MSH2 degradation result in undetectable levels of MSH2 protein in leukaemia cells, MMR deficiency and drug resistance [69].

Another agent, etoposide, is a topoisomerase II alpha (*TOP2A*) inhibitor, which is used in the treatment of breast cancer. Alterations in the expression of drug targets or DNA repair genes are among the important resistance mechanisms against TOP2A inhibitors. Decrease in the expression levels of *TOP2A*, and the MMR genes *MSH2* and *MLH1* may play significant roles in the development of chemotherapeutic resistance to etoposide in breast cancer. These genes may be considered for further development of new strategies to overcome resistance against topoisomerase II inhibitors [70].

MMR is also involved in repair of cross-linking agents such as platinum based chemotherapeutics. Increased tolerance to platinum-induced DNA damage can occur through loss of function of the MMR pathway. During MMR, cisplatin-induced DNA adducts are recognized by the MMR pathway, but are not repaired, giving rise to successive repair cycles, ultimately triggering apoptosis. Thus in MMR deficient cells, cell death is not as efficient, promoting tolerance to platinum agents [71].

MMR-deficient cells are also more tolerant to 6-thioguanine treatment, used to treat leukaemias, than MMR-proficient cells. The anti-metabolite 6-thioguanine is incorporated into DNA, where it can be methylated by *S*-adenosylmethionine to 6-methylthioguanine (Me6-thioguanine), which has similar miscoding properties as methylguanine [68].

Nevertheless, although many preclinical studies suggest MMR-deficient cells are resistant to alkylating agents, few clinical studies have been published regarding MMR deficiency and response to alkylating agents. On the contrary, for example, Maxwell *et al.*, [72] found that MMR deficiency does not seem to be responsible for mediating TMZ resistance in adult malignant glioma. Coupled with the lack of substantial data linking polymorphisms within the MMR genes and resistance to chemotherapy or radiotherapy, published work suggests that the MMR pathway has low priority in the quest for new cancer therapies. However, ongoing research on the role of microRNAs and cancer drug resistance could increase interest in this pathway. Published work has suggested that for example miR-21 targets *MSH2* and consequently induces resistance to 5-Fluorouracil (5-FU) in colorectal cancer [73] (see the section of microRNAs and drug resistance).

5. Targeting BER in cancer drug resistance

BER is the main pathway for removing small, non-helix-distorting base lesions from the genome. Thus, BER targets predominantly base lesions that arise due to oxidative, alkylation, deamination, and depurination/depyrimidination damage. Some examples of chemotherapeutic agents that generate lesions that are targeted by BER include TMZ, melphalan, dacarbazine/procarbazine, and streptozotocin [33]. Some chemotherapeutic agents also generate ROS as a "by-product" such as platinum-based drugs (*i.e.* oxaliplatin and cisplatin), anthracyclines, (*i.e.* epirubicin, daunorubicin, doxorubicin) and paclitaxel [31, 33]. ROS induce DNA lesions that are also repaired by the BER pathway. Additionally, IR produces a number of DNA lesions that are repaired by the BER pathway. Endogenous production of ROS also gives rise to several lesions, which are variable in number and consequence. For instance the highly mutagenic 8-hydroxyguanine (8-oxoG) is formed in large quantities as a consequence of the high oxidation potential of this base, and has a miscoding effect, due to DNA polymerase activity which inserts adenine opposite to 8-oxoG, resulting in G:C to A:T transition mutations.

The BER pathway is initiated by one of many DNA glycosylases, which recognize and catalyze the removal of different damaged bases. After recognition of the damaged base by the appropriate DNA glycosylase, it catalyzes the cleavage of an *N*-glycosidic bond, thus removing the damaged base and creating an apurinic or apyrimidinic site (AP site). The DNA backbone is cleaved by either a DNA AP endonuclease or a DNA AP lyase, activity present in some glycosylases. This creates a single-stranded DNA nick 5' to the AP site. The newly created nick is processed by the AP endonuclease, creating a single-nucleotide gap in the DNA. At this point BER can proceed through a short-patch BER, where polymerase β (pol β) introduces a single nucleotide past the abasic site and Ligase IIIα seals the DNA nick, or through a long-patch BER, where Polymerase δ/ε introduces two to eight nucleotides past the abasic site. The resulting overhang DNA is excised by FEN1 endonuclease and the nick sealed by DNA ligase I [74]. In addition to these enzymes, a number of accessory proteins are involved in BER, including the X-ray cross-complementation group 1 protein (XRCC1), PARP1, the proliferating cell nuclear antigen (PCNA), and the heterotrimer termed 9-1-1, which function in scaffolds for the core BER enzymes [75].

Preclinical evidences have implied the BER pathway in the repair of DNA lesions induced by antimetabolites, monofunctional alkylating drugs, radiotherapy and radiomimetic agents. Moreover, BER modulation may also sensitize cancer cells to the effect of chemotherapeutic drugs that are able to generate ROS [31, 33]. Therefore, targeting BER with inhibitors of the multifunctional AP Endonuclease 1 and DNA pol β is an attractive field to the development of novel therapeutic compounds.

Some studies have found deregulation of BER genes in tumours. For example pol β has been shown to be overexpressed in a variety of tumour cells [76]. N-methylpurine DNA glycosylase (MPG) overexpression, together with inhibition of BER, sensitizes glioma cells to the alkylating agent TMZ in a DNA pol β - dependent manner, suggesting that the expression level of both MPG and pol β might be used to predict the effectiveness of BER inhibition and PARP-mediated potentiation of TMZ in cancer treatment [77]. We recently observed an increase in expression of the BER genes *MDB4* and *NTHL1* in Imatinib resistant K562 leukaemia cells, and knockdown of their expression in resistant cells using siRNA decreased cell survival after treatment with doxorubicin [78]. Nevertheless, the involvement of deregulated BER components in chemotherapy resistance is not completely evident at present, except for PARP, and the AP endonucleases. The following text shall describe ongoing research targeting these components of the BER pathway.

The major AP endonuclease in mammalian cells is apurinic/apyrimidinic endonuclease 1/redox-factor-1 (APE1/Ref-1, also called APEX1), and has been found to be elevated in a number of cancers such as ovarian [79], prostate [80], osteosarcoma [81] and testicular cancer [82]. Over-expression of APE1 *in vitro* led to increased protection against bleomycin [82]. Thus elevated levels of APE1 in cancer cells have been postulated to be a reason for chemotherapeutic resistance [81, 83, 84]. Inhibition of APE1 has been shown to increase cell killing and apoptosis and also to sensitize cancer cells to chemotherapeutic agents, and thus APE1 is considered as a molecular target in therapeutics [85, 86].

APE1 endonuclease activity is indirectly inhibited by blocked AP sites that result from the binding of the small molecule methoxyamine (MX) to the DNA. With the APE1's substrate unavailable, BER cannot proceed and the cytotoxic abasic sites accumulate in the cell, eventually leading to cell death. The promising results from *in vitro* and *in vivo* experiments showing MX sensitization to the cytotoxic effect of TMZ [87-90], carmustine [91], pemetrexed [92] and 5-iodo-2'-deoxyuridine (IdUrd) as well as a potentiation of IdUrd-mediated radiosensitization [93, 94], in multiple solid tumours models, provided the proof-of-concept to conduct clinical trials with MX as adjuvant therapy of anticancer agents. A Phase I clinical trial of pemetrexed and oral methoxyamine hydrochloride (TRC102) in patients with advanced refractory cancer is already completed [95]. According to the authors, this drug is well tolerated after daily oral administration and potentiates the activity of chemotherapy. Safety, pharmacokinetic and pharmacodynamic profile of MX was also evaluated in combination with TMZ in a Phase I clinical trial for patients with advanced solid tumours [96]. Currently, two clinical trials (Phase I) are recruiting patients to study the side effects and the best dose of MX to be administered in combination with TMZ and fludarabine phosphate in patients with advanced solid tumours and relapsed or refractory hematologic malignancies, respectively.

In view of the emerging roles of APE1, many efforts have been made to develop small molecule inhibitors that can be translated to the clinic. *In silico* based approaches with design of pharmacophore models [97, 98] and high-throughput screening of several commercially available libraries of compounds have been performed to identify a pharmacologically active inhibitor for APE1 [86, 99-102]. Lucanthone acts as a direct inhibitor of APE1 but also interacts with other cellular targets and the associated toxicity hinders their therapeutic use [103, 104]. CRT0044876 was identified by a fluorescence-based high-throughput assay and showed promising results in *in vitro* studies [105]. However, some authors were not able to reproduce the reported effects of this compound [85].

Hypersensitivity of DNA pol β-null cells to methyl methanesulfonate (MMS), a DNA-methylating agent, displayed another potential target in BER [106]. Several small-molecule inhibitors of DNA pol β have been identified and many of these compounds are natural products, such as koetjapic acid (KJA), a triterpenoid. Pamoic acid was one of the first synthetic small molecule inhibitors of DNA pol β to be characterized and is more active than the former compound [107]. Nevertheless, the actually known inhibitors of DNA pol β have low potency and specificity that make them weak candidates to drug development (for a comprehensive review see [108]). In view of the preclinical data that suggest an important role of DNA Pol β in the repair of chemotherapeutic-induced DNA damage, the design of effective DNA Pol β inhibitors is an attractive research area.

In what concerns PARP1, this enzyme is a DNA damage sensor that binds to DNA breaks to activate the repair pathways. PARP1 is not directly involved in the repair of the lesions but is essential to signal the damage and to coordinate the functions of several BER and DSB repair proteins. PARP inhibitors have been thoroughly developed and several reviews papers published under this topic. For a recent comprehensive review on PARP inhibitors see Javle *et al* [109]. PARP inhibitors were first evaluated in clinical trials as chemosensitizers. After AG014699 combination with TMZ [110], other PARP inhibitors, specifically INO-1001, ABT-888 and AZD2281 were also tested as adjuvant therapy of multiple anticancer agents such as gemcitabine, carboplatin, TMZ or chemotherapeutic combinations (e.g. cisplatin plus gemcitabine) [111]. Currently, several PARP inhibitors are being evaluated in clinical trials, either in combination with chemotherapeutic drugs or in monotherapy [28, 109, 112-117].

Some of these chemicals showed an enhancement of the toxicity in normal tissues that required dose adjustments and optimization of the therapeutic schedule. Interestingly, preclinical and clinical data revealed that PARP inhibitors as single agents could be less toxic to the normal cells and are more effective in killing *BRCA1*- and *BRCA2*-mutated cancer cells since these cells are defective in HR, the backup pathway responsible for the repair of DSBs generated after PARP chemical inhibition. Similarly, mutations in other proteins related to the DNA damage response, such as ATM and PTEN have also been associated to defects in DSB repair and may be involved in an increased sensitivity to PARP inhibitors [118-120]. These findings led to a novel potential therapeutic indication of the DNA repair inhibitors as single agents in cancer therapy which is currently being evaluated in clinical trials [121]. This synthetic lethal approach was also reported in an *in vitro* study with APE1 inhibitors in BRCA and ATM deficient cells [116, 122].

Recently, negative results from the first phase III clinical trial in breast cancer patients with a combination of iniparib (BSI-201) and gemcitabine/carboplatin were reported [123]. The mechanism of action of this inhibitor is not fully understood, an issue that should be further clarified. Nonetheless, promising positive outcomes have already been suggested with other PARP inhibitors [124, 125]. A further understanding of the complex PARP interactome, the discovery of PARP1 specific small molecule inhibitors and an accurate selection of the best candidates to the treatment is still needed to improve the quality of information obtained from preclinical and clinical trials and to promote the development of currently known PARP inhibitors as well to discover novel compounds.

6. Targeting NER in drug resistance

NER repairs DNA lesions which alter the helical structure of the DNA molecule and interfere with DNA replication and transcription, such as bulky adducts and cross-linking agents [2]. Briefly, NER consists of the recognition of DNA damage and demarcation of the specific area affected, followed by the formation of a complex to unwind the damaged portion and excise a 24-32 oligonucleotide section that contains the lesion. Finally, the excised nucleotides are resynthesized and ligated. Two NER sub-pathways exist with partly distinct substrate specificity: global genome nucleotide excision repair (GGR) surveys the entire genome for distorting lesions and transcription-coupled repair (TCR) focuses specifically in the transcribed strand of expressed genes, by targeting damage that blocks elongating RNA polymerases. In total more than 30 proteins participate in NER [126]. The genes involved in GGR are DNA damage recognition by XPC-HR23B complex, lesion demarcation and verification by a TFIIH complex, assembly of a pre-incision complex (RPA, XPA and XPG), DNA opening by XPB and XPD helicases, dual incision by ERCC1-XPF and XPG endonucleases, release of the excised oligomer, repair synthesis to fill in the resulting gap, and ligation by ligase I. Defects in the proteins involved in NER result in three autosomal recessive disorders XP, CS, and TTD.

The most relevant class of chemotherapeutics associated with NER is the platinum-based group of agents. Platinum-based chemotherapy has been used for the treatment of a wide variety of solid tumours including lung, head and neck, ovarian, cervical, and testicular cancer for many years [127]. These agents interact with DNA to form predominantly intra-strand cross-link DNA adducts that trigger a series of intracellular events that ultimately result in cell death. The most studied platinum based cancer therapeutics are cisplatin and the less toxic carboplatin and oxaliplatin, but there has been a resurgence in the development of platinum based drugs, and more platinum based chemotherapeutics are in clinical trials [128].

The basic mechanism of action of cisplatin (and carboplatin) involves covalent binding to purine DNA bases: platinum binding to the N7 position of the imidazole ring of the purine bases of DNA — guanine (G) and adenine (A) — to form either monofunctional or bifunctional adducts. In the case of cisplatin, most occur on the same DNA strand and involve bases adjacent to one another, and are therefore known as intra-strand adducts or crosslinks, namely GpG 1,2 intra-strand (60–65% of all adducts) and ApG 1,2 intra-strand (20–25%) which primarily leads to cellular apoptosis [128]. These DNA lesions are repaired by the NER pathway.

Cisplatin has been used successfully as therapy to treat metastatic testicular cancer with >90 % cure rate. The high sensitivity of testicular tumour cells is attributed to reduced DNA-repair capacity in response to platinum–DNA adducts [129]. Extracts from testicular cancer cells had low constitutive NER capacity and, in particular, low levels of the protein XPA [130]. Further studies have shown low levels of XPA and other NER proteins (XPF and ERCC1), in testicular cancers. This suggested that reducing NER capacity in a cancer holds the potential to sensitize the cancer to cisplatin. Parallel studies revealed that increased DNA repair capacity was a common function in cancers that were inherently resistant to cisplatin or that acquired resistance following treatment [130].

Clinical studies in ovarian cancer patients have correlated increased excision repair cross-complementation group 1 – (ERCC1) mRNA levels with clinical resistance to platinum based chemotherapy [131, 132]. In metastatic colorectal cancer patients, higher ERCC1 expression levels were considered as predictive for lower survival rates when treated with oxaliplatin in combination with 5-fluorouracil, suggesting that enhanced DNA repair decreases the efficacy of platinum-based treatment [133]. In another study a subgroup of 761 patients with metastatic lung cancer treated with a platinum based compound were retrospectively evaluated by immunohistochemical analysis of ERCC1. This study showed a statistically significant survival benefit in patients with low levels of ERCC1 who had received platinum based chemotherapy, compared to patients with low levels of ERCC1 who did not receive chemotherapy and patients with high levels of ERCC1 who received cisplatin chemotherapy [134]. Also, low ERCC1 expression correlated with prolonged survival after cisplatin plus gemcitabine chemotherapy in non-small cell lung cancer (NSCLC) [135].

Hence, it is hypothesized that high expression of the ERCC1 gene might be a positive prognostic factor, and could predict decreased sensitivity to platinum-based chemotherapy. Expression of ERCC1 has been used to stratify patients treated with platinum based chemotherapeutics with some success, and also to predict improved survival in platinum treated patients [136]. Nonetheless, results from the published data are inconsistent. To derive a more precise estimation of the relationship between ERCC1 and the prognosis and predictive response to chemotherapy of NSCLC, a meta-analysis was performed and results indicated that high ERCC1 expression might indeed be a favourable prognostic and a drug resistance predictive factor for NSCLC [137].

Other studies with different tumour/chemotherapy associations have shown that ERCC1 mRNA expression in tumours may be a predictive marker of survival for Irinotecan-resistant metastatic colorectal cancer receiving 5-FU and Oxaliplatin combination chemotherapy [133]. In this study patients whose tumours had low ERCC1 mRNA expression had a significantly longer median survival than those with high ERCC1 expression.

Other genes involved in NER have been shown to influence drug resistance. For example, increased expression of excision repair cross-complementation group 4 (ERCC4 or XPF) was observed in hydroxycamptothecin (HCPT) treated bladder cancer tissue compared to untreated samples. Complementary *in vitro* studies showed that enhanced *ERCC4* expression decreased the sensitivity of bladder T24 cells and 5637 cells to HCPT, whereas after gene silencing of *ERCC4* the chemotherapeutic resistance of bladder cancer cells to HCPT was significantly decreased [138].

Since the NER pathway is crucial for the repair of bulky adducts and cross-linking agents in normal cells, the development and application of NER inhibitors in clinical settings is scarce, although preclinical data show that the manipulation of this pathway could be a relevant strategy in cancer chemotherapy. For example, preclinical studies have demonstrated that the chemotherapeutic action of the platinum agent oxaliplatin is improved when combined with cetuximab, a chimeric IgG1 monoclonal antibody targeting the epidermal growth factor receptor. This antibody has been shown to reduce the expression of *ERCC4* and *ERCC1*. A concomitant increase in the accumulation of platinum and apurinic/apyrimidinic sites on DNA during oxaliplatin treatment was observed, thus leading to an increase in apoptosis [139, 140]. These interesting results are suggestive that targeting other pathways that regulate expression of DNA repair genes could be a promising strategy.

7. HR and drug resistance

HR repairs DSBs, which occur through exposure to various chemotherapeutic agents, including IR, topoisomerase inhibitors and DNA crosslinking agents (e.g. mitomycin, campto-thecins, etoposide, doxorubicin, daunorubicin and bleomycin). HR is also recruited to restart stalled replication forks and to repair ICL, the repair of which also involves the FA protein complex. HR ensures the accurate repair of DSBs by using a homologous undamaged DNA strand from an intact sister chromatid as a template for DNA polymerase to extend past the break, and is thus restricted to late S and G2 of the cell cycle. Components of HR include the RAD group of proteins (including RAD50, RAD51, RAD52, and RAD54), RPA, XRCC2, XRCC3, and the BRCA proteins. Briefly, HR occurs through pre-synapsis, preparation of a recombination proficient DNA end; synapsis, formation of a joint molecule between the recombination proficient DNA end and a double-stranded homologous template DNA; post-synapsis and resolution, repair of DNA strands and separation of the recombined DNA molecules [19]. DSBs can also be repaired by NHEJ that do not utilize significant homology at the broken ends. In NHEJ, DSBs are recognized by the Ku protein that then binds and activates the protein kinase DNA-PKcs, leading to recruitment and activation of end-processing enzymes, polymerases and DNA ligase IV. Whereas HR is restricted to late S and G2, NHEJ functions in all phases of the cell cycle and ligates broken DNA ends without the need of an undamaged template.

Following DNA lesions initial checkpoint signalling is performed by the kinases ATR and ATM, two phosphatidylinositol 3-kinase family members. Activation of these kinases leads to activation of the effector kinases, checkpoint kinases 1 and 2 (Chk1 and Chk2; serine/threonine kinases). The activated effector kinases are then able to transiently delay cell cycle progression through the G1, S, or the G2 phases so that DNA can be efficiently repaired. The

ATM/Chk2 pathway predominantly regulates the G1 checkpoint and the ATR/Chk1 pathway the S and G2 checkpoints. However, there is cross-talk between the pathways implying a role for both ATR and ATM pathways in all cell cycle checkpoints. In addition to directly regulating the cell cycle, the pathways also affect DNA repair, transcription, chromatin regulation, and cell death. Many details of these pathways are not fully known.

One consequence of DSBs is the localized alteration of chromatin adjacent to DSBs in order to facilitate recruitment of repair proteins. For examples, ATM not only phosphorylates DNA repair proteins recruited to DNA ends but also the histone variant H2AX in nucleosomes adjacent to DSBs, which is also phosphorylated by DNA-dependent protein kinase (DNA-PK), another protein kinase activated by DSBs. Phosphorylated H2AX (known as γ-H2AX) around DSBs facilitates the recruitment of a number of DNA repair proteins and chromatin modulating factors. The presence of large patches of γ-H2AX around a DSB has made its detection by fluorescent tagged antibodies a biomarker for DSBs [141, 142].

There is accumulating evidence for the existence of HR defects not only in familial cancers but also in sporadic cancers. Mutations or epigenetic alterations have been observed in several genes known to be involved in HR regulation and repair, such as BRCA1 and BRCA2. Functional analysis of human cancer tissues and cancer cell lines has revealed HR deficiency, chromatid-type chromosomal aberrations, severe ICL hypersensitivity, and impaired formation of damage-induced RAD51 foci. For example, although genetic mutations in BRCA1 or BRCA2 are only rarely found in sporadic tumors, in contrast to familial breast and ovarian cancers, epigenetic gene inactivation of the BRCA1 promoter is a fairly common event in sporadic breast cancers, with aberrant methylation being detected in 11 to 14% of cases [143]. Non-triple-negative sporadic breast cancers may also harbor HR defects. It has been suggested that ~20% of these cancers are defective in HR as measured by an impaired ability to mount RAD51 foci in response to chemotherapy [144]. There is emerging evidence that approximately up to one fifth of non-familial breast cancers harbour HR defects that may be useful targets for therapy.

The BRCA1 and BRCA2 proteins are involved in HR, in association with FA proteins, forming a complex DNA damage response network [145]. BRCA1 expression levels have been demonstrated to be a biomarker of survival following cisplatin-based chemotherapy for NSCLC and ovarian cancer, suggesting that this gene could be involved in response to platinum therapy [146, 147]. In vitro studies indicate that loss of BRCA1 or BRCA2 increases sensitivity to agents that cause DSBs such as bleomycin and/or ICLs including platinum agents. Conversely, loss of BRCA1 or BRCA2 may increase resistance to microtubule interfering agents such as taxanes and vincristine [148, 149]. In contrast, BRCA1 may increase sensitivity to spindle poisons by activating the mitotic spindle checkpoint and signalling through a proapoptotic pathway. This dual role of increasing apoptosis and therefore sensitivity to spindle poisons and also promoting DNA repair and cell survival after treatment with DNA-damaging drugs may influence the response of breast and ovarian cancer cells to treatment [150]. Chemotherapy in breast and ovarian cancers is attained by treatment with platinum based compounds and anthracyclines and also taxanes, all of which induce both

SSBs and DSBs. Efforts are underway to use *BCRA1* as a predictive marker for chemotherapy customization and response [151].

Regarding other types of cancer, *BRCA1* promoter hypermethylation is also found in approximately 5-30% of sporadic ovarian cancers. Also, mutations in *BRCA1* and *BRCA2* have recently been found in up to 20% of unselected ovarian cancers [152]. Thus, these HR deficient cancers are viable targets for synthetic lethality approaches with PARP inhibitors. Defects in the FA/BRCA pathway as well as ATM defects have been described in a variety of other malignancies, such as prostatic adenocarcinoma, colorectal cancer, leukaemia, lymphoma, and medulloblastoma [153, 154]. However, it remains to be seen whether these defects can be targeted effectively in the clinic.

Single-agent chemotherapy with a nitrogen mustard, usually Chlorambucil, is the standard initial therapy for Chronic lymphocytic leukaemia (CLL) and at least 60–80% of patients respond but eventually all patients become resistant to these agents. XRCC3 protein levels and DNA-damage induced RAD51 foci correlates with chlorambucil drug resistance in lymphocytes from CLL patients and with melphalan and cisplatin resistance in epithelial tumor cell lines, indicating that increased HR can be involved in drug resistance to these agents [155].

Another component of the HR pathway, *RAD51,* has been found to be increased in expression in a wide range of human tumors, most likely contributing to drug resistance of these tumors. Over-expression of *RAD51* in different cell types leads to increased homologous recombination and increased resistance to DNA damaging agents to disruption of the cell cycle and apoptotic cell death. *RAD51* expression is increased in p53-negative cells, and since *TP53* is often mutated in tumor cells, there is a tendency for *RAD51* to be overexpressed in tumor cells, leading to increased resistance to DNA damage and drugs used in chemotherapies [156].

Chronic myeloid leukaemia (CML) cell lines expressing the fusion protein BCR-ABL1 utilize an alternative non-homologous end-joining pathway (ALT NHEJ) to repair DSBs. The expression levels of PARP1 and DNA ligase IIIα served as biomarkers to identify a subgroup of CML patients who may be candidates for therapies that target the ALT NHEJ pathway when treatment with TKIs has failed [157]. Tamoxifen- and aromatase-resistant derivatives of MCF7 cells and Estrogen Receptor⁻/Progesterone Receptor⁻ (ER⁻/PR⁻) cells have higher steady-state levels of DNA ligase IIIα and increased levels of PARP1, another ALT NHEJ component. Notably, therapy-resistant derivatives of MCF7 cells and ER⁻/PR⁻ cells exhibited significantly increased sensitivity to a combination of PARP and DNA ligase III inhibitors that increased the number of DSBs. Thus, ALT NHEJ may be a novel therapeutic target in breast cancers that are resistant to frontline therapies and changes in NHEJ protein levels may serve as biomarkers to identify tumors that are candidates for this therapeutic approach [158].

Another interesting approach in this field is to target components of the DNA damage response, namely DNA damage signalling and cell-cycle checkpoints [34]. The members of the phosphatidylinositol (PI) 3-kinase-like (PIKK) family perform crucial roles in the activation of DSB repair pathways, namely in HR and NHEJ. ATM, a PIKK family member, is a DSB signalling protein mainly implicated in the phosphorylation of effector proteins from HR. ATM has been also involved in the regulation of NHEJ. KU55933, 2-morpholin-4-yl-6-thianthren-1-yl-pyran-4-one is a specific and potent small-molecule inhibitor of ATM identified by screening of a combinatorial library. Preclinical studies have shown an increase in the cytotoxicity of multiple chemotherapeutic drugs as doxorubicin, etoposide, camptothecin and ionizing radiation [159, 160] while the UV-induced cellular effects were not modified. More recently, KU60019, an improved analogue of KU55933, was developed. Besides its radiosensitizing properties, *in vitro* studies revealed that KU60019 may also impair the migration and invasion of tumor cells by inhibiting ATM-mediated AKT phosphorylation [161].

DNA-PK is also a target to the development of chemo- and radiosensitizers [162]. In fact, the identification of specific small molecule modulators of DNA-PK [163-165], namely NU7441 and NU7026, was shown to potentiate the effects of ionizing radiation as well as chemotherapeutic agents in human tumor cell lines and in *in vivo* xenograft models.

Another example is the development of AZD7762, which potently inhibits Chk1 and Chk2, abrogates DNA damage-induced S and G2 checkpoints, enhances the efficacy of gemcitabine and topotecan, and modulates downstream checkpoint pathway proteins [166]. This agent has been evaluated in clinical trials, however due to an inadequate response the drug has been discontinued in 2011 (http://www.astrazenecaclinicaltrials.com).

8. MicroRNAs and chemotherapy resistance

MicroRNAs (miRs) are small non-coding RNAs (19 to 25 nucleotides) that regulate gene expression by binding to 3' untranslated region (UTR) of several mRNAs, thus blocking translation. Recently, it was also shown that miRs can act by binding to open reading frames or 5'UTR of mRNAs, as revised by Iorio and Croce [167]. Due to small size and incomplete complementarity to mRNA, one miR can have a widespread effect on the transcriptome of a cell, acting as a hallmark of several diseases, including cancer. Numerous studies have been performed regarding biogenesis and function of miRs, being revised elsewhere [168-170]. *In vitro* and *in vivo* studies have suggested that miRNAs might be useful as diagnostic and prognostic markers, and recent data suggest that miRNA profiling can be used for tumor typing.

Although it is well established that miRs have an important role in cancer, the complexity of their action remains to be understood and questions regarding their use as cancer therapy need further investigation. The strong pleiotropy of miRs in deregulating normal cellular homeostasis due to misexpression, has led investigators to believe that they are valuable targets for cancer therapy and consequently for drug resistance. Two major approaches for using miRs as therapeutics can be described. First, miRs can be used as single molecules or combined in order to target one or multiple transcripts. In this approach, a miR or a set of miRs are antagonized or mimicked to alter miR levels and consequently change the protein outcome in a cancer cell. Second, miRs can act as modulators of cell sensitivity for cancer therapy [167, 171]. This second approach will be our focus.

Many studies regarding miRs expression patterns in cancer cells have been performed. These studies not only allow investigators to determine novel biomarkers for a better and easily prognostication of several types of cancer but also the functional role of the same miRs. These can give us the knowledge if the loss or gain of miR function interferes with the original balance of protein levels which may be important, but not only, in drug response and consequently lead to drug resistance. Since miRs expression seems to be tissue, grade and stage specific, the ectopic expression or repression of miRs in conjugation with cancer therapy seems promising. For that reason, recent studies that evaluate miR expression profiles of sensitive and resistant cell lines have been made in order to find the key miR signatures related to drug response, which not only promote further analysis of the mechanisms of cancer drug resistance, but also allow the discovery of new drug targets and individualized medicine.

Although the study of the therapeutic potential of miRs is still recent, several studies have been published and compiled. For example, Tian et al. [172] and Kutanzi et al. [173], published compilations of several studies reporting influence of miRs in mechanisms of drug resistance and how they can modulate drug response in breast cancer.

With regard to miRs and modulation of drug resistance through regulation of DNA damage and repair genes, studies are scarce. It is known that miRs have an important role in DNA damage response, which includes DNA repair [174, 175]. One example how miRs can influence drug resistance through DNA repair is demonstrated by Valerie et al. [73]. The authors showed that miR-21 targets *MSH2* and consequently induces resistance to 5-FU in colorectal cancer. Since miR-21 has a pleiotropic effect, it is possible that it could regulate other genes associated with drug resistance. However, the impact of *MSH2* seems to be of extreme importance on acquired 5-FU resistance since when knocked out cells for *MSH2* are transfected with miR-21, cell-cycle arrest or apoptosis is not altered. These results show that the inhibition of miR-21 action might represent an important treatment to overcome 5-FU resistance. A correlation between miR-21 and *MSH2* in breast cancer was also found [176]. It is recognized that TGF-β is a promoter of miR-21 processing through the interaction with the SMAD and DROSHA complex. On the other hand, *MSH2* is a proven target of miR-21. Thus, TGF-β inhibits *MSH2* gene expression and consequently increases drug resistance. Indeed, to find out if TGF-β contributes to drug resistance through *MSH2*, the authors tested the response of breast cancer MDA-MB-231 cell line to cisplatin, methyl methanesulfonate (MMS) and doxorubicin in the presence and absence of TGF-β. Exposure to TGF-β for 24 h increased cell viability upon treatment with these DNA damaging agents and knock down of *MSH2* induced resistance to both cisplatin and doxorubicin. In contrast, transfection of the anti-miR-21 enhanced the effect of cisplatin in MDA-MB-231 cells.

Another example of miR influence in DNA repair and consequent drug response is miR-182 that targets *BRCA1*. Moskwa and colleagues showed that ectopic expression of miR-182 represses BRCA1 protein expression and sensitizes breast cancer cells to PARP inhibitors [177]. However, PARP inhibitors are mostly used in patients with BRCA1 inherited mutations. Therefore, the question if PARP inhibitors are useful therapeutic drugs in sporadic breast cancer rises. Theoretically, if administered with BRCA1 repressors such as miR-182, PARP inhibitors can have the same effect as in inherited breast cancer. Further studies need to be done in order to clarify this issue.

As described previously, MGMT has DNA repair activity insofar as it can remove mutagenic O^6-alkylguanine induced by alkylating agents. Although TMZ has been widely used in glioblastoma multiforme (GBM), many patients become or are resistant to this chemotherapy agent, since MGMT can repair the DNA damage induced by TMZ. Epigenetic regulation mechanisms, such as methylation of the *MGMT* gene promoter can sensitize cancer cells to alkylating chemotherapeutic drugs. Glioblastoma patients with positive methylation status of *MGMT* gene promoter have been reported to present a better response to TMZ treatment [44], but these results have not been confirmed by other studies, and therefore results are ambiguous [178]. Indeed, some patients with unmethylated status of *MGMT* promoter gene also have good response to TMZ, which points out to other regulatory mechanisms of *MGMT* expression [179]. Thus, miRs appear as good alternative regulation candidates of *MGMT* expression levels. Recent evidence also suggests that the miR-181 family might be associated to drug response [180]. The authors found that glioblastoma patients with low expression of miR-181b and miR-181c have a better response to TMZ. On the contrary, miR-181d seems to post-transcriptionally regulate *MGMT* since both directly interact and inversely correlate in relation to expression levels [181]. This fact is important because it could be a predictive biomarker for chemotherapy response in GBM. Lakomy and collaborators found that high expression of miR-195 and miR-196b is significantly associated with longer survival of GBM patients and miR-21 and miR-181c with high risk GBM patients [182]. However none of these miRs were associated with *MGMT* gene promoter status.

Altogether the potential for use of miRs in cancer therapy is high, so are the challenges, since each miR can target up to hundreds of mRNA targets. The rapid elucidation of the role they play in cancer suggest that translation of this knowledge will rapidly reach the clinic.

9. Phytochemicals as alternative therapies against drug resistance

As discussed previously, frequently novel therapeutics that show promising results in preclinical assays reveal unacceptable toxicity in clinical trials. Since cancer cells frequently present deregulation of multiple cellular pathways, targeting multiple pathways seems more promising than using single agents that target single pathways. In recent years natural dietary compounds such as curcumin, resveratrol and soy isoflavones such as genistein, have received attention due to the fact that they frequently target multiple cell signalling pathways, including the cell cycle, apoptosis, proliferation, survival, invasion, angiogenesis, metastasis and inflammation. Thus their use in chemoprevention has gained attention [183, 184]. Additionally, since most of the cancer drugs developed have been deliberately directed toward specific molecular targets that are involved in one way or another in enabling particular cellular functions, in response to monotherapy cancer cells may reduce their dependence on a particular proficiency (e.g. a single repair pathway), becoming more dependent on another, thus contributing to acquire drug resistance. Thus, as an alternative approach, selective co-targeting of multiple core and emerging hallmark proficiencies in mechanism-guided combinations could result in more effective and durable therapies for human cancer [185]. Phytochemicals can be highly pleiotropic, modulating numerous targets, including the activation of transcription factors, receptors, kinases, cytokines, enzymes, and growth factors [186]. Therefore current efforts are highly engaged in discovering natural plant-based chemicals that could assist in the fight against drug resistance.

For example soy isoflavones inhibited APE1 expression in prostate cancer cells in a time- and dose-dependent manner, whereas IR up-regulated expression of this BER gene, in response to DNA damage [187-190]. Pretreatment of cancer cells with soy isoflavones inhibited the increase in expression of APE1, and enhanced the efficacy of chemotherapy and radiation therapy of multiple cancers models *in vitro* and *in vivo*, possibly through down-regulation of this DNA repair gene [188]. Another phytochemical, resveratrol, was also shown to inhibit APE1 endonuclease activity and render melanoma cells more sensitive to treatment with the alkylating agent dacarbazine [191]. Thus both resveratrol and isoflavones such as genistein can have therapeutic potential as an APE inhibitor. A series of analogs of resveratrol have been generated in recent years, which exhibit increased potency and/or a range of selective activities compared to the parental compound resveratrol, and possibly improved pharmacokinetic properties [192]. A clinical trial of resveratrol in colon cancer has recently been completed (http://www.clinicaltrials.gov).

Resveratrol can also increase *BRCA1* and *BRCA2* expression, although no effect is seen at the protein level [193]. An increase in BRAC1 expression can lead to increased arrest of cells in the G2 phase, thus making them much more sensitive to conventional therapy. One common chemotherapeutic drug is doxorubicin, which predominantly induces DNA damage in G2 phase cells [194]. Resveratrol, curcumin and the naturally occurring flavolignan deoxypodophyllotoxin [195] can induce G2/M cell cycle arrest, and alter the expression of cell cycle regulatory proteins, thus allowing doxorubicin to induce lesions and as a consequence enhance the apoptotic effect [186, 196, 197]. Le Corre *et al.*, also demonstrated that resveratrol has an effect on the expression of genes implicated in the regulation of BRCA1 protein functions and in multiple nuclear processes modulated by BRCA1 in human breast cancer and fibrocystic breast cells [198]. One of the mechanisms by which resveratrol can enhance *BRAC1* expression is by association with BRCA1, repressing the aromatic hydrocarbon receptor (AhR). AhR binds many natural dietary bioactive compounds therefore combination diets with AhR antagonists may offer the advantage of higher cancer prevention efficacies [199]. In HR-deficient tumours, patients with heterozygous mutations in the HR genes *BRCA1* and *BRCA2* develop breast and ovarian tumours with functional loss of HR activity, and deficiency in this pathway may dictate the sensitivity of tumours to certain DNA-damaging agents and this may be another possible approach to test natural compounds to overcome resistance, and once more enhance combinatory strategies to optimize treatment outcome [32].

Recently an extract of neem leaves was characterized and a significant up-regulation of genes associated with metabolism, inflammation and angiogenesis, such as *HMOX1* and

AKR was observed. However genes associated with cell cycle, DNA replication, recombination, and repair functions were down-regulated [200]. One study analysed 531 compounds derived from plants and found no correlation with genes involved in NER (*ERCC1, XPA, XPC, DDB2, ERCC4, ERCC5*) or BER (*MPG, APE1, OGG1, XRCC1, LIG3, POLB*). It is possible that natural compounds may target different molecular pathways from those of standard anti-tumor drugs, hence if DNA repair is involved in the development of resistance to established anticancer drugs, natural compounds may be attractive sources of novel drugs suitable to treat drug resistant tumours, with the advantage of having reduced side effects [201].

Likewise, most plant derivatives can act as antioxidants and some of them can increase human *MGMT* expression (e.g. curcumin, silymarin, sulforaphane and resveratrol) beyond its steady-state levels, having a role in cancer chemoprevention [202]. Additionally, both *BRCA1* and *MGMT* genes are susceptible to hypermethylation, and green tea polyphenols and bioflavonoids have been shown to reverse the effects of DNA hypermethylation [203].These results suggest that some dietary compounds may have a potential demethylating effect, and could be promising adjuvants to chemotherapy in drug resistant settings.

Another issue in cancer chemotherapy is the use of monotherapy *vs* combined therapy, and several studies have been performed regarding possible combinatory chemotherapy with natural compounds (less aggressive than the majority of chemotherapeutic drugs), albeit in pre-clinical settings, *e.g.* silibinin extract [204], ixabepilone [205] and curcumin [206]. Some of these agents are being evaluated in clinical trials. Silibinin strongly synergized the growth-inhibitory effect of doxorubicin in prostate carcinoma cells, which was associated with a strong G2-M arrest followed by apoptosis [204]. Ixabepilone, an analogue of the natural product epothilone B, is already indicated for the treatment of locally advanced or metastatic breast cancer in the US. In a phase III trial in women with locally advanced or metastatic breast cancer that were pretreated with, or resistant to, anthracyclines (e.g. doxorubicin) and resistant to taxanes, progression-free survival was significantly longer in ixabepilone plus capecitabine recipients compared with recipients of capecitabine monotherapy [205]. Combination therapy using curcumin with gemcitabine-based chemotherapy, in a phase I/II study, in patients with pancreatic cancer warrants further investigation into its efficacy [206].

Finally, an interesting recent development concerns the observation that miRs could be regulated by natural agents, leading to the inhibition of cancer cell growth, epithelial to mesenchymal transition (EMT), drug resistance, and metastasis [207]. For most epithelial tumors, progression toward malignancy is accompanied by a loss of epithelial differentiation and a shift toward mesenchymal phenotype [185]. During the acquisition of EMT characteristics, cancer cells lose the expression of proteins that promote cell-cell contact, such as E-cadherin and γ-catenin, and gain the expression of mesenchymal markers, such as vimentin, fibronectin, and N-cadherin, leading to enhanced cancer cell migration and invasion. It has been shown that down-regulation or the loss in the expression of the miR-200 family is associated with EMT. Gemcitabine-resistant pancreatic cells having EMT characteristics showed low expression of the miR-200 family and miR-200 is lost in invasive breast cancer cell lines with mesenchymal phenotype. Hence the interesting observation that isoflavone could induce miR-200 expression in gemcitabine-resistant pancreatic cells, resulting in altered cellular morphology from mesenchymal-to-epithelial appearance and induced E-cadherin distribution that is more similar to epithelial-like cells. Likewise, let-7 has been found to regulate cell proliferation and differentiation, and inhibit the expression of multiple oncogenes, including ras and myc, and again it was observed that isoflavone could significantly up-regulate the expression of let-7 family, suggesting that this phytochemical could reverse EMT characteristics in part due to the up-regulation of let-7 [207]. Other reports have shown that curcumin, isoflavone, indole-3-car-binol (I3C), 3,3'-diindolylmethane (DIM), (−)-epigallocatechin-3-gallate (EGCG) or resveratrol, can alter miRNA expression profiles, leading to the inhibition of cancer growth, induction of apoptosis, reversal of EMT phenotype, and increasing drug sensitivity [208].

It remains to be seen if phytochemicals can affect miRs that regulate DNA repair pathways, but since any given miR can target several transcripts, this regulation is highly likely. Overall, natural compounds, may have an important role in chemoprevention and in combined therapy, and may prevent resistance to chemotherapy [188, 189, 208-210].

10. Conclusion and future directions

As discussed in this chapter, the ultimate target of chemotherapy and radiotherapy is the cancer cell, and use of DNA damaging agents is justifiable since most of these cells are highly cycling cells. The targeting of DNA repair pathways is but one of the many strategies developed in the fight against cancer. Cancer cells frequently possess altered DNA repair capacities, and this can be put to use in the clinic. Thus the quest for specific therapies that target DNA repair has produced many potentially useful agents (Table 1). Using such agents can theoretically increase the efficacy of existing chemotherapy and/or radiotherapy. Nevertheless, the same difficulties encountered by all other alternative strategies are also arising when we disrupt DNA repair processes.

The success of these agents ultimately will depend on our basic knowledge of the various DNA repair processes present in a given cell type or tissue. Not all DNA repair pathways are present in all tissues, as evidenced by the fact that mutations in specific pathways give rise preferentially to certain tumour types and not others. Secondly, the success will also depend on the specific genomic and genetic landscape of each tumour, implying that different combinations of inhibitors and chemical agents shall have to be tailored to each tumour. We are still far from achieving this goal, but great strides have been taken in the past years. Thirdly, we shall have to redirect the strategy to discover a "cure for cancer" and instead follow strategies that allow us to accompany the inevitable and inexorable evolution of the cancer cell and consistently find and implement more and more targeted therapies, even if these strategies lead us to return to abandoned therapies. The resurgence of drug holidays, in which a therapy is abandoned temporarily to be taken up after a certain period, not unlike what can be adopted with antibiotics, is one such strategy. In this case the absence of a selective pressure imposed by a specific agent may lead cancer cells to lose resistance to this agent, making them again vulnerable to the same agent. This strategy has been followed in certain cancers and could be adapted in others, with the advantage of offering reduced time on chemotherapy, reduced cumulative toxic effects, and improved quality of life [211, 212].

Target	Drug	Condition or tumor	Combination therapy agent(s)	Phase of clinical trial planned, ongoing or recently completed*	Reference
MGMT	O^6-Benzylguanine	Multiple Myeloma and Plasma Cell Neoplasm	Carmustine	Phase II completed	www.cancer.gov
		Glioblastoma, Gliosarcoma	Temodar	Phase II completed	
		Melanoma	Carmustine	Phase II completed	
		Colorectal Cancer	Carmustine	Phase II completed	
PARP1	AZD-2281/ KU59436 (Olaparib)	Triple Negative Breast Cancer	Cisplatin	Phase I/II active	www.astrazeneca.com
		Triple Negative Metastatic Breast Cancer	Paclitaxel	Phase I/II completed	
		Known BRCA Ovarian Cancer or Known BRCA/ Triple Neg. Breast Cancer			
	AG014699/ PF-01367338 (Rucaparib)	Solid tumors	Temozolomide	Phase I completed	www.pfizer.com
		Melanoma	Various agents	Phase II ongoing	
	INO-1001	Melanoma	Temozolomide	Phase I terminated	www.inotekcorp.com
	BSI-201/ (Iniparib)	Uterine Carcinosarcoma	Carboplatin, Paclitaxel,	Phase II active	www.biparsciences.com www.sanofi.com
		Breast Cancer	Gemcitabine/ Carboplatin	Phase II completed Phase III active	
	ABT-888/ (Veliparib)	Breast cancer	Carboplatin Temozolomide	Phase II active	www.abbott.com
		Prostate Cancer	Temozolomide	Phase I active	
		Melanoma	Temozolomide	Phase II active	
		Various cancers	Various agents	Phase I/II active	
	MK4827	Solid BRCA Ovarian	Single agent Various agents	Phase I ongoing	www.merck.com
	CEP-9722	Solid tumours	TMZ Various agents	Phase I	www.cephalon.com www.tevapharm.com
	GPI 1016/ E7016	Solid tumours	TMZ Various agents	Phase I	www.eisai.com
	LT673	Hematological cancers Solid tumours	Various agents	Phase I ongoing	www.bmrn.com
	NMS-P118			Preclinical; highly selective against PARP-5 (tankyrase)	www.nervianoms.com
BER	Methoxyamine/ TRC-102	Advanced refractory solid cancers	Pemetrexed	Phase I active	www.traconpharma.com
		Hematological cancers	TMZ Fludarabine	Phase I ongoing	
ATM Kinase	KU55933			Preclinical	www.astrazeneca.com
CHK1	AZD7762				www.astrazeneca.com
	PF-00477736				www.pfizer.com
	XL844				www.exelixis.com
FA	Curcumin	Gastrointestinal cancers		Phase II	

Target	Drug	Condition or tumor	Combination therapy agent(s)	Phase of clinical trial planned, ongoing or recently completed*	Reference
Pathway					
c-ABL	Imatinib	Various solid tumours		Phase III	www.novartis.com
EGFR	Erlotinib	NSCLC	Monotherapy or combination	Phase II/III	www.gene.com
	Gefinitib				www.astrazeneca.com

* As of 10 September 2012, http://clinicaltrials.gov

Table 1. Targeted therapeutics in development, in clinical use or in clinical trials*.

This leads to the final and perhaps most challenging problem in the development of agents that modulate DNA repair, which is toxicity to normal cells, in particular to the hematopoietic system and the gastrointestinal epithelia. Various strategies are being followed to minimize toxicity, which include the intermittent administration during therapy, mentioned above, alternating with other therapies, using highly localized radiotherapy together with inhibitors to minimize collateral damage, and using inhibitors as single agents [213, 214]. Altogether, the combined use of the various weapons at our disposal in a coordinated, comprehensive fashion could effectively lead to improved patient treatment.

Acknowledgements

This work was supported by grants PTDC/SAUGMG/71720/2006 from Fundação de Ciência e Tecnologia (FCT), and PEst-OE/SAU/UI0009/2011-12 from FCT. M.G. was supported by CIENCIA 2008 (FCT). CIGMH is supported by FCT. B.G. (SFRH/BD/64131/2009), P.G. (SFRH/BD/70293/2010) and C.M. (SFRH/BD/ 81097/2011) are supported by Ph.D. grants from FCT.

Author details

António S. Rodrigues[1]*, Bruno Costa Gomes[1], Célia Martins[1], Marta Gromicho[1], Nuno G. Oliveira[2], Patrícia S. Guerreiro[2] and José Rueff[1]

*Address all correspondence to: sebastiao.rodrigues@fcm.unl.pt

1 CIGMH – Department of Genetics, Faculty of Medical Sciences, Universidade Nova de Lisboa, Lisboa, Portugal

2 Research Institute for Medicines and Pharmaceutical Sciences (iMed.UL), UL, Faculty of Pharmacy, Universidade de Lisboa, Lisboa, Portugal

References

[1] Friedberg EC. DNA Repair And Mutagenesis: ASM Press; 2006.

[2] Hoeijmakers J. Genome maintenance mechanisms for preventing cancer. Nature. 2001;411:366 - 74.

[3] Harper JW, Elledge SJ. The DNA Damage Response: Ten Years After. Molecular cell. 2007;28(5):739-45.

[4] Lin Z, Nei M, Ma H. The origins and early evolution of DNA mismatch repair genes —multiple horizontal gene transfers and co-evolution. Nucleic Acids Research. 2007;35(22):7591-603.

[5] Lehmann A, McGibbon D, Stefanini M. Xeroderma pigmentosum. Orphanet Journal of Rare Diseases. 2011;6(1):70.

[6] Paradiso A, Formenti S. Hereditary breast cancer: clinical features and risk reduction strategies. Annals of Oncology. 2011;22(suppl 1):i31-i6.

[7] Hoeijmakers JHJ. DNA Damage, Aging, and Cancer. New England Journal of Medicine. 2009;361(15):1475-85.

[8] Lengauer C, Kinzler KW, Vogelstein B. Genetic instabilities in human cancers. Nature. 1998;396(6712):643-9.

[9] Loeb LA. A Mutator Phenotype in Cancer. Cancer Research. 2001;61(8):3230-9.

[10] Lagerwerf S, Vrouwe MG, Overmeer RM, Fousteri MI, Mullenders LHF. DNA damage response and transcription. DNA Repair. 2011;10(7):743-50.

[11] Rodrigues AS, Dinis, J., Gromicho, M., Martins, M., Laires, A. and Rueff, J.. Genomics and Cancer Drug Resistance. Current Pharmaceutical Biotechnology. 2012;13(5): 651-73.

[12] Gillet J-P, Gottesman MM. Mechanisms of Multidrug Resistance in Cancer Multi-Drug Resistance in Cancer. In: Zhou J, editor.: Humana Press; 2010. p. 47-76.

[13] O'Connor MJ, Martin NMB, Smith GCM. Targeted cancer therapies based on the inhibition of DNA strand break repair. Oncogene. 2007;26(56):7816-24.

[14] Al-Ejeh F, Kumar R, Wiegmans A, Lakhani SR, Brown MP, Khanna KK. Harnessing the complexity of DNA-damage response pathways to improve cancer treatment outcomes. Oncogene. 2010;29(46):6085-98.

[15] Hurley LH. DNA and its associated processes as targets for cancer therapy. Nat Rev Cancer. 2002;2(3):188-200.

[16] Kokkinakis DM, Ahmed MM, Delgado R, Fruitwala MM, Mohiuddin M, Albores-Saavedra J. Role of O6-Methylguanine-DNA Methyltransferase in the Resistance of Pancreatic Tumors to DNA Alkylating Agents. Cancer Research. 1997;57(23):5360-8.

[17] Sancar A, Lindsey-Boltz LA, Ünsal-Kaçmaz K, Linn S. Molecular Mechanisms of Mammalian DNA repair and the DNA Damage Checkpoints. Annual Review of Biochemistry. 2004;73(1):39-85.

[18] Lieberman HB. DNA Damage Repair and Response Proteins as Targets for Cancer Therapy. Current Medicinal Chemistry. 2008;15(4):360-7.

[19] Jackson SP, Bartek J. The DNA-damage response in human biology and disease. Nature. 2009;461(7267):1071-8.

[20] Schofield M, Hsieh P. DNA mismatch repair: molecular mechanisms and biological function. Annu Rev Microbiol. 2003;57:579 - 608.

[21] Peltomäki P. Role of DNA Mismatch Repair Defects in the Pathogenesis of Human Cancer. Journal of Clinical Oncology. 2003;21(6):1174-9.

[22] Nguewa PA, Fuertes MA, Valladares B, Alonso C, Pérez JM. Poly(ADP-Ribose) Polymerases: Homology, Structural Domains and Functions. Novel Therapeutical Applications. Progress in Biophysics and Molecular Biology. 2005;88(1):143-72.

[23] Kass EM, Jasin M. Collaboration and competition between DNA double-strand break repair pathways. FEBS Letters. 2010;584(17):3703-8.

[24] Li SX, Sjolund A, Harris L, Sweasy JB. DNA repair and personalized breast cancer therapy. Environmental and Molecular Mutagenesis. 2010;51(8-9):897-908.

[25] Helleday T. Homologous recombination in cancer development, treatment and development of drug resistance. Carcinogenesis. 2010;31(6):955-60.

[26] Bryant HE, Schultz N, Thomas HD, Parker KM, Flower D, Lopez E, et al. Specific killing of BRCA2-deficient tumours with inhibitors of poly(ADP-ribose) polymerase. Nature. 2005;434(7035):913-7.

[27] Farmer H, McCabe N, Lord CJ, Tutt ANJ, Johnson DA, Richardson TB, et al. Targeting the DNA repair defect in BRCA mutant cells as a therapeutic strategy. Nature. 2005;434(7035):917-21.

[28] Chiarugi A. A snapshot of chemoresistance to PARP inhibitors. Trends in Pharmacological Sciences. 2012;33(1):42-8.

[29] Davar D, Beumer JH, Hamieh L, Tawbi H. Role of PARP Inhibitors in Cancer Biology and Therapy. Curr Med Chem. 2012;19(23):3907-21. Epub 2012/07/14.

[30] Fong PC, Boss DS, Yap TA, Tutt A, Wu P, Mergui-Roelvink M, et al. Inhibition of Poly(ADP-Ribose) Polymerase in Tumors from BRCA Mutation Carriers. New England Journal of Medicine. 2009;361(2):123-34.

[31] Kelley MR, Fishel ML. DNA repair proteins as molecular targets for cancer therapeutics. Anti-Cancer Agents in Medicinal Chemistry. 2008;8(4):417-25.

[32] Evers B, Helleday T, Jonkers J. Targeting homologous recombination repair defects in cancer. Trends in Pharmacological Sciences. 2010;31(8):372-80.

[33] Helleday T, Petermann E, Lundin C, Hodgson B, Sharma RA. DNA repair pathways as targets for cancer therapy. Nature Reviews Cancer. 2008;8(3):193-204.

[34] Bouwman P, Jonkers J. The effects of deregulated DNA damage signalling on cancer chemotherapy response and resistance. Nat Rev Cancer. 2012;12(9):587-98.

[35] Longley DB, Johnston PG. Molecular mechanisms of drug resistance. The Journal of Pathology. 2005;205(2):275-92.

[36] Stratton MR, Campbell PJ, Futreal PA. The cancer genome. Nature. 2009;458(7239): 719-24.

[37] Drabløs F, Feyzi E, Aas PA, Vaagbø CB, Kavli B, Bratlie MS, et al. Alkylation damage in DNA and RNA—repair mechanisms and medical significance. DNA Repair. 2004;3(11):1389-407.

[38] Kaina B, Christmann M, Naumann S, Roos WP. MGMT: Key node in the battle against genotoxicity, carcinogenicity and apoptosis induced by alkylating agents. DNA Repair. 2007;6(8):1079-99.

[39] Jacinto FV, Esteller M. MGMT hypermethylation: A prognostic foe, a predictive friend. DNA Repair. 2007;6(8):1155-60.

[40] Margison GP, Santibáñez Koref MF, Povey AC. Mechanisms of carcinogenicity/ chemotherapy by O6-methylguanine. Mutagenesis. 2002;17(6):483-7.

[41] Zaidi NH, Liu L, Gerson SL. Quantitative immunohistochemical estimates of O6-alkylguanine-DNA alkyltransferase expression in normal and malignant human colon. Clinical Cancer Research. 1996;2(3):577-84.

[42] Lee SM, Rafferty JA, Elder RH, Fan CY, Bromley M, Harris M, et al. Immunohistological examination of the inter- and intracellular distribution of O6-alkylguanine DNA-alkyltransferase in human liver and melanoma. Br J Cancer. 1992;66(2):355-60.

[43] Augustine CK, Yoo JS, Potti A, Yoshimoto Y, Zipfel PA, Friedman HS, et al. Genomic and Molecular Profiling Predicts Response to Temozolomide in Melanoma. Clinical Cancer Research. 2009;15(2):502-10.

[44] Hegi ME, Liu L, Herman JG, Stupp R, Wick W, Weller M, et al. Correlation of O6-Methylguanine Methyltransferase (MGMT) Promoter Methylation With Clinical Outcomes in Glioblastoma and Clinical Strategies to Modulate MGMT Activity. Journal of Clinical Oncology. 2008;26(25):4189-99.

[45] Hegi ME, Diserens A-C, Gorlia T, Hamou M-F, de Tribolet N, Weller M, et al. MGMT Gene Silencing and Benefit from Temozolomide in Glioblastoma. New England Journal of Medicine. 2005;352(10):997-1003.

[46] Hegi ME, Diserens A-C, Godard S, Dietrich P-Y, Regli L, Ostermann S, et al. Clinical Trial Substantiates the Predictive Value of O-6-Methylguanine-DNA Methyltransferase Promoter Methylation in Glioblastoma Patients Treated with Temozolomide. Clinical Cancer Research. 2004;10(6):1871-4.

[47] Hegi ME, Sciuscio D, Murat A, Levivier M, Stupp R. Epigenetic Deregulation of DNA Repair and Its Potential for Therapy. Clinical Cancer Research. 2009;15(16): 5026-31.

[48] Suzuki T, Nakada M, Yoshida Y, Nambu E, Furuyama N, Kita D, et al. The Correlation between Promoter Methylation Status and the Expression Level of O6-Methylguanine-DNA Methyltransferase in Recurrent Glioma. Japanese Journal of Clinical Oncology. 2011;41(2):190-6.

[49] Esteller M, Hamilton SR, Burger PC, Baylin SB, Herman JG. Inactivation of the DNA repair gene O6-methylguanine-DNA methyltransferase by promoter hypermethylation is a common event in primary human neoplasia. Cancer Res. 1999;59(4):793-7. Epub 1999/02/24.

[50] Nakada M, Furuta T, Hayashi Y, Minamoto T, Hamada J-i. The strategy for enhancing temozolomide against malignant glioma. Frontiers in Oncology. 2012;2.

[51] Quinn JA, Desjardins A, Weingart J, Brem H, Dolan ME, Delaney SM, et al. Phase I Trial of Temozolomide Plus O6-Benzylguanine for Patients With Recurrent or Progressive Malignant Glioma. Journal of Clinical Oncology. 2005;23(28):7178-87.

[52] Maki Y, Murakami J, Asaumi J-i, Tsujigiwa H, Nagatsuka H, Kokeguchi S, et al. Role of O6-methylguanine–DNA methyltransferase and effect of O6-benzylguanine on the anti-tumor activity of cis-diaminedichloroplatinum(II) in oral cancer cell lines. Oral oncology. 2005;41(10):984-93.

[53] Batts E, Maisel C, Kane D, Liu L, Fu P, O'Brien T, et al. O6-benzylguanine and BCNU in multiple myeloma: a phase II trial. Cancer Chemotherapy and Pharmacology. 2007;60(3):415-21.

[54] Ranson M, Hersey P, Thompson D, Beith J, McArthur GA, Haydon A, et al. Randomized Trial of the Combination of Lomeguatrib and Temozolomide Compared With Temozolomide Alone in Chemotherapy Naive Patients With Metastatic Cutaneous Melanoma. Journal of Clinical Oncology. 2007;25(18):2540-5.

[55] Khan OA, Ranson M, Michael M, Olver I, Levitt NC, Mortimer P, et al. A phase II trial of lomeguatrib and temozolomide in metastatic colorectal cancer. Br J Cancer. 2008;98(10):1614-8.

[56] Tawbi HA, Villaruz L, Tarhini A, Moschos S, Sulecki M, Viverette F, et al. Inhibition of DNA repair with MGMT pseudosubstrates: phase I study of lomeguatrib in combination with dacarbazine in patients with advanced melanoma and other solid tumours. Br J Cancer. 2011;105(6):773-7.

[57] Ranson M, Middleton MR, Bridgewater J, Lee SM, Dawson M, Jowle D, et al. Lomeguatrib, a potent inhibitor of O6-alkylguanine-DNA-alkyltransferase: phase I safety, pharmacodynamic, and pharmacokinetic trial and evaluation in combination with temozolomide in patients with advanced solid tumors. Clin Cancer Res. 2006;12(5): 1577-84. Epub 2006/03/15.

[58] Caporaso P, Turriziani M, Venditti A, Marchesi F, Buccisano F, Tirindelli MC, et al. Novel role of triazenes in haematological malignancies: Pilot study of Temozolomide, Lomeguatrib and IL-2 in the chemo-immunotherapy of acute leukaemia. DNA Repair. 2007;6(8):1179-86.

[59] Sabharwal A, Corrie PG, Midgley RS, Palmer C, Brady J, Mortimer P, et al. A phase I trial of lomeguatrib and irinotecan in metastatic colorectal cancer. Cancer Chemother Pharmacol. 2010;66(5):829-35. Epub 2009/12/30.

[60] Kaina B, Margison GP, Christmann M. Targeting O(6)-methylguanine-DNA methyltransferase with specific inhibitors as a strategy in cancer therapy. Cell Mol Life Sci. 2010;67(21):3663-81. Epub 2010/08/19.

[61] Koch D, Hundsberger T, Boor S, Kaina B. Local intracerebral administration of O⁶-benzylguanine combined with systemic chemotherapy with temozolomide of a patient suffering from a recurrent glioblastoma. Journal of Neuro-Oncology. 2007;82(1): 85-9.

[62] Jiricny J. The multifaceted mismatch-repair system. Nat Rev Mol Cell Biol. 2006;7:335 - 46.

[63] Li G. Mechanisms and functions of DNA mismatch repair. Cell Res. 2008;18:85 - 98.

[64] Aaltonen L, Peltomaki P, Leach F, Sistonen P, Pylkkanen L, Mecklin J, et al. Clues to the pathogenesis of familial colorectal cancer. Science. 1993;260(5109):812-6.

[65] Boland CR, Thibodeau SN, Hamilton SR, Sidransky D, Eshleman JR, Burt RW, et al. A National Cancer Institute Workshop on Microsatellite Instability for Cancer Detection and Familial Predisposition: Development of International Criteria for the Determination of Microsatellite Instability in Colorectal Cancer. Cancer Research. 1998;58(22):5248-57.

[66] Aebi S, Kurdi-Haidar B, Gordon R, Cenni B, Zheng H, Fink D, et al. Loss of DNA Mismatch Repair in Acquired Resistance to Cisplatin. Cancer Research. 1996;56(13): 3087-90.

[67] Hawn MT, Umar A, Carethers JM, Marra G, Kunkel TA, Boland CR, et al. Evidence for a Connection between the Mismatch Repair System and the G2 Cell Cycle Checkpoint. Cancer Research. 1995;55(17):3721-5.

[68] Lage H, Dietel M. Involvement of the DNA mismatch repair system in antineoplastic drug resistance. Journal of cancer research and clinical oncology. 1999;125(3):156-65.

[69] Diouf B, Cheng Q, Krynetskaia NF, Yang W, Cheok M, Pei D, et al. Somatic deletions of genes regulating MSH2 protein stability cause DNA mismatch repair deficiency and drug resistance in human leukemia cells. Nat Med. 2011;17(10):1298-303.

[70] Kaplan E, Gunduz U. Expression analysis of TOP2A, MSH2 and MLH1 genes in MCF7 cells at different levels of etoposide resistance. Biomed Pharmacother. 2012;66(1):29-35. Epub 2012/01/31.

[71] Povey AC, Badawi AF, Cooper DP, Hall CN, Harrison KL, Jackson PE, et al. DNA Alkylation and Repair in the Large Bowel: Animal and Human Studies. The Journal of Nutrition. 2002;132(11):3518S-21S.

[72] Maxwell JA, Johnson SP, McLendon RE, Lister DW, Horne KS, Rasheed A, et al. Mismatch Repair Deficiency Does Not Mediate Clinical Resistance to Temozolomide in Malignant Glioma. Clinical Cancer Research. 2008;14(15):4859-68.

[73] Valeri N, Gasparini P, Braconi C, Paone A, Lovat F, Fabbri M, et al. MicroRNA-21 induces resistance to 5-fluorouracil by down-regulating human DNA MutS homolog 2 (hMSH2). Proceedings of the National Academy of Sciences. 2010;107(49): 21098-103.

[74] Robertson A, Klungland A, Rognes T, Leiros I. DNA Repair in Mammalian Cells. Cellular and Molecular Life Sciences. 2009;66(6):981-93.

[75] Wallace SS, Murphy DL, Sweasy JB. Base excision repair and cancer. Cancer Letters. 2012(0).

[76] Srivastava DK, Husain I, Arteaga CL, Wilson SH. DNA polymerase β expression differences in selected human tumors and cell lines. Carcinogenesis. 1999;20(6):1049-54.

[77] Tang J-b, Svilar D, Trivedi RN, Wang X-h, Goellner EM, Moore B, et al. N-methylpurine DNA glycosylase and DNA polymerase β modulate BER inhibitor potentiation of glioma cells to temozolomide. Neuro-Oncology. 2011;13(5):471-86.

[78] Dinis J, Silva V, Gromicho M, Martins C, Laires A, Tavares P, et al. DNA damage response in imatinib resistant chronic myeloid leukemia K562 cells. Leuk Lymphoma. 2012;53(10):2004-14. Epub 2012/04/06.

[79] Moore DH, Michael H, Tritt R, Parsons SH, Kelley MR. Alterations in the Expression of the DNA Repair/Redox Enzyme APE/ref-1 in Epithelial Ovarian Cancers. Clinical Cancer Research. 2000;6(2):602-9.

[80] Kelley MR, Cheng L, Foster R, Tritt R, Jiang J, Broshears J, et al. Elevated and Altered Expression of the Multifunctional DNA Base Excision Repair and Redox Enzyme Ape1/ref-1 in Prostate Cancer. Clinical Cancer Research. 2001;7(4):824-30.

[81] Wang D, Luo M, Kelley MR. Human apurinic endonuclease 1 (APE1) expression and prognostic significance in osteosarcoma: enhanced sensitivity of osteosarcoma to DNA damaging agents using silencing RNA APE1 expression inhibition. Mol Cancer Ther. 2004;3(6):679-86. Epub 2004/06/24.

[82] Robertson KA, Bullock HA, Xu Y, Tritt R, Zimmerman E, Ulbright TM, et al. Altered Expression of Ape1/ref-1 in Germ Cell Tumors and Overexpression in NT2 Cells Confers Resistance to Bleomycin and Radiation. Cancer Research. 2001;61(5):2220-5.

[83] Bobola MS, Finn LS, Ellenbogen RG, Geyer JR, Berger MS, Braga JM, et al. Apurinic/ Apyrimidinic Endonuclease Activity Is Associated with Response to Radiation and

Chemotherapy in Medulloblastoma and Primitive Neuroectodermal Tumors. Clinical Cancer Research. 2005;11(20):7405-14.

[84] Silber JR, Bobola MS, Blank A, Schoeler KD, Haroldson PD, Huynh MB, et al. The Apurinic/Apyrimidinic Endonuclease Activity of Ape1/Ref-1 Contributes to Human Glioma Cell Resistance to Alkylating Agents and Is Elevated by Oxidative Stress. Clinical Cancer Research. 2002;8(9):3008-18.

[85] Bapat A, Fishel ML, Kelley MR. Going ape as an approach to cancer therapeutics. Antioxidants & redox signaling. 2009;11(3):651-68. Epub 2008/08/22.

[86] Bapat A, Glass LS, Luo M, Fishel ML, Long EC, Georgiadis MM, et al. Novel small-molecule inhibitor of apurinic/apyrimidinic endonuclease 1 blocks proliferation and reduces viability of glioblastoma cells. J Pharmacol Exp Ther. 2010;334(3):988-98. Epub 2010/05/28.

[87] Taverna P, Liu L, Hwang HS, Hanson AJ, Kinsella TJ, Gerson SL. Methoxyamine potentiates DNA single strand breaks and double strand breaks induced by temozolomide in colon cancer cells. Mutat Res. 2001;485(4):269-81. Epub 2001/10/05.

[88] Rinne M, Caldwell D, Kelley MR. Transient adenoviral N-methylpurine DNA glycosylase overexpression imparts chemotherapeutic sensitivity to human breast cancer cells. Mol Cancer Ther. 2004;3(8):955-67. Epub 2004/08/10.

[89] Yan L, Bulgar A, Miao Y, Mahajan V, Donze JR, Gerson SL, et al. Combined treatment with temozolomide and methoxyamine: blocking apurininc/pyrimidinic site repair coupled with targeting topoisomerase IIalpha. Clin Cancer Res. 2007;13(5): 1532-9. Epub 2007/03/03.

[90] Fishel ML, He Y, Smith ML, Kelley MR. Manipulation of base excision repair to sensitize ovarian cancer cells to alkylating agent temozolomide. Clin Cancer Res. 2007;13(1):260-7. Epub 2007/01/04.

[91] Liu L, Yan L, Donze JR, Gerson SL. Blockage of abasic site repair enhances antitumor efficacy of 1,3-bis-(2-chloroethyl)-1-nitrosourea in colon tumor xenografts. Mol Cancer Ther. 2003;2(10):1061-6. Epub 2003/10/28.

[92] Bulgar AD, Weeks LD, Miao Y, Yang S, Xu Y, Guo C, et al. Removal of uracil by uracil DNA glycosylase limits pemetrexed cytotoxicity: overriding the limit with methoxyamine to inhibit base excision repair. Cell Death Dis. 2012;3:e252. Epub 2012/01/13.

[93] Taverna P, Hwang HS, Schupp JE, Radivoyevitch T, Session NN, Reddy G, et al. Inhibition of base excision repair potentiates iododeoxyuridine-induced cytotoxicity and radiosensitization. Cancer Research. 2003;63(4):838-46.

[94] Yan T, Seo Y, Schupp JE, Zeng XH, Desai AB, Kinsella TJ. Methoxyamine potentiates iododeoxyuridine-induced radiosensitization by altering cell cycle kinetics and enhancing senescence. Molecular Cancer Therapeutics. 2006;5(4):893-902.

[95] Weiss G, Gordon M, Rosen L, Savvides P, Adams B, Alvarez D, et al. Final results from a phase I study of oral TRC102 (methoxyamine HCl), an inhibitor of base-excision repair, to potentiate the activity of pemetrexed in patients with refractory cancer. [abstract]. J Clin Oncol. 2010;28(15s):abstract 2576.

[96] Sawides P, Xu Y, Liu L, Bokar J, Silverman P, Dowlati A, et al. Pharmacokinetic profile of the base-excision repair inhibitor methoxyamine-HCl (TRC102; MX) given as an one-hour intravenous infusion with temozolomide (TMZ) in the first-in-human phase I clinical trial. [abstract]. J Clin Oncol 2010;28(15s):abstract e13662.

[97] Mohammed MZ, Vyjayanti VN, Laughton CA, Dekker LV, Fischer PM, Wilson DM, 3rd, et al. Development and evaluation of human AP endonuclease inhibitors in melanoma and glioma cell lines. Br J Cancer. 2011;104(4):653-63. Epub 2011/01/27.

[98] Zawahir Z, Dayam R, Deng J, Pereira C, Neamati N. Pharmacophore guided discovery of small-molecule human apurinic/apyrimidinic endonuclease 1 inhibitors. J Med Chem. 2009;52(1):20-32. Epub 2008/12/17.

[99] Simeonov A, Kulkarni A, Dorjsuren D, Jadhav A, Shen M, McNeill DR, et al. Identification and characterization of inhibitors of human apurinic/apyrimidinic endonuclease APE1. PLoS One. 2009;4(6):e5740. Epub 2009/06/02.

[100] Seiple LA, Cardellina JH, 2nd, Akee R, Stivers JT. Potent inhibition of human apurinic/apyrimidinic endonuclease 1 by arylstibonic acids. Mol Pharmacol. 2008;73(3): 669-77. Epub 2007/11/29.

[101] Rai G, Vyjayanti VN, Dorjsuren D, Simeonov A, Jadhav A, Wilson DM, et al. Synthesis, Biological Evaluation, and Structure-Activity Relationships of a Novel Class of Apurinic/Apyrimidinic Endonuclease 1 Inhibitors. Journal of Medicinal Chemistry. 2012;55(7):3101-12.

[102] Srinivasan A, Wang L, Cline C, Xie Z, Sobol RW, Xie X, et al. Identification and Characterization of Human Apurinic/Apyrimidinic Endonuclease-1 Inhibitors. Biochemistry. 2012;[Epub ahead of print].

[103] Luo M, Kelley MR. Inhibition of the human apurinic/apyrimidinic endonuclease (APE1) repair activity and sensitization of breast cancer cells to DNA alkylating agents with lucanthone. Anticancer Res. 2004;24(4):2127-34. Epub 2004/08/28.

[104] Naidu MD, Agarwal R, Pena LA, Cunha L, Mezei M, Shen M, et al. Lucanthone and its derivative hycanthone inhibit apurinic endonuclease-1 (APE1) by direct protein binding. PLoS One. 2011;6(9):e23679. Epub 2011/09/22.

[105] Madhusudan S, Smart F, Shrimpton P, Parsons JL, Gardiner L, Houlbrook S, et al. Isolation of a small molecule inhibitor of DNA base excision repair. Nucleic Acids Res. 2005;33(15):4711-24. Epub 2005/08/23.

[106] Horton JK, Joyce-Gray DF, Pachkowski BF, Swenberg JA, Wilson SH. Hypersensitivity of DNA polymerase beta null mouse fibroblasts reflects accumulation of cytotoxic

repair intermediates from site-specific alkyl DNA lesions. DNA Repair. 2003;2(1): 27-48.

[107] Hu HY, Horton JK, Gryk MR, Prasad R, Naron JM, Sun DA, et al. Identification of small molecule synthetic inhibitors of DNA polymerase beta by NMR chemical shift mapping. J Biol Chem. 2004;279(38):39736-44. Epub 2004/07/20.

[108] Barakat K, Gajewski M, Tuszynski JA. DNA Repair Inhibitors: The Next Major Step to Improve Cancer Therapy. Curr Top Med Chem. 2012;12(12):1376-90.

[109] Javle M, Curtin NJ. The role of PARP in DNA repair and its therapeutic exploitation. Brit J Cancer. 2011;105(8):1114-22.

[110] Plummer R, Jones C, Middleton M, Wilson R, Evans J, Olsen A, et al. Phase I Study of the Poly (ADP-Ribose) Polymerase Inhibitor, AG014699, in Combination with Temozolomide in Patients with Advanced Solid Tumors. Clin Cancer Res. 2008;14(23): 7917-23.

[111] Underhill C, Toulmonde M, Bonnefoi H. A review of PARP inhibitors: from bench to bedside. Annals of Oncology. 2011;22(2):268-79.

[112] Ashworth A. A Synthetic Lethal Therapeutic Approach: Poly(ADP) Ribose Polymerase Inhibitors for the Treatment of Cancers Deficient in DNA Double-Strand Break Repair. Journal of Clinical Oncology. 2008;26(22):3785-90.

[113] de Bono JS, Ashworth A. Translating cancer research into targeted therapeutics. Nature. 2010;467(7315):543-9.

[114] Rouleau M, Patel A, Hendzel MJ, Kaufmann SH, Poirier GG. PARP inhibition: PARP1 and beyond. Nat Rev Cancer. 2010;10(4):293-301.

[115] Calvert H, Azzariti A. The clinical development of inhibitors of poly(ADP-ribose) polymerase. Annals of Oncology. 2011;22(suppl 1):i53-i9.

[116] Yap TA, Sandhu SK, Carden CP, de Bono JS. Poly(ADP-Ribose) polymerase (PARP) inhibitors: Exploiting a synthetic lethal strategy in the clinic. CA: A Cancer Journal for Clinicians. 2011;61(1):31-49.

[117] Kummar S, Chen A, Parchment R, Kinders R, Ji J, Tomaszewski J, et al. Advances in using PARP inhibitors to treat cancer. BMC medicine. 2012;10(1):25.

[118] Bryant HE, Helleday T. Inhibition of poly (ADP-ribose) polymerase activates ATM which is required for subsequent homologous recombination repair. Nucleic Acids Res. 2006;34(6):1685-91. Epub 2006/03/25.

[119] Mendes-Pereira AM, Martin SA, Brough R, McCarthy A, Taylor JR, Kim JS, et al. Synthetic lethal targeting of PTEN mutant cells with PARP inhibitors. EMBO Mol Med. 2009;1(6-7):315-22. Epub 2010/01/06.

[120] Leung M, Rosen D, Fields S, Cesano A, Budman DR. Poly(ADP-ribose) polymerase-1 inhibition: preclinical and clinical development of synthetic lethality. Mol Med. 2011;17(7-8):854-62. Epub 2011/03/23.

[121] Tutt A, Robson M, Garber JE, Domchek SM, Audeh MW, Weitzel JN, et al. Oral poly(ADP-ribose) polymerase inhibitor olaparib in patients with BRCA1 or BRCA2 mutations and advanced breast cancer: a proof-of-concept trial. Lancet. 2010;376(9737):235-44. Epub 2010/07/09.

[122] Sultana R, McNeill DR, Abbotts R, Mohammed MZ, Zdzienicka MZ, Qutob H, et al. Synthetic lethal targeting of DNA double-strand break repair deficient cells by human apurinic/apyrimidinic endonuclease inhibitors. Int J Cancer. 2012. Epub 2012/03/02.

[123] Guha M. PARP inhibitors stumble in breast cancer. Nat Biotechnol. 2011;29(5):373-4.

[124] Dent R, Lindeman G, Clemons M, Wildiers H, Chan A, McCarthy N, et al. Safety and efficacy of the oral PARP inhibitor olaparib (AZD2281) in combination with paclitaxel for the first- or second-line treatment of patients with metastatic triple-negative breast cancer: Results from the safety cohort of a phase I/II multicenter trial [abstract]. J Clin Oncol 2010;28(15s):abstr 1018.

[125] Gelmon KA, Tischkowitz M, Mackay H, Swenerton K, Robidoux A, Tonkin K, et al. Olaparib in patients with recurrent high-grade serous or poorly differentiated ovarian carcinoma or triple-negative breast cancer: a phase 2, multicentre, open-label, non-randomised study. The Lancet Oncology. 2011;12(9):852-61.

[126] Wood RD, Mitchell M, Lindahl T. Human DNA repair genes, 2005. Mutation Research/Fundamental and Molecular Mechanisms of Mutagenesis. 2005;577(1–2): 275-83.

[127] Rabik CA, Dolan ME. Molecular mechanisms of resistance and toxicity associated with platinating agents. Cancer Treatment Reviews. 2007;33(1):9-23.

[128] Kelland L. The resurgence of platinum-based cancer chemotherapy. Nat Rev Cancer. 2007;7(8):573-84.

[129] Köberle B, Grimaldi KA, Sunters A, Hartley JA, Kelland LR, Masters JRW. DNA Repair capacity and cisplatin sensitivity of human testis tumour cells. International Journal of Cancer. 1997;70(5):551-5.

[130] Masters JRW, Koberle B. Curing metastatic cancer: lessons from testicular germ-cell tumours. Nat Rev Cancer. 2003;3(7):517-25.

[131] Dabholkar M, Bostick-Bruton F, Weber C, Bohr VA, Egwuagu C, Reed E. ERCC1 and ERCC2 Expression in Malignant Tissues From Ovarian Cancer Patients. Journal of the National Cancer Institute. 1992;84(19):1512-7.

[132] Reed E. ERCC1 and Clinical Resistance to Platinum-Based Therapy. Clinical Cancer Research. 2005;11(17):6100-2.

[133] Shirota Y, Stoehlmacher J, Brabender J, Xiong Y-P, Uetake H, Danenberg KD, et al. ERCC1 and Thymidylate Synthase mRNA Levels Predict Survival for Colorectal Cancer Patients Receiving Combination Oxaliplatin and Fluorouracil Chemotherapy. Journal of Clinical Oncology. 2001;19(23):4298-304.

[134] Olaussen KA, Dunant A, Fouret P, Brambilla E, André F, Haddad V, et al. DNA Repair by ERCC1 in Non–Small-Cell Lung Cancer and Cisplatin-Based Adjuvant Chemotherapy. New England Journal of Medicine. 2006;355(10):983-91.

[135] Lord RVN, Brabender J, Gandara D, Alberola V, Camps C, Domine M, et al. Low ERCC1 Expression Correlates with Prolonged Survival after Cisplatin plus Gemcitabine Chemotherapy in Non-Small Cell Lung Cancer. Clinical Cancer Research. 2002;8(7):2286-91.

[136] Weberpals J, Garbuio K, O'Brien A, Clark-Knowles K, Doucette S, Antoniouk O, et al. The DNA repair proteins BRCA1 and ERCC1 as predictive markers in sporadic ovarian cancer. International Journal of Cancer. 2009;124(4):806-15.

[137] Jiang J, Liang X, Zhou X, Huang R, Chu Z, Zhan Q. ERCC1 expression as a prognostic and predictive factor in patients with non-small cell lung cancer: a meta-analysis. Molecular Biology Reports. 2012;39(6):6933-42.

[138] Li J, Zhang J, Liu Y, Ye G. Increased expression of DNA repair gene XPF enhances resistance to hydroxycamptothecin in bladder cancer. Medical science monitor : international medical journal of experimental and clinical research. 2012;18(4):BR156-62. Epub 2012/03/31.

[139] Prewett M, Deevi DS, Bassi R, Fan F, Ellis LM, Hicklin DJ, et al. Tumors Established with Cell Lines Selected for Oxaliplatin Resistance Respond to Oxaliplatin if Combined with Cetuximab. Clinical Cancer Research. 2007;13(24):7432-40.

[140] Balin-Gauthier D, Delord JP, Pillaire MJ, Rochaix P, Hoffman JS, Bugat R, et al. Cetuximab potentiates oxaliplatin cytotoxic effect through a defect in NER and DNA replication initiation. Br J Cancer. 2008;98(1):120-8.

[141] Rogakou EP, Pilch DR, Orr AH, Ivanova VS, Bonner WM. DNA Double-stranded Breaks Induce Histone H2AX Phosphorylation on Serine 139. Journal of Biological Chemistry. 1998;273(10):5858-68.

[142] Paull TT, Rogakou EP, Yamazaki V, Kirchgessner CU, Gellert M, Bonner WM. A critical role for histone H2AX in recruitment of repair factors to nuclear foci after DNA damage. Current biology : CB. 2000;10(15):886-95.

[143] Esteller M, Silva JM, Dominguez G, Bonilla F, Matias-Guiu X, Lerma E, et al. Promoter Hypermethylation and BRCA1 Inactivation in Sporadic Breast and Ovarian Tumors. Journal of the National Cancer Institute. 2000;92(7):564-9.

[144] Graeser M, McCarthy A, Lord CJ, Savage K, Hills M, Salter J, et al. A Marker of Homologous Recombination Predicts Pathologic Complete Response to Neoadjuvant

Chemotherapy in Primary Breast Cancer. Clinical Cancer Research. 2010;16(24): 6159-68.

[145] Wang W. Emergence of a DNA-damage response network consisting of Fanconi anaemia and BRCA proteins. Nat Rev Genet. 2007;8(10):735-48.

[146] Taron M, Rosell R, Felip E, Mendez P, Souglakos J, Ronco MS, et al. BRCA1 mRNA expression levels as an indicator of chemoresistance in lung cancer. Human Molecular Genetics. 2004;13(20):2443-9.

[147] Quinn JE, Carser JE, James CR, Kennedy RD, Harkin DP. BRCA1 and implications for response to chemotherapy in ovarian cancer. Gynecologic Oncology. 2009;113(1): 134-42.

[148] Quinn JE, Kennedy RD, Mullan PB, Gilmore PM, Carty M, Johnston PG, et al. BRCA1 Functions as a Differential Modulator of Chemotherapy-induced Apoptosis. Cancer Research. 2003;63(19):6221-8.

[149] Husain A, He G, Venkatraman ES, Spriggs DR. BRCA1 Up-Regulation Is Associated with Repair-mediated Resistance to cis-Diamminedichloroplatinum(II). Cancer Research. 1998;58(6):1120-3.

[150] Kennedy RD, Quinn JE, Mullan PB, Johnston PG, Harkin DP. The Role of BRCA1 in the Cellular Response to Chemotherapy. Journal of the National Cancer Institute. 2004;96(22):1659-68.

[151] Margeli M, Cirauqui B, Castella E, Tapia G, Costa C, Gimenez-Capitan A, et al. The Prognostic Value of BRCA1 mRNA Expression Levels Following Neoadjuvant Chemotherapy in Breast Cancer. PLoS ONE. 2010;5(3):e9499.

[152] Hennessy BTJ, Timms KM, Carey MS, Gutin A, Meyer LA, Flake DD, et al. Somatic Mutations in BRCA1 and BRCA2 Could Expand the Number of Patients That Benefit From Poly (ADP Ribose) Polymerase Inhibitors in Ovarian Cancer. Journal of Clinical Oncology. 2010;28(22):3570-6.

[153] Weston VJ, Oldreive CE, Skowronska A, Oscier DG, Pratt G, Dyer MJS, et al. The PARP inhibitor olaparib induces significant killing of ATM-deficient lymphoid tumor cells in vitro and in vivo. Blood. 2010;116(22):4578-87.

[154] Offit K, Levran O, Mullaney B, Mah K, Nafa K, Batish SD, et al. Shared Genetic Susceptibility to Breast Cancer, Brain Tumors, and Fanconi Anemia. Journal of the National Cancer Institute. 2003;95(20):1548-51.

[155] Panasci L, Paiement J-P, Christodoulopoulos G, Belenkov A, Malapetsa A, Aloyz R. Chlorambucil Drug Resistance in Chronic Lymphocytic Leukemia. Clinical Cancer Research. 2001;7(3):454-61.

[156] Klein HL. The consequences of Rad51 overexpression for normal and tumor cells. DNA Repair. 2008;7(5):686-93.

[157] Tobin LA, Robert C, Rapoport AP, Gojo I, Baer MR, Tomkinson AE, et al. Targeting abnormal DNA double-strand break repair in tyrosine kinase inhibitor-resistant chronic myeloid leukemias. Oncogene. 2012. Epub 2012/05/30.

[158] Tobin LA, Robert C, Nagaria P, Chumsri S, Twaddell W, Ioffe OB, et al. Targeting abnormal DNA repair in therapy-resistant breast cancers. Molecular cancer research : MCR. 2012;10(1):96-107. Epub 2011/11/25.

[159] Hickson I, Zhao Y, Richardson CJ, Green SJ, Martin NMB, Orr AI, et al. Identification and Characterization of a Novel and Specific Inhibitor of the Ataxia-Telangiectasia Mutated Kinase ATM. Cancer Research. 2004;64(24):9152-9.

[160] Cowell IG, Durkacz BW, Tilby MJ. Sensitization of breast carcinoma cells to ionizing radiation by small molecule inhibitors of DNA-dependent protein kinase and ataxia telangiectsia mutated. Biochemical Pharmacology. 2005;71(1–2):13-20.

[161] Golding SE, Rosenberg E, Valerie N, Hussaini I, Frigerio M, Cockcroft XF, et al. Improved ATM kinase inhibitor KU-60019 radiosensitizes glioma cells, compromises insulin, AKT and ERK prosurvival signaling, and inhibits migration and invasion. Molecular Cancer Therapeutics. 2009;8(10):2894-902.

[162] Shinohara ET, Geng L, Tan J, Chen H, Shir Y, Edwards E, et al. DNA-Dependent Protein Kinase Is a Molecular Target for the Development of Noncytotoxic Radiation–Sensitizing Drugs. Cancer Research. 2005;65(12):4987-92.

[163] Leahy JJJ, Golding BT, Griffin RJ, Hardcastle IR, Richardson C, Rigoreau L, et al. Identification of a highly potent and selective DNA-dependent protein kinase (DNA-PK) inhibitor (NU7441) by screening of chromenone libraries. Bioorganic & Medicinal Chemistry Letters. 2004;14(24):6083-7.

[164] Collis SJ, DeWeese TL, Jeggo PA, Parker AR. The life and death of DNA-PK. Oncogene. 2004;24(6):949-61.

[165] Hardcastle IR, Cockcroft X, Curtin NJ, El-Murr MD, Leahy JJJ, Stockley M, et al. Discovery of Potent Chromen-4-one Inhibitors of the DNA-Dependent Protein Kinase (DNA-PK) Using a Small-Molecule Library Approach. Journal of Medicinal Chemistry. 2005;48(24):7829-46.

[166] Zabludoff SD, Deng C, Grondine MR, Sheehy AM, Ashwell S, Caleb BL, et al. AZD7762, a novel checkpoint kinase inhibitor, drives checkpoint abrogation and potentiates DNA-targeted therapies. Molecular Cancer Therapeutics. 2008;7(9):2955-66.

[167] Iorio MV, Croce CM. MicroRNA dysregulation in cancer: diagnostics, monitoring and therapeutics. A comprehensive review. EMBO Molecular Medicine. 2012;4(3):143-59.

[168] Bartel D. MicroRNAs: genomics, biogenesis, mechanism, and function. Cell. 2004;116(2):281 - 97.

[169] Bartel D. MicroRNAs: target recognition and regulatory functions. Cell. 2009;136:215 - 33.

[170] Garzon R, Calin GA, Croce CM. MicroRNAs in Cancer. Annual Review of Medicine. 2009;60(1):167-79.

[171] Di Leva G, Briskin D, Croce CM. MicroRNA in cancer: New hopes for antineoplastic chemotherapy. Upsala Journal of Medical Sciences. 2012;117(2):202-16.

[172] Tian W, Chen J, He H, Deng Y. MicroRNAs and drug resistance of breast cancer: basic evidence and clinical applications. Clinical and Translational Oncology. 2012:1-8.

[173] Kutanzi KR, Yurchenko OV, Beland FA, Checkhun VF, Pogribny IP. MicroRNA-mediated drug resistance in breast cancer. Clin Epigenetics. 2011;2(2):171-85. Epub 2011/09/29.

[174] Hu H, Gatti RA. MicroRNAs: new players in the DNA damage response. Journal of Molecular Cell Biology. 2010.

[175] Wouters MD, van Gent DC, Hoeijmakers JHJ, Pothof J. MicroRNAs, the DNA damage response and cancer. Mutation Research/Fundamental and Molecular Mechanisms of Mutagenesis. 2011;717(1–2):54-66.

[176] Yu Y, Wang Y, Ren X, Tsuyada A, Li A, Liu LJ, et al. Context-Dependent Bidirectional Regulation of the MutS Homolog 2 by Transforming Growth Factor β Contributes to Chemoresistance in Breast Cancer Cells. Molecular Cancer Research. 2010;8(12): 1633-42.

[177] Moskwa P, Buffa FM, Pan Y, Panchakshari R, Gottipati P, Muschel RJ, et al. miR-182-Mediated Downregulation of BRCA1 Impacts DNA Repair and Sensitivity to PARP Inhibitors. Molecular cell. 2011;41(2):210-20.

[178] Sadones J, Michotte A, Veld P, Chaskis C, Sciot R, Menten J, et al. MGMT promoter hypermethylation correlates with a survival benefit from temozolomide in patients with recurrent anaplastic astrocytoma but not glioblastoma. European Journal of Cancer. 2009;45(1):146-53.

[179] Stupp R, Hegi ME, Mason WP, van den Bent MJ, Taphoorn MJB, Janzer RC, et al. Effects of radiotherapy with concomitant and adjuvant temozolomide versus radiotherapy alone on survival in glioblastoma in a randomised phase III study: 5-year analysis of the EORTC-NCIC trial. The Lancet Oncology. 2009;10(5):459-66.

[180] Slaby O, Lakomy R, Fadrus P, Hrstka R, Kren L, Lzicarova E, et al. MicroRNA-181 family predicts response to concomitant chemoradiotherapy with temozolomide in glioblastoma patients. Neoplasma. 2010;57(3):264-9. Epub 2010/04/01.

[181] Zhang W, Zhang J, Hoadley K, Kushwaha D, Ramakrishnan V, Li S, et al. miR-181d: a predictive glioblastoma biomarker that downregulates MGMT expression. Neuro-Oncology. 2012;14(6):712-9.

[182] Lakomy R, Sana J, Hankeova S, Fadrus P, Kren L, Lzicarova E, et al. MiR-195, miR-196b, miR-181c, miR-21 expression levels and O-6-methylguanine-DNA methyl-transferase methylation status are associated with clinical outcome in glioblastoma patients. Cancer Science. 2011;102(12):2186-90.

[183] Surh Y-J. Cancer chemoprevention with dietary phytochemicals. Nat Rev Cancer. 2003;3(10):768-80.

[184] Khan N, Afaq F, Mukhtar H. Apoptosis by dietary factors: the suicide solution for delaying cancer growth. Carcinogenesis. 2006;28(2):233-9.

[185] Hanahan D, Weinberg Robert A. Hallmarks of Cancer: The Next Generation. Cell. 2011;144(5):646-74.

[186] Anand P, Sundaram C, Jhurani S, Kunnumakkara AB, Aggarwal BB. Curcumin and cancer: An "old-age" disease with an "age-old" solution. Cancer Letters. 2008;267(1): 133-64.

[187] Raffoul JJ, Banerjee S, Che M, Knoll ZE, Doerge DR, Abrams J, et al. Soy isoflavones enhance radiotherapy in a metastatic prostate cancer model. Int J Cancer. 2007;120(11):2491-8. Epub 2007/02/17.

[188] Raffoul JJ, Banerjee S, Singh-Gupta V, Knoll ZE, Fite A, Zhang H, et al. Down-regulation of Apurinic/Apyrimidinic Endonuclease 1/Redox Factor-1 Expression by Soy Isoflavones Enhances Prostate Cancer Radiotherapy In vitro and In vivo. Cancer Research. 2007;67(5):2141-9.

[189] Raffoul JJ, Heydari AR, Hillman GG. DNA Repair and Cancer Therapy: Targeting APE1/Ref-1 Using Dietary Agents. Journal of oncology. 2012;2012:11.

[190] Raffoul JJ, Sarkar FH, Hillman GG. Radiosensitization of prostate cancer by soy iso-flavones. Curr Cancer Drug Targets. 2007;7(8):759-65. Epub 2008/01/29.

[191] Yang S, Irani K, Heffron SE, Jurnak F, Meyskens FL. Alterations in the expression of the apurinic/apyrimidinic endonuclease-1/redox factor-1 (APE/Ref-1) in human mel-anoma and identification of the therapeutic potential of resveratrol as an APE/Ref-1 inhibitor. Molecular Cancer Therapeutics. 2005;4(12):1923-35.

[192] Fulda S. Resveratrol and derivatives for the prevention and treatment of cancer. Drug Discovery Today. 2010;15(17–18):757-65.

[193] Fustier P, Le Corre L, Chalabi N, Vissac-Sabatier C, Communal Y, Bignon YJ, et al. Resveratrol increases BRCA1 and BRCA2 mRNA expression in breast tumour cell lines. Br J Cancer. 2003;89(1):168-72.

[194] Potter AJ, Gollahon KA, Palanca BJA, Harbert MJ, Choi YM, Moskovitz AH, et al. Flow cytometric analysis of the cell cycle phase specificity of DNA damage induced by radiation, hydrogen peroxide and doxorubicin. Carcinogenesis. 2002;23(3): 389-401.

[195]   Shin SY, Yong Y, Kim CG, Lee YH, Lim Y. Deoxypodophyllotoxin induces G2/M cell cycle arrest and apoptosis in HeLa cells. Cancer Letters. 2010;287(2):231-9.

[196]   Park C, Kim GY, Kim GD, Choi BT, Park YM, Choi YH. Induction of G2/M arrest and inhibition of cyclooxygenase-2 activity by curcumin in human bladder cancer T24 cells. Oncol Rep. 2006;15(5):1225-31. Epub 2006/04/06.

[197]   Sa G, Das T. Anti cancer effects of curcumin: cycle of life and death. Cell Division. 2008;3(1):14.

[198]   Le Corre L, Fustier P, Chalabi N, Bignon Y-J, Bernard-Gallon D. Effects of resveratrol on the expression of a panel of genes interacting with the BRCA1 oncosuppressor in human breast cell lines. Clinica Chimica Acta. 2004;344(1–2):115-21.

[199]   Papoutsis AJ, Lamore SD, Wondrak GT, Selmin OI, Romagnolo DF. Resveratrol Prevents Epigenetic Silencing of BRCA-1 by the Aromatic Hydrocarbon Receptor in Human Breast Cancer Cells. The Journal of Nutrition. 2010;140(9):1607-14.

[200]   Mahapatra S, Karnes R, Holmes M, Young C, Cheville J, Kohli M, et al. Novel Molecular Targets of <i>Azadirachta indica Associated with Inhibition of Tumor Growth in Prostate Cancer. The AAPS Journal. 2011;13(3):365-77.

[201]   Konkimalla VSB, Wang G, Kaina B, Efferth T. Microarray-based Expression of DNA Repair Genes Does not Correlate with Growth Inhibition of Cancer Cells by Natural Products Derived from Traditional Chinese Medicine. Cancer Genomics - Proteomics. 2008;5(2):79-83.

[202]   Niture SK, Velu CS, Smith QR, Bhat GJ, Srivenugopal KS. Increased expression of the MGMT repair protein mediated by cysteine prodrugs and chemopreventative natural products in human lymphocytes and tumor cell lines. Carcinogenesis. 2006;28(2):378-89.

[203]   Aggarwal BB, Shishodia S. Molecular targets of dietary agents for prevention and therapy of cancer. Biochemical Pharmacology. 2006;71(10):1397-421.

[204]   Tyagi AK, Singh RP, Agarwal C, Chan DCF, Agarwal R. Silibinin Strongly Synergizes Human Prostate Carcinoma DU145 Cells to Doxorubicin-induced Growth Inhibition, G2-M Arrest, and Apoptosis. Clinical Cancer Research. 2002;8(11):3512-9.

[205]   Moen MD. Ixabepilone: In Locally Advanced or Metastatic Breast Cancer. Drugs. 2009;69(11):1471-81 10.2165/00003495-200969110-00006.

[206]   Kanai M, Yoshimura K, Asada M, Imaizumi A, Suzuki C, Matsumoto S, et al. A phase I/II study of gemcitabine-based chemotherapy plus curcumin for patients with gemcitabine-resistant pancreatic cancer. Cancer Chemotherapy and Pharmacology. 2011;68(1):157-64.

[207]   Li Y, VandenBoom TG, Kong D, Wang Z, Ali S, Philip PA, et al. Up-regulation of miR-200 and let-7 by Natural Agents Leads to the Reversal of Epithelial-to-Mesen-

chymal Transition in Gemcitabine-Resistant Pancreatic Cancer Cells. Cancer Research. 2009;69(16):6704-12.

[208] Li Y, Kong D, Wang Z, Sarkar F. Regulation of microRNAs by Natural Agents: An Emerging Field in Chemoprevention and Chemotherapy Research. Pharmaceutical research. 2010;27(6):1027-41.

[209] Sarkar FH, Li Y. Harnessing the fruits of nature for the development of multi-targeted cancer therapeutics. Cancer Treatment Reviews. 2009;35(7):597-607.

[210] Sarkar FH, Li Y, Wang Z, Kong D, Ali S. Implication of microRNAs in drug resistance for designing novel cancer therapy. Drug resistance updates : reviews and commentaries in antimicrobial and anticancer chemotherapy. 2010;13(3):57-66.

[211] Adams RA, Meade AM, Seymour MT, Wilson RH, Madi A, Fisher D, et al. Intermittent versus continuous oxaliplatin and fluoropyrimidine combination chemotherapy for first-line treatment of advanced colorectal cancer: results of the randomised phase 3 MRC COIN trial. The Lancet Oncology. 2011;12(7):642-53.

[212] Becker A, Crombag L, Heideman DAM, Thunnissen FB, van Wijk AW, Postmus PE, et al. Retreatment with erlotinib: Regain of TKI sensitivity following a drug holiday for patients with NSCLC who initially responded to EGFR-TKI treatment. European journal of cancer (Oxford, England : 1990). 2011;47(17):2603-6.

[213] Kelley MR. Chapter 14 - Future Directions with DNA Repair Inhibitors: A Roadmap for Disruptive Approaches to Cancer Therapy. In: Mark RK, editor. DNA Repair in Cancer Therapy. San Diego: Academic Press; 2012. p. 301-10.

[214] Kelley MR. DNA repair inhibitors: where do we go from here? DNA Repair (Amst). 2011;10(11):1183-5. Epub 2011/10/04.

Chapter 4

DNA Base Excision Repair: Evolving Biomarkers for Personalized Therapies in Cancer

Vivek Mohan and Srinivasan Madhusudan

Additional information is available at the end of the chapter

1. Introduction

DNA repair is critical for maintaining genomic integrity. The DNA damage such as those induced by endogenous processes (methylation, hydroxylation, oxidation by free radicals) or by exogenous agents such as ionizing radiation, environmental toxins, and chemotherapy is processed through the DNA repair machinery in cells. At least six distinct DNA repair pathways have been described. A detailed discussion of individual pathways is beyond the scope of this chapter as several recent excellent reviews on DNA repair are available [1-6]. Briefly, direct repair is involved in the repair of alkylated bases (such as O^6 methyl guanine) by MGMT (O^6 methyl guanine DNA methyl transferases [7-10]. DNA mismatch repair (MMR) corrects base-base mismatches and insertion-deletion loops (IDLs) erroneously generated during DNA replication and by exogenous DNA damage [11-13]. Bulky DNA adducts are processed through the nucleotide excision repair pathway (NER) [14-16]. DNA double strand breaks are repaired through the homologous recombination pathway (predominantly during S-phase of cell cycle) [17-19] or the non-homologous end joining pathway (NHEJ), that operates outside the S-phase of the cell cycle [20-22]. DNA base damage is processed by the base excision repair (BER) machinery. In the current chapter we focus on BER. Evolving preclinical and clinical data suggests that BER factors are likely to be important prognostic, predictive and therapeutic targets in cancer.

2. Base excision repair pathway (BER)

Exogenous as well as endogenously derived reactive metabolites cause DNA damage such as base oxidation, deamination and alkylation. If the damaged bases are left unrepaired, then during replication or transcription misincorporation of erroneous complementary bases usher mutagenesis. For example, reactive oxygen species (ROS) generated during cellular respiration, phagocytosis, inflammation and in tumour hypoxia milieu can lead to base oxidation and generation of oxidised bases such as 8-hydroxyguanine (8-oxoG) [23]. DNA polymerase inserts adenine opposite to 8-oxoG, resulting in GC to AT transversion mutations after replication. Similarly, pyrimidine oxidation leads to the formation of 5-hydroxycytosine (5-OHC) which leads to the insertion of a thymine creating a potential mutagenic lesion [24]. Purine deamination products such as hypoxanthine and xanthine generated from adenine and guanine respectively are highly mutagenic. Hypoxanthine in DNA can cause AT to GC mutations, whereas xanthine generate GC to AT mutations [25]. Deamination of cytosine generates uracil which can occur in DNA at a frequency of upto 100–500 per cell per day. Uracil misincorporation can induce CG to TA transition mutations [26]. Although endogenous S-adenosyl methionine (SAM) participates in targeted enzymatic DNA base methylation, non-enzymatic methylation of ring nitrogen of purine base adenine can be cytotoxic[26]. Exogenous agents that cause base alkylation are common chemotherapeutic agents and include mono functional alkylating agents [27] (e.g. temozolomide, nitrosurea compounds, alkylsulfonates) and bifunctional alkylating agents (e.g. cisplatin, mitomycin C, nitrogen mustards). DNA bases damaged by oxidation, deamination and alkylation produce a non-helix distorting, non-bulky base lesion. Such lesions are the prime repair target of BER [6, 28-30].

BER is a complex process and utilizes a number of enzymes and accessory scaffold proteins (Figure 1). DNA glycosylases, AP endonuclease (APE-1) also called REF-1(Redox Effector Factor-1), DNA Polymerases, flap endonuclease (FEN-1), poly (ADP-ribose) polymerase 1(PARP-1) and DNA ligases are the key enzymes involved in BER. The core enzymes depend on accessory proteins such as X-ray cross complementation group 1 protein (XRCC1), proliferating cell nuclear antigen (PCNA), and protein 9-1-1 for coordinated action. DNA glycosylases initiate BER by excising the damaged base from DNA and generating an abasic site. APE1 hydrolyzes the phosphate bond 5′ to the AP site leaving a 3′-OH group and a 5′-dRP flanking the nucleotide gap. Polymerase β (pol β) excises the 5′-dRP moiety generating a 5′-P. Members of the poly (ADP-ribose) polymerase (PARP) family of proteins get activated by single strand DNA breaks induced by APE1 and catalyze the addition of poly (ADP-ribose) polymers to target proteins, affecting protein-protein interactions. PARP may also be involved in the coordination of BER. At this point, BER can proceed through the short-patch (SP-BER) where pol β introduces a single nucleotide with the help of XRCC1. Ligase-IIIα subsequently seals the DNA nick establishing the phosphodiester DNA backbone. The long patch (LP-BER) processes those lesions that cannot be handled by the short patch such as oxidised AP sites. PCNA mediated Polymerase δ/ε introduces two to eight nucleotides past the abasic site. The resulting overhang DNA is excised by FEN1 endonuclease and the nick is then sealed by DNA ligase I.[28-32]

2.1. BER factors are promising biomarkers in cancer

Prognostic factors are defined as patient and/or cancer characteristics that help to estimate patient survival independent of treatment. Conventionally these include patient age, fitness to withstand treatment toxicity (usually measured as performance status), tumour stage, histological grade, neuro-lymphovascular invasion by cancer cells, presence or absence of certain signal protein expression (for example Her-2 in breast cancer is a poor prognostic marker). Predictive factors are those factors that help estimate the probability of a patient responding to a specific treatment. For example BRAF V 600 gene mutation in patients with metastatic melanoma predicts the response to treatment with Vemurafenib [33].

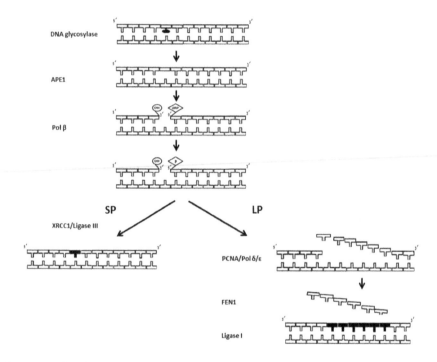

Figure 1. DNA glycosylase initiates BER by excising the damaged base from DNA and generating an abasic site. APE 1 nicks the phosphodiester bond and hydrolyzes the phosphate bond 5' to the AP site leaving a 3'-OH group and a 5'-dRP flanking the nucleotide gap. Pol β excises the 5'-dRP moiety generating a 5'-P. The short-patch (SP-BER) where pol β introduces a single nucleotide with the help of XRCC1. Ligase-IIIα subsequently seals the DNA nick establishing the phosphodiester DNA backbone. The long patch (LP-BER) processes those lesions that cannot be handled by the short patch such as oxidised AP sites. PCNA mediated Polymerase δ/ε introduces two to eight nucleotides past the abasic site. The resulting overhang DNA is excised by FEN1 endonuclease and the nick is then sealed by DNA ligase I.

Chemotherapeutic agents and ionizing radiation achieve cellular cytotoxicity by inducing DNA base damages [34]. However proficient BER in cancer cells results in therapeutic resistance and adversely impact patient outcomes. BER factors, therefore, are emerging as important prognostic factors as well as predictors of response to cytotoxic therapy in patients. For example, Temozolomide is an effective treatment for patients with high grade brain tumours. It induces O6-meG, N3-meA and N7-meG base alkylation lesions which are processed by BER. [35]. Similarly, Melphalan which is used in the treatment of multiple myeloma induces N3-meA lesions that is processed through BER [36]. Thiotepa is used with or without total body irradiation as a conditioning treatment prior to allogeneic or autologous haematopoietic progenitor cell transplantation in haematological diseases in adult and paediatric patients. Thiotepa produces formamidopyrimidine, 7-Methyl-formamidopyrimidine base lesions [37] which is repaired by BER. Dacarbazine is used in the treatment of patients with advanced malignant melanoma and Procarbazine is used in the treatment of Hodgkin's disease. They both produce O6-meG, N7-meG alkylation lesions which are targets of BER [38]. Streptozotocin generates O6-meG, N3-meA, N7-meG metabolites and is used in the treatment of neuroendocrine tumours of the gastro-intestinal (GI) tract.[39]. Platinating agents usually cause DNA inter-strand lesions which are repaired via NER, MMR and HR pathways (see table 1). Cisplatin is used in the treatment of advanced and metastatic non-small cell lung cancer (NSCLC), small cell carcinoma (SCLC), head and neck squamous cell carcinoma (HNSCC), germ cell tumour (GCT), gastric, pancreatic, bladder and cervical cancer. In addition to the DNA inter-strand lesions it also generates reactive oxygen species (ROS) that results in oxidative base damages. ROS derived base damages are also seen in patients with colorectal cancer (CRC) treated with Oxaliplatin. ROS induced base damages are also seen with anthracyclines (epirubicin and doxorubicin), used in the treatment of breast, gastric, ovarian, sarcoma and in haematological malignancy. The antimetabolite gemcitabine used in the treatment of NSCLC, pancreatico-biliary, bladder, breast and ovarian cancer also causes DNA base damage. Given the essential role of BER in cytotoxic therapy induced base damage, it is perhaps not surprising to note that several components of BER are promising prognostic and predictive factors. The following section will review individual markers and their relevance to cancer therapy.

2.2 APE1

Human apurinic / apyrimidinic endonuclease 1 (APE1) is a major endonuclease accounting for > 95% of the cellular AP endonuclease activity in most of the human cell lines [40]. It is also involved in redox regulation of transcription factors [41-43]. APE1 may be expressed in the cytoplasm and/ or in the nucleus of cancer cells. Although the precise sub-cellular localization and regulation is not clearly known, altered localization may have prognostic or predictive significance in patients. Table 2 summarizes the current knowledge regarding the association between APE1 and its role as a biomarker. We recently demonstrated that APE1 is over expressed in Ovarian, Gastro-oesophageal and pancreatico-biliary cancers [44]. In ovarian cancers, nuclear APE1 expression was seen in 71.9% of tumours and correlated with tumour type (P 0.006), optimal debulking (P 0.009), and overall survival (P 0.05). In gastro-oesophageal cancers previously exposed to neoadjuvant chemotherapy, 34.8% of tumours were positive in the nucleus and this correlated with shorter overall survival (P 0.005), whereas cytoplasmic localisation correlated with tumour de-differentiation (P 0.034). In pancreatico-biliary cancer, nuclear staining was seen in 44% of tumours. Absence of cytoplasmic staining was associated with perineural invasion (P 0.007), vascular invasion (P 0.05), and poorly differentiated tumours (P 0.068). [45]. In another study, a cohort of ninety one NSCLC patients treated with radical resection, tumour samples were analyzed for expression of APE1 protein. In patients with adenocarcinoma, cytoplasmic expression of APE1 was significantly associated with poor survival rate in univariate (P 0.01) and multivariate (P 0.07) analyses. In addition, a cytoplasmic expression was also predictive of worse prognosis (log-rank test, P 0.02) in NSCLC patients with lymph node involvement, regardless of the histology [46]. In another study, high nuclear and cytoplasmic APE1 expression was demonstrated in prostate cancer biopsy samples [47].

	DNA damaging agents	DNA Lesions	DNA repair pathways [104, 105]
1	Mono-functional alkylators: temozolomide, nitrosurea, alkylsulfonates	Small alkyl base adducts	Direct reversal
		Non-bulky alkyl adducts, base oxidation, deamination, AP sites	BER
		Bulky alkyl adducts, helix distorting lesions	NER
		Mismatched base pairs, insertion deletion loops	MMR
		DS DNA break	HR
2	Bi-functional alkylators: cisplatin, mitomycin C, nitrogen mustards, psoralen	DNA cross-links	NER
		DS DNA break	HR
		Bulky adducts	NER, MMR
		Replication fork arrest	BER
3	Anti-metabolites: 5-Fluorouracil (5FU) Thiopurines Folate analogues	Base damages, replication fork arrest	BER
4	Topoisomerase inhibitors: Etoposide	Double-strand breaks Single-strand breaks Replication lesions	HR, NHEJ
5	Replication inhibitors: Hydroxyrea	Double-strand breaks, Replication lesions	HR, NHEJ
6	Ionising Radiation and Radiomimetics: Bleomycin	Single-strand breaks Double-strand breaks Base damage	NHEJ, HR,BER

Abbreviations: BER : base excision repair pathway, NER: Nucleotide excision repair pathway, MMR: mis match repair pathway, HR: homologous repair pathway, NHEJ: non homologous end joining repair pathway.

Table 1. Cytotoxic agents and DNA Repair pathways

	BER factor	Key findings	Year of publication	Ref
1	APE1	Profound deregulation of APE1 acetylation status in triple negative breast cancer	2012	[52]
2	APE1	Ape1 expression elevated by p53 aberration may be used to predict poor survival and relapse in patients with NSCLC.	2012	[53]
3	APE1, XRCC1,HOGG1	APE1 genetic variants may be associated with endometrial cancer in Turkish women.	2012	[54]
4	APE1	APE1 T1349G polymorphism may be a marker for the development of gastric cancer in the Chinese population	2012	[55]
5	APE1, XRCC1	APE1 allele and the 399Gln XRCC1 allele apparently increased the risk of colon cancer	2012	[57]
6	APE1	APE1-656 T "/> G polymorphism has a possible protective effect on cancer risk particularly among Asian populations	2011	[106]
7	APE1, XRCC, OOG1	Polymorphisms within BER genes may contribute to the tumorigenesis of lung cancer.	2011	[59]
8	APE1	Loss of APE1 expression causes cell growth arrest, mitochondrial impairment and apoptosis	2011	[107]
9	APE1	Genetic variant rs1760944 in APE1 was associated with gastric cancer survival in a Chinese population.	2011	[56]
10	APE1, OGG1, XRCC1	APE1 Asp148Glu and hOGG1 Ser326Cys polymorphisms might be associated with increasing risk of CRC in a Turkish population.	2011	[58]
11	APE1	Polymorphisms of APE1 may confer susceptibility to RCC.	2011	[60]
12	APE1	Cytoplasmic localization of APE1 is associated with tumor progression and might be a valuable prognostic marker for EOC	2011	[51]
13	APE1	Genetic variant in the APE1 promoter may modulate risk of glioblastoma.	2011	[61]
14	APE1	Changes in the expression of APE1 might contribute to lip carcinogenesis.	2011	[108]
15	APE1	APE1 inhibitors potentiated the cytotoxicity of alkylating agents in melanoma and glioma cell lines	2011	[109]
16	APE1	Ape1 promotes radiation resistance in pediatric ependymomas	2011	[110]
17	APE1	The APE1 expression had significant correlation with osteosarcoma local recurrence and/or metastasis.	2010	[111]
18	APE1	APE1 may be a potential therapeutic target of MM.	2010	[112]
19	APE1	APE1 is a potential drug target in ovarian, gastro-oesophageal, and pancreatico-biliary cancers.	2010	[44]

	BER factor	Key findings	Year of publication	Ref
20	APE1	Nuclear expression of APE1 in gastro-oesophageal cancer patients treated with neo-adjuvant chemotherapy is associated with poor prognosis.	2010	[45]
21	APE1	Polymorphism in APE1 gene may affect response to palliative chemotherapy in NSCLC.	2009	[113]
22	APE1	Altered APE1 expression found in platinum resistant ovarian cancer patients	2009	[114]
23	APE1	APE1 is up-regulated in the NSCLC	2008	[115]
24	APE1	APE1 activity promotes resistance to radiation plus chemotherapy in Medulloblastomas and primitive neuroectodermaltumours	2005	[116]
25	APE1, XRCC1	High APE1 and XRCC1 protein expression levels predict better cancer-specific survival following radical radiotherapy in bladder cancer.	2005	[77]
26	APE1	APE1 over expression corresponds to poor prognosis in osteosarcoma	2004	[117]
27	APE1	APE 1 activity mediates resistance to alkylating agents and radiation and may be a useful predictor of progression after adjuvant therapy in a subset of gliomas.	2004	[118]
28	APE1	Cytoplasmic localization of APE1 seems to confer a poor survival outcome in patients with lung adenocarcinoma. Cytoplasmic expression of APE1 is a poor prognostic marker in node positive NSCLC regardless of the Histology.	2002	[46]
29	APE1	Increased expression of APE1 is seen in GCT and may be responsible for resistance to treatment with chemotherapy and IR	2001	[119]
30	APE1	APE1 nuclear expression in HNSCC is directly related to resistance to chemoradiotherapy and poor survival	2001	[120]
31	APE1	Increased APE1 cytoplasmic staining in prostate carcinoma as compared to BPH	2001	[47]
32	APE1	APE1 expression in carcinompa of the cervix is a marker of radio-resistance	1998	[121]

Abbreviations: NSCLC: non small cell lung cancer, CRC: colorectal cancer, RCC: renal cell carcinoma, EOC: epithelial ovarian carcinoma, GCT: germ cell tumour, HNSCC: Head and neck Squamous cell carcinoma, BPH: benign prostatic hypertrophy.

Table 2. APE1

The commonly reported APE1 polymorphisms include Asp148Glu, Leu104Arg, Glu126Asp, Arg237Ala, Asp283Gly, Gln51His, Ile64Val, Gly306Glu and Thr141Gly [48-50]. In a cohort of epithelial ovarian cancer patients, cytoplasmic APE1 positivity was significantly associated with higher grade of tumour (P = 0.002), advanced stage (III + IV) compared to early stage (I + II) patients (40.7% vs. 11.8%; P = 0.002) and a lower survival rate compared to patients with cytoplasmic negative localization (P < 0.05) of APE1 [51]. Profound deregulation of APE1 acetylation status in triple negative breast cancer patients has recently been demonstrated. This may be a potential biomarker for breast cancer aggressiveness [52]. In another study, one hundred and twenty five lung tumour samples were analysed for APE1 protein and mRNA expression by immunohistochemistry and real-time RT-PCR respectively. Cytoplasmic APE1 overexpression and p53 aberration was shown to be a potential predictor of poor survival and relapse in patients with NSCLC[53]. In a case control study of one hundred and four endometrial cancer patients with aged matched normal controls, APE1 Asp148Glu genotypes were determined by PCR-RFLP assays. Frequencies of Glu+ and Asp/Glu genotypes of APE1 were found to be more prevalent in patients than controls. This may represent a future diagnostic biomarker in endometrial cancer [54]. In a study involving three hundred and thirty eight newly diagnosed gastric cancer patients and matched control, APE1 genotype T1349G polymorphism was assessed. Compared with the APE1 TT genotype, individuals with the variant TG/GG genotypes had a significantly increased risk of gastric cancer (OR 1.69, 95% CI 1.19-2.40). Further analyses revealed that the variant genotypes were associated with an increased risk for diffuse-type, low depth of tumour infiltration (T1 and T2), and lymph node metastatic gastric cancer. The APE1 T1349G polymorphism may be a biomarker for the development of aggressive gastric cancer [55]. Another cohort of nine hundred and twenty five gastric cancer patients was evaluated for the genetic variant rs1760944 in APE1. Survival analyses showed a statistically significant (P 0.025, log-rank test) differences in median survival time between gastric cancer patients with APE1 rs1760944 TT (55 months) versus those with GT/GG (78 months). These studies suggest that APE1 polymorphism is a potential biomarker in patients with Gastric cancer [56]. In another study, significant differences in the distribution of APE1 genotype were found between colon cancer patients and healthy individuals. The 148Asp APE1 allele apparently increased the risk of colon cancer (OR 1.9-2.3), suggesting it to be a biomarker in colorectal cancer (CRC) [57]. Polymorphisms of APE1 Asp148Glu (rs3136820) were determined by polymerase chain reaction (PCR) and restriction fragment length polymorphism (RFLP) methods in blood samples of seventy nine CRC patients at their initial staging and two hundred and forty seven healthy controls. Frequency of Glu allele of APE1 Asp148Glu was higher in CRC patients than in controls (P 0.006, OR 3.43; 95% CI 1.76-6.70)[58]. In a hospital-based case-control study of four hundred and fifty five lung cancer patients and four hundred and forty three controls, the single nucleotide polymorphisms (SNPs) of APE1 (Asp148Glu and -141T/G) were genotyped and analyzed. In a multivariate logistic regression model, individuals homozygous for the variants APE1 -141GG showed a protective effect for lung cancer (OR 0.62; 95% CI 0.42-0.91; p 0.02). This study indirectly suggests that polymorphism in APE1 genes may be a biomarker and contribute in the pathogenesis of lung cancer [59]. In a case-control study of six hundred and twelve renal cell carcinoma (RCC) patients and six hundred and thirty two age and sex matched healthy controls, APE1 polymorphisms (-656 T>G, rs1760944 and 1349 T>G, rs1130409) were assessed. Compared with 1349 TT/TG genotypes, the variant genotype 1349 GG had a significantly increased risk of RCC (adjusted odds ratio 1.47; 95% CI 1.10-1.95), suggesting a role for APE1 polymorphism as a biomarker in RCC [60]. In a case-control study of seven hundred and sixty six glioma patients and eight hundred and twenty four cancer-free controls APE1/Ref-1 promoter -141T/G variant (rs1760944) was evaluated. Allele G was associated with significant decreased glioblastoma risk (OR 0.80; 95% CI 0.65-0.98; P 0.032) [61].In conclusion emerging studies of APE1 in tumours suggest that APE1 is a promising biomarker in cancer. However, large prospective studies are required to confirm these observations.

2.3. XRCC1

X-ray repair cross-complementing group 1 (XRCC1) is a scaffolding protein and coordinates BER [62]. Cells deficient in XRCC1 are hypersensitive to DNA damaging agents such as ionizing radiation and alkylating agents. Pre-clinically XRCC1 deficiency can induce mutagenesis [63]. Embryonic knock out of XRCC1 is lethal. The most extensively studied polymorphisms of XRCC1 are Arg194Trp, Arg280His, Arg399Gln, Arg399Gln, Pro161Leu and Tyr576S. Ensembl data base records ten somatic mutations and six genetic variations of human XRCC1gene. Table 3 summarizes the current knowledge regarding the association between APE1 and its role as a biomarker. XRCC1 SNPs rs1799782 and rs25487 were investigated using the TaqMan assay in one hundred and eighty five pancreatic cancer cases and one thousand four hundred and sixty five controls. The minor allele, rs25487 was significantly associated with pancreatic cancer risk in the per-allele model (OR 1.29; CI 1.01-1.65; P 0.043). Haplotype analysis of XRCC1 also showed a statistically significant association with pancreatic cancer risk [64]. Endometrial biopsy samples in a case control study assessed the polymorphisms Arg399Gln. Gln/Gln genotype of XRCC1was more prevalent in patients than in controls suggesting XRCC1 polymorphisms as a biomarker in endometrial cancer [54]. In a case control study, polymorphisms of XRCC1 Arg399Gln allele increased the risk of colon cancer (OR 1.5-2.1)[57]. In a cohort of ninety nine advanced colorectal cancer patients treated with oxaliplatin based chemotherapy, polymorphisms of XRCC1 Arg399Gln (G-->A) genotypes were detected by TaqMan-MGB probe allelic discrimination method. Cox proportional hazards model, adjusted for stage, performance status, and chemotherapy regimen, showed that XRCC1 G/G genotype increased the OR significantly (OR 3.555; 95 % CI, 2.119 - 5.963; P < 0.01). The result suggests that XRCC1 Arg399Gln polymorphism is associated with response to chemotherapy and time to progression in advanced colorectal cancer patients. This study pointed XRCC1 polymorphism as a predictive biomarker in advanced CRC patients treated with oxaliplatin based chemotherapy [65].

	BER factor	Key findings	Year of publication	Ref
1	XRCC1	XRCC1 polymorphisms affect pancreatic cancer risk in Japanese.	2012	[64]
2	XRCC1	Elevated cancer risk associated with XRCC1 polymorphism.	2012	[66]
3	XRCC1	XRCC1 polymorphism might influence the risk of developing glioma	2012	[68]
4	XRCC1, XRCC3	Polymorphisms in DNA repair genes have roles in the susceptibility and survival of ovarian cancer patients.	2012	[69]
5	XRCC1	XRCC1 polymorphism is associated with significantly increased risk of gastric cancer	2012	[70]
6	XRCC1	High XRCC1 and low ATM were independently associated with poor survival in gastric cancer	2012	[76]
7	XRCC1	XRCC1 polymorphisms affect pancreatic cancer risk in Japanese.	2012	[64]
8	XRCC1	Genetic variations in XRCC1 exhibit variation in the sensitivity to platinum based chemotherapy in NSCLC	2012	[71]
9	XRCC1	Polymorphisms of XRCC1 gene might have contributed to individual susceptibility to lung cancer.	2012	[72]
10	XRCC1	Arg194Trp polymorphism could be associated with nonmelanoma skincancer and extramammary Paget's disease risk in a Japanese population.	2012	[Chiyomaru, 2012 #1047] [122]
11	XRCC1	Polymorphism of XRCC1 Arg399Gln may be a candidate for contributing to the difference in the OS of gemcitabine/platinum-treated advanced NSCLC patients.	2012	[73]
12	XRCC1	XRCC1 Arg399Gln polymorphisms is associated with a response to oxaliplantin-based chemotherapy and time to progression in advanced colorectal cancer in Chinese population.	2012	[65]

	BER factor	Key findings	Year of publication	Ref
13	XRCC1	The 751 Lys/Gln polymorphism of the ERCC2 gene may be linked to endometrial cancer	2012	[Sobczuk, 2012 #1050][123]
14	XRCC1, XRCC3	XRCC1 and XRCC3 gene polymorphisms for risk of colorectal cancer in the Chinese population.	2012	[124]
15	XRCC1	XRCC1 399Gln is an independent unfavourable prognostic factor in unresected NSCLC treated with radiotherapy and chemoradiotherapy	2012	[125]
16	XRCC1	XRCC1-Arg399Cln polymorphism is associated with susceptibility to HCC, and XRCC1 Gln allele genotype showed significant prognostic associations.	2012	[126]
17	XRCC1	XRCC1 -77T"/>C polymorphism is associated with cancer risk, and individuals with XRCC1-77C variant have a significantly higher cancer risk, particularly in the Asian population	2012	[67]
18	XRCC1	XRCC1 protein expressions in tumor is novel candidate prognostic markers and predictive factor for benefit from adjuvant platinum-based chemotherapy in resectable gastric carcinoma.	2012	[75]
19	XRCC1	Genetic polymorphisms in XRCC1 gene might be associated with overall survival and response to platinum-based chemotherapy in lung cancer patients.	2012	[127]
20	XRCC1	XRCC1 T-77C and eNOS G874T may confer an increased risk of acute skin reactions to radiotherapy in breast cancer patients	2012	[128]
21	XRCC1	XRCC1 399Gln/Gln genotype have an increased risk of colorectal cancer	2012	[129]
22	XRCC1	XRCC1 Arg399Gln allele is a risk factor for the development breast cancer, especially among Asian and African populations.	2011	[130]
23	XRCC1	genetic polymorphisms in XRCC1 may affect survival post radiotherapy for localized prostate cancer.	2010	[131]
24	XRCC1	Combined polymorphisms of ERCC1 and XRCC1 may predict OS and response to palliative chemotherapy with FOLFOX / XELOX in metastatic CRC patients	2010	[132]

	BER factor	Key findings	Year of publication	Ref
25	XRCC1	XRCC1 194 CT genotype associated with inferior overall survival in advanced gastric cancer patients treated with Cisplatin-Taxane combined chemotherapy.	2010	[133]
26	XRCC1	XRCC1 codon 194 and codon 399 polymorphisms may predict the sensitivity of advanced NSCLC to palliative chemotherapy treatment with vinorelbine and Cisplatin.	2009	[134]
27	XRCC1	XRCC 1 polymorphism may predict higher response rate to palliative Cisplatin based chemotherapy in NSCLC patients	2009	[135]
28	XRCC1	XRCC1 polymorphism in clinical stage III may be a prognostic survival marker in HNSCC.	2009	[136]
29	XRCC1	SNP of XRCC1 gene at codon 399 influences the response to platinum based neo-adjuvant chemotherapy treatment in patients with cervical cancer.	2009	[137]
30	XRCC1, APE1	Polymorphism in APE1 and XRCC1 may represent prognostic factors in metastatic melanoma.	2009	[138]
31	XRCC1	A rarely occurring XRCC1 variant may predict response to Neoadjuvant chemo-radiotherapy for the treatment of oesophageal cancer.	2009	[139]
32	XRCC1	XRCC1 variant alleles may be associated with shorter overall survival in lung cancer patients	2008	[140]
33	XRCC1	Genotypes of XRCC1 Arginine194Tryptophan and GGH-401Cytosine/Thymine associated with the response to platinum based neo-adjuvant chemotherapy treatment in patients with cervical cancer	2008	[141]
34	XRCC1	XRCC1 variant may predict the risk of recurrence of bladder TCC post BCG treatment.	2008	[142]
35	XRCC1	XRCC1 gene polymorphism may predict survival in good PS advanced Gastric cancer patients treated with Oxalipaltin based palliative chemotherapy.	2007	[143]
36	XRCC1	Polymorphism in XRCC1 gene is a potential prognostic and predictive marker in breast cancer patients treated with adjuvant CMF chemotherapy	2007	[144]

	BER factor	Key findings	Year of publication	Ref
37	XRCC1	XRCC1 polymorphism may predict survival advantage for SCLC and NSCLC patients after platinum based treatment	2007	[145]
38	XRCC1	XRCC1 polymorphism may predict response to palliative FOLFOX and can also be a prognostic survival factor in metastatic colorectal cancer.	2006	[146]
39	XRCC1	Variant alleles of XRCC1 associated with the absence of pathologic complete response and poor survival in oesophageal cancer	2006	[147]
40	XRCC1	XRCC1 polpmorphism may represent a prognostic factor in advanced NSCLC patients treated with palliative Cisplatin and Gemcitabine.	2006	[148]
41	OGG1, LIG3, APE1, POLB, XRCC1, PCNA	XRCC1 polymorphism may be a prognostic factor in patients with CRC	2006	[95]
42	XRCC1	XRCC1-01 may predict survival outcome in patients with MBC treated with high dose chemotherapy.	2006	[74]
43	XRCC1	Combined XPD and XRCC1 genotypes might be prognostic factors in muscle-invasive bladder cancer patients treated with CRT.	2006	[149]
44	XRCC1	Polymorphism of XRCC1 R399Q is associated with response to platinum-based NAC in bulky cervical cancer	2006	[150]
45	XRCC1	Polymorphisms in the XRCC1 gene may impact the response rate to platinum based palliative chemotherapy in NSCLC patients.	2004	[151]
46	XRCC1	Polymorphism of XRCC1 gene may be associated with resistance to oxaliplatin/5-FU chemotherapy in advanced colorectal cancer.	2001	[152]

Abbreviations:ATM: ataxia telangiectasia mutated protein, FOLFOX: oxaliplatin and 5FU based chemotherapy, XELOX: oxliplatin and Capecitabine based chemotherapy, TCC: transitional cell carcinoma, BCG: Bacillus Calmette–Guérin, CRT: chemoradiotherapy, MBC: metastatic breast cancer.

Table 3. XRCC1

Meta-analysis of fifty three case-control studies with twenty one thousand three hundred and forty nine cases and twenty three thousand six hundred forty nine controls for XRCC1 Arg280His polymorphism and its cancer risk were estimated using fixed or random effect models. Minor variant His allele and Arg-His/His-His genotypes showed a statistical association with the risk of cancer (OR 1.16; 95% CI 1.08-1.25) [66]. Meta-analysis of thirteen studies involving a total of eleven thousand six hundred and seventy eight individuals showed that there was significant association between the C variant of XRCC1-77T>C polymorphism and cancer risk in all four genetic comparison models (OR C vs. T 1.19; 95% CI 1.07-1.31; P 0.001; OR homozygote model 1.28; 95% CI 1.07-1.52; P 0.007; OR recessive genetic model 1.22; 95% CI 1.04-1.44; P 0.015; OR dominant model 1.21; 95% CI 1.07-1.35, P 0.001). XRCC1 -77T>C polymorphism is associated with cancer risk, and individuals with XRCC1 -77C variant have a significantly higher cancer risk, particularly in the Asian population [67]. Using a PCR-RFLP method, XRCC1 Arg194Trp, Arg280His and Arg399Gln were genotyped in six hundred and twenty four glioma patients and five hundred and eighty healthy controls. Significant differences in the distribution of the Arg399Gln allele were detected between glioma patients and healthy controls by a logistic regression analysis (OR 1.35; 95% CI 1.17-1.68; P 0.001). Arg399Gln variant (allele A) carriers had an increased glioma risk compared to the wild-type (allele G) homozygous carriers (OR 1.40, 95%CI 1.12-1.76, P 0.003)[68]. In a prospective follow-up study, a cohort of three hundred and ten ovarian cancer patients treated with platinum-based chemotherapy between January 2005 to January 2007 were followed up to 2010. Genotyping of XRCC1 and XRCC3 polymorphisms was conducted by TaqMan Gene Expression assays. Lower survival rate in XRCC1 399 Arg/Arg genotype than in Gln/ Gln, with a significant increased risk of death (HR 1.69; 95% CI 1.07-2.78) were observed. However no significant association between XRCC1 Arg194Trp and XRCC1 Arg280His gene polymorphisms and ovarian cancer death was observed. [69]. A multicenter 1:1 matched case- control study of three hundred and seven pairs of gastric cancers patients and controls between October 2010 and August 2011 was undertaken. XRCC1 Arg194Trp and ADPRT Val762Ala were sequenced. Demographic data collected using a self-designed questionnaire. Individuals carrying XRCC1 Trp/Trp or Arg/Trp variant genotype had a significantly increased risk of gastric cancer (OR 1.718; 95% CI, 1.190-2.479). [70]. In a cohort of advanced NSCLC patients treated with platinum based chemotherapy, XRCC1 polymorphism was evaluated. XRCC1 Arg194Arg, FAS-1377GG, and FASL-844T allele displayed no response to platinum, whereas patients with XRCC1 194Trp allele and XPC PAT +/+ had 68.8% response rate to platinum. In Logistic Regression analysis, a significant gene-dosage effect was detected along with the increasing number of favourable genotypes of these four polymorphisms (P 0.00002). Multi-loci analysis showed the importance of genetic variations involved in BER repair and apoptotic pathways in sensitivity of platinum-based chemotherapy in NSCLC [71]. In a meta-analysis of forty four published case-control studies demonstrated that codon 194, codon 399 and -77 T > C polymorphisms of XRCC1 gene might have contributed to individual susceptibility to lung cancer [72]. In a another study, sixty two advanced NSCLC patients in a training set and forty five patients in a validation set treated with gemcitabine/platinum were genotyped for XRCC1 polymorphism. Wild-type genotype of XRCC1 Arg399Gln (G/G) was associated with decreased median overall survival than those carrying variant genotypes (G/A+A/A). In addition, there was a statistically significant longer median OS in patients carrying wild-type ERCC2 Asp312Asn genotype (G/G) (51 months, 95% CI, 19-82 months versus 10 months, log-rank test, P < 0.001) than those carrying heterozygous variant genotypes (G/A). This points out the predictive biomarker status of XRCC1 in platinum treated NSCLC patients[73]. XRCC1 polymorphism is a potential predictive marker of platinum based treatment response in non-small cell lung carcinoma, col-

orectal carcinoma, advanced gastric, advanced cervical, advanced operable oesophageal cancer. It may also predict response to adjuvant CMF chemotherapy and high dose chemotherapy
in breast cancer [74].

In a training and validating cohort of Gastric cancer patients, XRCC1 protein levels were significantly downregulated in gastric cancers compared to adjacent non-cancerous tissues. Low tumour XRCC1 expression significantly correlated with shorter overall survival as well as with
clinic-pathologic characteristics in patients without adjuvant treatment. Multivariate regression analysis showed that low XRCC1 expressions, separately and together, were independent
negative markers of OS. Adjuvant fluorouracil-leucovorin-oxaliplatin (FLO) significantly improved OS compared with surgery alone (log-rank test, P 0.01). However, this effect was evident only in the XRCC1 low expression group (HR 0.44, 95% CI 0.26-0.75; P 0.002); Adjuvant
fluorouracil-leucovorin-platinum (FLP) did not improve OS, except in the patients with low
XRCC1 expressions (P 0.024). XRCC1 protein expressions in tumour are novel candidate prognostic markers and predictive factors for benefit from adjuvant platinum-based chemotherapy
(FLO or FLP) in patients with resectable gastric carcinoma [75]. SMUG1, FEN1, XRCC1 and
ATM are involved in ROS induced oxidative DNA damage repair in gastric cancer patients.
High expression of SMUG1, FEN1 and XRCC1 correlated to high T-stage (T3/T4) (P 0.001, 0.005
& 0.02 respectively). High expression of XRCC1 and FEN1 also correlated to lymph node positive disease (P 0.009 and 0.02 respectively). High expression of XRCC1, FEN1 & SMUG1 correlated with poor disease specific survival (P 0.001, 0.006 and 0.05 respectively) and poor disease
free survival (P 0.001, 0.001 & 0.02 respectively) [76]. Muscle-invasive transitional cell carcinoma tumour samples from ninety patients treated with radical radiotherapy was evaluated for
XRCC1 protein expression. Nuclear staining of XRCC1 was 96.5% (range, 0.6-99.6%). High expression levels of XRCC1 (> or = 95% positivity) were associated with improved patient cancer-
specific survival (log-rank, P 0.006) [77].

XRCC1 has shown to be a promising prognostic biomarker in a majority of cancer groups
including HNSCC, breast, ovarian, endometrial, cervical, lung, gastric, oesophageal, pancreatic, glial, colorectal, hepatocellular, bladder transitional cell carcinoma, metastatic melanoma
and non melanomatous skin cancer.

2.4. FEN1

FEN1 is a structure-specific 5' endo/exonuclease with a range of functions during DNA repair
and replication. It is a BER long patch protein. FEN1 also has a role in the processing of the
okazaki lagging DNA strand synthesis. As an endonuclease, FEN1 recognizes double-stranded
DNA with a 5'-unannealed flap and makes an endonucleolytic cleavage at the base of the flap.
As a 5' exonuclease, it degrades nucleotides from a nick or a gap. It may also be involved in
maintaining stability of telomeres, inhibiting repeat sequence expansion and involved in
creation of double-stranded DNA breaks when mammalian cells are subjected to X-ray
irradiation[78]. Human flap endonuclease 1 gene has been shown to have 4 somatic mutations,
one polymorphism, and two transcripts in the Ensembl data base. In the following section we
will review the potential of FEN1 as a prognostic, predictive biomarker and its feasibility as a
drug target in cancer treatment (See Table 4).

	BER factor	Key findings	Year of publication	Ref
1	FEN1	Polymorphisms in FEN1 confer susceptibility to gastrointestinal cancers	2012	[79]
2	FEN1	High expression of XRCC1, FEN1 & SMUG1 correlated with poor disease free survival	2012	[76]
3	FEN1	FEN1 protein expression was also associated with poor prognosis in prostatectomy-treated patients. Knock-down of FEN1 with small interfering RNA inhibited the growth of LNCaP cells.	2012	[83]
4	FEN1	Genetic polymorphisms in FEN1 confer susceptibility to lung cancer.	2009	[153]
5	FEN1	FEN1 overexpression is common in testis, lung and brain tumors. Down-regulation of FEN1 by siRNA increased sensitivity to methylating agents (temozolomide, MMS) and cisplatin in LN308 glioma cells	2009	[81]
6	FEN1	RAD54B-deficient human colorectal cancer cells are sensitive to SL killing by reduced FEN1 expression	2009	[85]
7	FEN1	FEN1 is significantly up-regulated in multiple cancers. The overexpression and promoter hypomethylation of FEN1 may serve as biomarkers for monitoring the progression of cancers	2008	[82]
8	FEN1	FEN-1 is overexpressed in prostate cancer and is associated with higher Gleason score.	2006	[84]

Abbreviations: SMUG1: Single-strand selective monofunctional uracil-DNA glycosylase

Table 4. FEN1

Human germ line variants (-69G >A and 4150G > T) in the FEN1 gene have been associated with DNA damage in coke oven workers and lung cancer risk in general populations. This was studied in one thousand eight hundred and fifty gastrointestinal cancer (hepatocellular carcinoma, esophageal cancer, gastric cancer and colorectal cancer) patients and two thousand two hundred and twenty two healthy controls. It was found that the FEN1 -69GG genotypes were significantly correlated to increased risk for developing gastrointestinal cancer compared with the -69AA genotype highlighting FEN1 as an important gene in human gastrointestinal oncogenesis and a potential biomarker [79]. We recently investigated this relationship in a cohort of gastric cancer patients and found high expression of FEN1 correlated to lymph node positive disease with poor disease specific survival and poor disease free survival [76]. In promyelocytic leukemia cell line HL-60, gene expression of FEN-1 has been shown to be higher in cells during mitotic phase as compared to cells in the resting phase. FEN1 expression markedly decreases when these cells reach maturity upon induction of terminal differentiation [80]. This study pointed out the relationship between increased FEN1 expression and proliferating cancer cells. Subsequent studies showed increased FEN 1 expression in testis, lung and brain cancer specimens as studied by Western blot analysis and compared with the normal tissue from the same patient. FEN1 over expression was observed in nineteen samples from testicular tumours (mostly seminomas), four samples from NSCLC, nine samples from glioblastoma multiforme and in five samples from astrocytomas. Down regulation of FEN1 expression in LN308 glioblastoma cell line by siRNA resulted in hypersensitivity to cisplatin, temozolomide, nimustine and methyl methanesulfonate (MMS)[81]. Statistically significant increased amount of FEN1 expression has been demonstrated in breast tumor tissue (~2.4 fold, P< 0.0001, n = 50), uterine tumor tissue (~2.3 fold, P = 0.0006, n = 42), colon tumor tissue (~1.5 fold, P < 0.0001, n = 35), stomach tumor tissue (~1.5 fold, P= 0.0005, n = 28), lung tumor tissue (~1.9 fold, P = 0.0066, n=21) and kidney tumor tissue (~2.3 fold, P = 0.0063, n = 20), compared to matched normal tissues[82]. FEN1 also found to be increased in castration refractory prostate cancer (CRPC) cells. The knock-down of FEN1 with si RNA inhibited the growth of these LNCaP cells [83] pointing it as a potential drug target in prostate cancer. In primary prostate cancer from two hundred and forty six patients who had had a radical prostatectomy, FEN-1 nuclear expression correlated with Gleason score. These results suggest that FEN-1 might be a potential marker for selecting patients at high risk and therapy [84]. Interestingly, synthetic lethality (SL) has been observed in RAD54B-deficient human colorectal cancer cell line by iatrogenic reduction of FEN1 expression thus demonstrating it to be a potential novel therapeutic biological target [85].

2.5. Polymerase beta, PCNA

Polymerase beta (pol β) is essential for short patch BER. It is present in all tissues at a lower level [86] and has no cell-cycle dependence. Majority of BER proceeds through the short-patch whereby a single nucleotide is removed and replaced. Unlike other DNA polymerases, pol β has no proof reading capability[87] hence its over expression has the potential for mutagenesis[88, 89]. Proliferating cell nuclear antigen(PCNA) is an accessory protein required for replication by DNA polymerase δ, and as a consequence, PCNA is required during the long patch BER [90]. Lesions left unrepaired by the short patch BER is facilitated by PCNA to switch to the long patch BER. PCNA then helps polymerase δ to excise and replace 2-8 nucleotide patch in the long path of BER. Table 5 summarizes recent insight into the prognostic and predictive significance of pol β and PCNA.

Twenty somatic pol β mutations in prostate tumors are already known. The somatic missense pol β mutations (p.K27N, p.E123K, p.E232K, p.P242R, p.E216K, p.M236L, and the triple mutant p.P261L/T292A/I298T) were assessed *in vitro* for the biochemical properties of the polymerase. Experiments suggest that interfering with normal polymerase beta function may be a frequent mechanism of prostate tumour progression [91].Three non-synonymous single nucleotide substitutions, Gln8Arg, Arg137Gln and Pro242Arg have been identified as polymorphisms in DNA Pol β. The Arg137Gln variant demonstrates significantly reduced polymerase activity and impaired interaction with PCNA, and reduced BER efficiency when assayed in a reconstitution assay or with cellular extracts. Other polymorphisms within DNA Pol β include *A165G* and *T2133C*, which were associated with overall survival in a study of patients with pancreatic cancer [92, 93]. One hundred and fifty two ovarian cancer samples subjected to RT-PCR and sequencing, a variant of polymerase beta (deletion of exon 4-6 and 11-13, comprising of amino acid 63-123, and 208-304) was detected in heterozygous condition. Statistical analysis showed this variant to be associated with risk of stage IV, endometrioid type ovarian carcinoma[94]. In a case-control study (three hundred and seventy seven cases along with three hundred and twenty nine controls) designed to assess gene-environment interactions, samples were genotyped by use of an oligonucleotide microarray and the arrayed primer extension technique. Twenty-eight single nucleotide polymorphisms in 15 DNA repair genes including pol β P242R were evaluated. It was demonstrated that pol β polymorphism is associated with a decreased risk of colorectal cancer [95]. Pol β over expression reduces the efficacy of anticancer drug therapies including ionizing radiation, bleomycin, monofunctional alkylating agents and cisplatin. Small-scale studies in different cancers showed that pol β is mutated in approximately 30% of tumours. These mutations further lower pol β fidelity in DNA synthesis exposing the genome to serious mutations. These findings suggested pol β to be a promising therapeutic target in cancer treatment [96].

	BER factor	Key findings	Year of publication	Ref
1	Pol beta	variant form of Pol β cDNA is associated with edometriod type, stage IV ovarian carcinoma	2012	[94]
2	Pol beta	A proportion of prostate cancer patients express functionally important somatic mutations of pol β.	2011	[91]
3	Pol Beta	Over expression of pol β reduces the efficacy of anticancer drug therapies including, Cisplatin, bleomycin, monofunctional alkylating agents and ionizing radiation.	2011	[96]
4	Pol beta, PCNA	More than 30% of human tumors characterized to date express DNA pol β variants, a polymorphism encoding an arginine to glutamine substitution, R137Q, has lower polymerase activity	2009	[92]
4	Pol beta, PCNA	Pancreatic cancer patients carrying at least 1 of the 2 homozygous variant pol β GG or CC genotypes have a significantly better overall survival	2007	[93]
5	OGG1, LIG3, APE1, POLB, XRCC1, PCNA	pol β P242R was also associated with decreased risk of colorectal cancer	2006	[95]

Table 5. Other BER factors

3. Summary and the future developments

Numerous DNA base excision repair proteins are currently under development as potential biomarkers and therapeutic targets. Studies presented above provide compelling evidence that BER factors are promising prognostic and predictive biomarkers in cancer. More recent evidence also suggests that BER is an attractive target for drug discovery. APE1 inhibitors, for example, are currently in development and may have therapeutic application in the near future. [34, 97, 98]. Moreover, DNA polymerase beta inhibitor is also currently under developmental stage and early reports reveal the ability of DNA pol β inhibitors to potentiate the cytotoxicity of alkylating agents [99]. In contrast, several other studies demonstrate that pol β-null cells, although sensitive to temozolomide, are not sensitive to other chemotherapeutic agents such as melphalan, mitozolomide, BCNU, and IR [34, 100, 101]. Therefore further research is warranted to confirm pol β as a drug target in cancer.The principles of synthetic lethality has been transferred from the bench to the bedside with PARP-1 inhibitors in BRCA-deficient (HR-defective) cancer cells [102, 103]. Recent evidence suggests that other factors in BER are also important synthetic lethality targets for personalized cancer therapy.

Author details

Vivek Mohan and Srinivasan Madhusudan*

*Address all correspondence to: srinivasan.madhusudan@nottingham.ac.uk

Translational DNA Repair Group, Academic Unit of Oncology, School of Molecular Medical Sciences, University of Nottingham, Nottingham University Hospitals, Nottingham NG 1PB, UK

References

[1] Friedberg, E. C. A brief history of the DNA repair field. Cell Res, (2008). , 3-7.

[2] Hoeijmakers, J. H. DNA damage, aging, and cancer. N Engl J Med, (2009). , 1475-1485.

[3] Hanawalt, P. C. Subpathways of nucleotide excision repair and their regulation. Oncogene, (2002). , 8949-8956.

[4] Fleck, O, & Nielsen, O. DNA repair. J Cell Sci, (2004). Pt 4): , 515-517.

[5] Essers, J, et al. Homologous and non-homologous recombination differentially affect DNA damage repair in mice. EMBO J, (2000). , 1703-1710.

[6] Kelley, M. R. DNA repair in cancer therapy : molecular targets and clinical applications. 1st ed. (2012). London ; Waltham, MA: Elsevier/Academic Press. xiii, 316 p.

[7] Kokkinakis, D. M, et al. Role of O6-methylguanine-DNA methyltransferase in the resistance of pancreatic tumors to DNA alkylating agents. Cancer Res, (1997). , 5360-5368.

[8] Milsom, M. D, & Williams, D. A. Live and let die: in vivo selection of gene-modified hematopoietic stem cells via MGMT-mediated chemoprotection. DNA Repair (Amst), (2007). , 1210-1221.

[9] Cai, S, et al. Mitochondrial targeting of human O6-methylguanine DNA methyltransferase protects against cell killing by chemotherapeutic alkylating agents. Cancer Res, (2005). , 3319-3327.

[10] Bobola, M. S, et al. O6-Methylguanine-DNA methyltransferase in pediatric primary brain tumors: relation to patient and tumor characteristics. Clin Cancer Res, (2001). , 613-619.

[11] Modrich, P, & Lahue, R. Mismatch repair in replication fidelity, genetic recombination, and cancer biology. Annu Rev Biochem, (1996). , 101-133.

[12] Brown, K. D, et al. The mismatch repair system is required for S-phase checkpoint activation. Nat Genet, (2003). , 80-84.

[13] Charara, M, et al. Microsatellite status and cell cycle associated markers in rectal cancer patients undergoing a combined regimen of 5-FU and CPT-11 chemotherapy and radiotherapy. Anticancer Res, (2004). B): , 3161-3167.

[14] Dip, R, Camenisch, U, & Naegeli, H. Mechanisms of DNA damage recognition and strand discrimination in human nucleotide excision repair. DNA Repair (Amst), (2004). , 1409-1423.

[15] Nouspikel, T. Nucleotide excision repair and neurological diseases. DNA Repair (Amst), (2008). , 1155-1167.

[16] Koberle, B, Roginskaya, V, & Wood, R. D. XPA protein as a limiting factor for nucleotide excision repair and UV sensitivity in human cells. DNA Repair (Amst), (2006). , 641-648.

[17] San FilippoJ., P. Sung, and H. Klein, Mechanism of eukaryotic homologous recombination. Annu Rev Biochem, (2008). , 229-257.

[18] Zou, L. DNA repair: A protein giant in its entirety. Nature, (2010). , 667-668.

[19] Heyer, W. D, et al. Rad54: the Swiss Army knife of homologous recombination? Nucleic Acids Res, (2006). , 4115-4125.

[20] Hartlerode, A. J, & Scully, R. Mechanisms of double-strand break repair in somatic mammalian cells. Biochem J, (2009). , 157-168.

[21] Lieber, M. R, et al. Nonhomologous DNA end joining (NHEJ) and chromosomal translocations in humans. Subcell Biochem, (2010). , 279-296.

[22] Adachi, N, et al. Hypersensitivity of nonhomologous DNA end-joining mutants to VP-16 and ICRF-193: implications for the repair of topoisomerase II-mediated DNA damage. J Biol Chem, (2003). , 35897-35902.

[23] Kasai, H, et al. Hydroxyguanine, a DNA adduct formed by oxygen radicals: its implication on oxygen radical-involved mutagenesis/carcinogenesis. J Toxicol Sci, (1991). Suppl 1: , 95-105.

[24] Lindahl, T. Instability and decay of the primary structure of DNA. Nature, (1993). , 709-715.

[25] Kow, Y. W. Repair of deaminated bases in DNA. Free Radic Biol Med, (2002). , 886-893.

[26] Huffman, J. L, Sundheim, O, & Tainer, J. A. DNA base damage recognition and removal: new twists and grooves. Mutat Res, (2005). , 55-76.

[27] Beranek, D. T. Distribution of methyl and ethyl adducts following alkylation with monofunctional alkylating agents. Mutat Res, (1990). , 11-30.

[28] Sancar, A, et al. Molecular mechanisms of mammalian DNA repair and the DNA damage checkpoints. Annu Rev Biochem, (2004). , 39-85.

[29] Wilson, D. M, & Rd, V. A. Bohr, The mechanics of base excision repair, and its relationship to aging and disease. DNA Repair (Amst), (2007). , 544-559.

[30] Dianov, G. L, et al. Repair of abasic sites in DNA. Mutat Res, (2003). , 157-163.

[31] Almeida, K. H, & Sobol, R. W. A unified view of base excision repair: lesion-dependent protein complexes regulated by post-translational modification. DNA Repair (Amst), (2007). , 695-711.

[32] Fortini, P, & Dogliotti, E. Base damage and single-strand break repair: mechanisms and functional significance of short- and long-patch repair subpathways. DNA Repair (Amst), (2007). , 398-409.

[33] Sosman, J. A, et al. Survival in BRAF advanced melanoma treated with vemurafenib. N Engl J Med, (2012). , 600-mutant, 707-714.

[34] Kelley, M. R, & Fishel, M. L. DNA repair proteins as molecular targets for cancer therapeutics. Anticancer Agents Med Chem, (2008). , 417-425.

[35] Denny, B. J, et al. NMR and molecular modeling investigation of the mechanism of activation of the antitumor drug temozolomide and its interaction with DNA. Biochemistry, (1994). , 9045-9051.

[36] Mchugh, P. J, et al. Excision repair of nitrogen mustard-DNA adducts in Saccharomyces cerevisiae. Nucleic Acids Res, (1999). , 3259-3266.

[37] Xu, Y, et al. Protection of mammalian cells against chemotherapeutic agents thiotepa, 1,3-N,N'-bis(2-chloroethyl)-N-nitrosourea, and mafosfamide using the DNA base excision repair genes Fpg and alpha-hOgg1: implications for protective gene therapy applications. J Pharmacol Exp Ther, (2001). , 825-831.

[38] Pletsa, V, et al. DNA damage and mutagenesis induced by procarbazine in lambda lacZ transgenic mice: evidence that bone marrow mutations do not arise primarily through miscoding by O6-methylguanine. Carcinogenesis, (1997). , 2191-2196.

[39] Drablos, F, et al. Alkylation damage in DNA and RNA--repair mechanisms and medical significance. DNA Repair (Amst), (2004). , 1389-1407.

[40] Demple, B, Herman, T, & Chen, D. S. Cloning and expression of APE, the cDNA encoding the major human apurinic endonuclease: definition of a family of DNA repair enzymes. Proc Natl Acad Sci U S A, (1991). , 11450-11454.

[41] Evans, A. R, Limp-foster, M, & Kelley, M. R. Going APE over ref-1. Mutat Res, (2000). , 83-108.

[42] Mitra, S, et al. Intracellular trafficking and regulation of mammalian AP-endonuclease 1 (APE1), an essential DNA repair protein. DNA Repair (Amst), (2007). , 461-469.

[43] Tell, G, et al. The intracellular localization of APE1/Ref-1: more than a passive phenomenon? Antioxid Redox Signal, (2005). , 367-384.

[44] Al-attar, A, et al. Human apurinic/apyrimidinic endonuclease (APE1) is a prognostic factor in ovarian, gastro-oesophageal and pancreatico-biliary cancers. Br J Cancer, (2010). , 704-709.

[45] Fareed, K. R, et al. Tumour regression and ERCC1 nuclear protein expression predict clinical outcome in patients with gastro-oesophageal cancer treated with neoadjuvant chemotherapy. Br J Cancer, (2010). , 1600-1607.

[46] Puglisi, F, et al. Prognostic significance of Ape1/ref-1 subcellular localization in non-small cell lung carcinomas. Anticancer Res, (2001). A): , 4041-4049.

[47] Kelley, M. R, et al. Elevated and altered expression of the multifunctional DNA base excision repair and redox enzyme Ape1/ref-1 in prostate cancer. Clin Cancer Res, (2001). , 824-830.

[48] Hung, R. J, et al. Genetic polymorphisms in the base excision repair pathway and cancer risk: a HuGE review. Am J Epidemiol, (2005). , 925-942.

[49] Hadi, M. Z, et al. Functional characterization of Ape1 variants identified in the human population. Nucleic Acids Res, (2000). , 3871-3879.

[50] Lu, J, et al. Functional characterization of a promoter polymorphism in APE1/Ref-1 that contributes to reduced lung cancer susceptibility. FASEB J, (2009). , 3459-3469.

[51] Sheng, Q, et al. Prognostic significance of APE1 cytoplasmic localization in human epithelial ovarian cancer. Med Oncol, (2012). , 1265-1271.

[52] Poletto, M, et al. Acetylation on critical lysine residues of Apurinic/apyrimidinic endonuclease 1 (APE1) in triple negative breast cancers. Biochem Biophys Res Commun, (2012). , 34-39.

[53] Wu, H. H, et al. Cytoplasmic Ape1 Expression Elevated by Aberration May Predict Survival and Relapse in Resected Non-Small Cell Lung Cancer. Ann Surg Oncol, (2012). , 53.

[54] Cincin, Z. B, et al. DNA repair gene variants in endometrial carcinoma. Med Oncol, (2012).

[55] Gu, D, et al. The DNA repair gene APE1 T1349G polymorphism and risk of gastric cancer in a Chinese population. PLoS One, (2011). , e28971.

[56] Zhao, Q, et al. A genetic variation in APE1 is associated with gastric cancer survival in a Chinese population. Cancer Sci, (2011). , 1293-1297.

[57] Jelonek, K, et al. Association between single-nucleotide polymorphisms of selected genes involved in the response to DNA damage and risk of colon, head and neck, and breast cancers in a Polish population. J Appl Genet, (2010). , 343-352.

[58] Canbay, E, et al. Association of APE1 and hOGG1 polymorphisms with colorectal cancer risk in a Turkish population. Curr Med Res Opin, (2011). , 1295-1302.

[59] Li, Z, et al. Genetic polymorphism of DNA base-excision repair genes (APE1, OGG1 and XRCC1) and their correlation with risk of lung cancer in a Chinese population. Arch Med Res, (2011). , 226-234.

[60] Cao, Q, et al. Genetic polymorphisms in APE1 are associated with renal cell carcinoma risk in a Chinese population. Mol Carcinog, (2011). , 863-870.

[61] Zhou, K, et al. A genetic variant in the APE1/Ref-1 gene promoter-141T/G may modulate risk of glioblastoma in a Chinese Han population. BMC Cancer, (2011). , 104.

[62] Caldecott, K. W. XRCC1 and DNA strand break repair. DNA Repair (Amst), (2003). , 955-969.

[63] Thompson, L. H, & West, M. G. XRCC1 keeps DNA from getting stranded. Mutat Res, (2000). , 1-18.

[64] Nakao, M, et al. Selected Polymorphisms of Base Excision Repair Genes and Pancreatic Cancer Risk in Japanese. J Epidemiol, (2012).

[65] Lv, H, et al. Genetic Polymorphism of XRCC1 Correlated with Response to Oxaliplatin-Based Chemotherapy in Advanced Colorectal Cancer. Pathol Oncol Res, (2012).

[66] Zhang, K, et al. The XRCC1 Arg280His polymorphism contributes to cancer susceptibility: an update by meta-analysis of 53 individual studies. Gene, (2012).

[67] Wang, Y. G, Zheng, T. Y, & Xrcc, T. C polymorphism and cancer risk: a meta- analysis. Asian Pac J Cancer Prev, (2012). , 111-115.

[68] Wang, D, et al. Genetic polymorphisms in the DNA repair gene XRCC1 and suscepti-bility to glioma in a Han population in northeastern China: A case-control study. Gene, (2012).

[69] Cheng, C. X, et al. Predictive Value of XRCC1 and XRCC3 Gene Polymorphisms for Risk of Ovarian Cancer Death After Chemotherapy. Asian Pac J Cancer Prev, (2012). , 2541-2545.

[70] Wen, Y. Y, et al. ADPRT Val762Ala and XRCC1 Arg194Trp Polymorphisms and Risk of Gastric Cancer in Sichuan of China. Asian Pac J Cancer Prev, (2012). , 2139-2144.

[71] Liu, L, et al. Multi-loci analysis reveals the importance of genetic variations in sensitivity of platinum-based chemotherapy in non-small-cell lung cancer. Mol Carcinog, (2012).

[72] Dai, L, et al. XRCC1 gene polymorphisms and lung cancer susceptibility: a meta-analysis of 44 case-control studies. Mol Biol Rep, (2012). , 9535-9547.

[73] Liao, W. Y, et al. Genetic polymorphism of XRCC1 Arg399Gln is associated with survival in non-small-cell lung cancer patients treated with gemcitabine/platinum. Journal of thoracic oncology : official publication of the International Association for the Study of Lung Cancer, (2012). , 973-981.

[74] Bewick, M. A, Conlon, M. S, & Lafrenie, R. M. Polymorphisms in XRCC1, XRCC3, and CCND1 and survival after treatment for metastatic breast cancer. J Clin Oncol, (2006). , 5645-5651.

[75] Wang, S, et al. Prognostic and predictive role of JWA and XRCC1 expressions in gastric cancer. Clin Cancer Res, (2012). , 2987-2996.

[76] Abdel-fatah, T, et al. Are DNA repair factors promising biomarkers for personalized therapy in gastric cancer? Antioxid Redox Signal, (2012).

[77] Sak, S. C, et al. APE1 and XRCC1 protein expression levels predict cancer-specific survival following radical radiotherapy in bladder cancer. Clin Cancer Res, (2005). , 6205-6211.

[78] Liu, Y, Kao, H. I, & Bambara, R. A. Flap endonuclease 1: a central component of DNA metabolism. Annu Rev Biochem, (2004). , 589-615.

[79] Liu, L, et al. Functional FEN1 genetic variants contribute to risk of hepatocellular carcinoma, esophageal cancer, gastric cancer and colorectal cancer. Carcinogenesis, (2012). , 119-123.

[80] Kim, I. S, et al. Gene expression of flap endonuclease-1 during cell proliferation and differentiation. Biochim Biophys Acta, (2000). , 333-340.

[81] Nikolova, T, Christmann, M, & Kaina, B. FEN1 is overexpressed in testis, lung and brain tumors. Anticancer Res, (2009). , 2453-2459.

[82] Singh, P, et al. Overexpression and hypomethylation of flap endonuclease 1 gene in breast and other cancers. Mol Cancer Res, (2008). , 1710-1717.

[83] Urbanucci, A, et al. Overexpression of androgen receptor enhances the binding of the receptor to the chromatin in prostate cancer. Oncogene, (2012). , 2153-2163.

[84] Lam, J. S, et al. Flap endonuclease 1 is overexpressed in prostate cancer and is associated with a high Gleason score. BJU Int, (2006). , 445-451.

[85] Mcmanus, K. J, et al. Specific synthetic lethal killing of RAD54B-deficient human colorectal cancer cells by FEN1 silencing. Proc Natl Acad Sci U S A, (2009). , 3276-3281.

[86] Hirose, F, et al. Difference in the expression level of DNA polymerase beta among mouse tissues: high expression in the pachytene spermatocyte. Exp Cell Res, (1989). , 169-180.

[87] Zhang, Q. M, & Dianov, G. L. DNA repair fidelity of base excision repair pathways in human cell extracts. DNA Repair (Amst), (2005). , 263-270.

[88] Chan, K. K, Zhang, Q. M, & Dianov, G. L. Base excision repair fidelity in normal and cancer cells. Mutagenesis, (2006). , 173-178.

[89] Chan, K, et al. Overexpression of DNA polymerase beta results in an increased rate of frameshift mutations during base excision repair. Mutagenesis, (2007). , 183-188.

[90] Klungland, A, & Lindahl, T. Second pathway for completion of human DNA base excision-repair: reconstitution with purified proteins and requirement for DNase IV (FEN1). EMBO J, (1997). , 3341-3348.

[91] An, C. L, Chen, D, & Makridakis, N. M. Systematic biochemical analysis of somatic missense mutations in DNA polymerase beta found in prostate cancer reveal alteration of enzymatic function. Hum Mutat, (2011). , 415-423.

[92] Guo, Z, et al. Human DNA polymerase beta polymorphism, Arg137Gln, impairs its polymerase activity and interaction with PCNA and the cellular base excision repair capacity. Nucleic Acids Res, (2009). , 3431-3441.

[93] Li, D, et al. Effects of base excision repair gene polymorphisms on pancreatic cancer survival. Int J Cancer, (2007). , 1748-1754.

[94] Khanra, K, Bhattacharya, C, & Bhattacharyya, N. Association of a Newly Identified Variant of DNA Polymerase Beta (polbeta63-123, 208-304) with the Risk Factor of Ovarian Carcinoma in India. Asian Pac J Cancer Prev, (2012). , 1999-2002.

[95] Moreno, V, et al. Polymorphisms in genes of nucleotide and base excision repair: risk and prognosis of colorectal cancer. Clin Cancer Res, (2006). Pt 1): , 2101-2108.

[96] Barakat, K, & Tuszynski, J. Relaxed complex scheme suggests novel inhibitors for the lyase activity of DNA polymerase beta. J Mol Graph Model, (2011). , 702-716.

[97] Srinivasan, A, et al. Identification and characterization of human apurinic/apyrimidinic endonuclease-1 inhibitors. Biochemistry, (2012). , 6246-6259.

[98] Rai, G, et al. Synthesis, biological evaluation, and structure-activity relationships of a novel class of apurinic/apyrimidinic endonuclease 1 inhibitors. J Med Chem, (2012). , 3101-3112.

[99] Jaiswal, A. S, et al. DNA polymerase beta as a novel target for chemotherapeutic intervention of colorectal cancer. PLoS One, (2011). , e16691.

[100] Horton, J. K, et al. Hypersensitivity of DNA polymerase beta null mouse fibroblasts reflects accumulation of cytotoxic repair intermediates from site-specific alkyl DNA lesions. DNA Repair (Amst), (2003). , 27-48.

[101] Sobol, R. W, et al. Requirement of mammalian DNA polymerase-beta in base-excision repair. Nature, (1996). , 183-186.

[102] Bryant, H. E, et al. Specific killing of BRCA2-deficient tumours with inhibitors of poly(ADP-ribose) polymerase. Nature, (2005). , 913-917.

[103] Farmer, H, et al. Targeting the DNA repair defect in BRCA mutant cells as a therapeutic strategy. Nature, (2005). , 917-921.

[104] Tell, G, & Wilson, D. M. rd, Targeting DNA repair proteins for cancer treatment. Cell Mol Life Sci, (2010). , 3569-3572.

[105] Helleday, T, et al. DNA repair pathways as targets for cancer therapy. Nat Rev Cancer, (2008). , 193-204.

[106] Zhou, B, et al. The association of APE1-656T > G and 1349 T > G polymorphisms and cancer risk: a meta-analysis based on 37 case-control studies. BMC Cancer, (2011). , 521.

[107] Vascotto, C, et al. Knock-in reconstitution studies reveal an unexpected role of Cys-65 in regulating APE1/Ref-1 subcellular trafficking and function. Mol Biol Cell, (2011). , 3887-3901.

[108] Souza, L. R, et al. Immunohistochemical analysis of APE1, hMSH2 and ERCC1 proteins in actinic cheilitis and lip squamous cell carcinoma. Histopathology, (2011). p. 352-60., 53.

[109] Mohammed, M. Z, et al. Development and evaluation of human AP endonuclease inhibitors in melanoma and glioma cell lines. Br J Cancer, (2011). , 653-663.

[110] Bobola, M. S, et al. Apurinic/apyrimidinic endonuclease is inversely associated with response to radiotherapy in pediatric ependymoma. Int J Cancer, (2011). , 2370-2379.

[111] Yang, J, et al. APEX1 gene amplification and its protein overexpression in osteosarcoma: correlation with recurrence, metastasis, and survival. Technol Cancer Res Treat, (2010). , 161-169.

[112] Xie, J. Y, et al. Elevated expression of APE1/Ref-1 and its regulation on IL-6 and IL-8 in bone marrow stromal cells of multiple myeloma. Clin Lymphoma Myeloma Leuk, (2010). , 385-393.

[113] Su, D, et al. Genetic polymorphisms and treatment response in advanced non-small cell lung cancer. Lung Cancer, (2007). , 281-288.

[114]   Zhang, Y, et al. Alterations in the expression of the apurinic/apyrimidinic endonu-
        clease-1/redox factor-1 (APE1/Ref-1) in human ovarian cancer and indentification of
        the therapeutic potential of APE1/Ref-1 inhibitor. Int J Oncol, (2009). , 1069-1079.

[115]   Yoo, D. G, et al. Alteration of APE1/ref-1 expression in non-small cell lung cancer: the
        implications of impaired extracellular superoxide dismutase and catalase antioxidant
        systems. Lung Cancer, (2008). , 277-284.

[116]   Bobola, M. S, et al. Apurinic/apyrimidinic endonuclease activity is associated with
        response to radiation and chemotherapy in medulloblastoma and primitive neuroec-
        todermal tumors. Clin Cancer Res, (2005). , 7405-7414.

[117]   Wang, D, Luo, M, & Kelley, M. R. Human apurinic endonuclease 1 (APE1) expression
        and prognostic significance in osteosarcoma: enhanced sensitivity of osteosarcoma to
        DNA damaging agents using silencing RNA APE1 expression inhibition. Mol Cancer
        Ther, (2004). , 679-686.

[118]   Bobola, M. S, et al. Apurinic endonuclease activity in adult gliomas and time to tumor
        progression after alkylating agent-based chemotherapy and after radiotherapy. Clin
        Cancer Res, (2004). , 7875-7883.

[119]   Robertson, K. A, et al. Altered expression of Ape1/ref-1 in germ cell tumors and
        overexpression in NT2 cells confers resistance to bleomycin and radiation. Cancer Res,
        (2001). , 2220-2225.

[120]   Koukourakis, M. I, et al. Nuclear expression of human apurinic/apyrimidinic endonu-
        clease (HAP1/Ref-1) in head-and-neck cancer is associated with resistance to chemo-
        radiotherapy and poor outcome. Int J Radiat Oncol Biol Phys, (2001). , 27-36.

[121]   Herring, C. J, et al. Levels of the DNA repair enzyme human apurinic/apyrimidinic
        endonuclease (APE1, APEX, Ref-1) are associated with the intrinsic radiosensitivity of
        cervical cancers. Br J Cancer, (1998). , 1128-1133.

[122]   Chiyomaru, K, Nagano, T, & Nishigori, C. XRCC1 Arg194Trp polymorphism, risk of
        nonmelanoma skin cancer and extramammary Paget's disease in a Japanese popula-
        tion. Arch Dermatol Res, (2012). , 363-370.

[123]   Sobczuk, A, Poplawski, T, & Blasiak, J. Polymorphisms of DNA Repair Genes in
        Endometrial Cancer. Pathol Oncol Res, (2012).

[124]   Zhao, Y, et al. Genetic polymorphisms of DNA repair genes XRCC1 and XRCC3 and
        risk of colorectal cancer in Chinese population. Asian Pac J Cancer Prev, (2012). ,
        665-669.

[125]   Butkiewicz, D, et al. Influence of DNA repair gene polymorphisms on prognosis in
        inoperable non-small cell lung cancer patients treated with radiotherapy and platinum-
        based chemotherapy. Int J Cancer, (2012). , E1100-E1108.

[126]   Li, Q. W, et al. Evaluation of DNA repair gene XRCC1 polymorphism in prediction and
        prognosis of hepatocellular carcinoma risk. Asian Pac J Cancer Prev, (2012). , 191-194.

[127] Cui, Z, et al. Association between polymorphisms in XRCC1 gene and clinical outcomes of patients with lung cancer: a meta-analysis. BMC Cancer, (2012). , 71.

[128] Terrazzino, S, et al. Common variants of eNOS and XRCC1 genes may predict acute skin toxicity in breast cancer patients receiving radiotherapy after breast conserving surgery. Radiother Oncol, (2012). , 199-205.

[129] Yin, G, et al. Genetic polymorphisms of XRCC1, alcohol consumption, and the risk of colorectal cancer in Japan. J Epidemiol, (2012). , 64-71.

[130] Wu, Z, et al. High risk of benzo[alpha]pyrene-induced lung cancer in E160D FEN1 mutant mice. Mutat Res, (2012). , 85-91.

[131] Gao, R, et al. Genetic polymorphisms in XRCC1 associated with radiation therapy in prostate cancer. Cancer Biol Ther, (2010). , 13-18.

[132] Liang, J, et al. The combination of ERCC1 and XRCC1 gene polymorphisms better predicts clinical outcome to oxaliplatin-based chemotherapy in metastatic colorectal cancer. Cancer Chemother Pharmacol, (2010). , 493-500.

[133] Shim, H. J, et al. BRCA1 and XRCC1 polymorphisms associated with survival in advanced gastric cancer treated with taxane and cisplatin. Cancer Sci, (2010). , 1247-1254.

[134] Hong, C. Y, et al. Correlation of the sensitivity of NP chemotherapy in non-small lung cancer with DNA repair gene XRCC1 polymorphism]. Ai Zheng, (2009). , 1291-1297.

[135] Sun, X, et al. Polymorphisms in XRCC1 and XPG and response to platinum-based chemotherapy in advanced non-small cell lung cancer patients. Lung Cancer, (2009). , 230-236.

[136] Csejtei, A, et al. Association between XRCC1 polymorphisms and head and neck cancer in a Hungarian population. Anticancer Res, (2009). , 4169-4173.

[137] Cheng, X. D, et al. The association of XRCC1 gene single nucleotide polymorphisms with response to neoadjuvant chemotherapy in locally advanced cervical carcinoma. J Exp Clin Cancer Res, (2009). , 91.

[138] Figl, A, et al. Single nucleotide polymorphisms in DNA repair genes XRCC1 and APEX1 in progression and survival of primary cutaneous melanoma patients. Mutat Res, (2009). , 78-84.

[139] Warnecke-eberz, U, et al. ERCC1 and XRCC1 gene polymorphisms predict response to neoadjuvant radiochemotherapy in esophageal cancer. J Gastrointest Surg, (2009). , 1411-1421.

[140] Sreeja, L, et al. Prognostic importance of DNA repair gene polymorphisms of XRCC1 Arg399Gln and XPD Lys751Gln in lung cancer patients from India. J Cancer Res Clin Oncol, (2008). , 645-652.

[141] Kim, K, et al. XRCC1 Arginine194Tryptophan and GGH-401Cytosine/Thymine
       polymorphisms are associated with response to platinum-based neoadjuvant chemo-
       therapy in cervical cancer. Gynecol Oncol, (2008). , 509-515.

[142] Mittal, R. D, et al. XRCC1 codon 399 mutant allele: a risk factor for recurrence of
       urothelial bladder carcinoma in patients on BCG immunotherapy. Cancer Biol Ther,
       (2008). , 645-650.

[143] Liu, B, et al. Polymorphism of XRCC1 predicts overall survival of gastric cancer patients
       receiving oxaliplatin-based chemotherapy in Chinese population. European journal of
       human genetics : EJHG, (2007). , 1049-1053.

[144] Jaremko, M, et al. Polymorphism of the DNA repair enzyme XRCC1 is associated with
       treatment prediction in anthracycline and cyclophosphamide/methotrexate/5-fluo-
       rouracil-based chemotherapy of patients with primary invasive breast cancer. Phar-
       macogenet Genomics, (2007). , 529-538.

[145] Giachino, D. F, et al. Prospective assessment of XPD Lys751Gln and XRCC1 Arg399Gln
       single nucleotide polymorphisms in lung cancer. Clin Cancer Res, (2007). , 2876-2881.

[146] Suh, K. W, et al. Which gene is a dominant predictor of response during FOLFOX
       chemotherapy for the treatment of metastatic colorectal cancer, the MTHFR or XRCC1
       gene? Ann Surg Oncol, (2006). , 1379-1385.

[147] Wu, X, et al. Genetic variations in radiation and chemotherapy drug action pathways
       predict clinical outcomes in esophageal cancer. J Clin Oncol, (2006). , 3789-3798.

[148] de las PenasR., et al., Polymorphisms in DNA repair genes modulate survival in
       cisplatin/gemcitabine-treated non-small-cell lung cancer patients. Ann Oncol, (2006). ,
       668-675.

[149] Sakano, S, et al. Single nucleotide polymorphisms in DNA repair genes might be
       prognostic factors in muscle-invasive bladder cancer patients treated with chemora-
       diotherapy. Br J Cancer, (2006). , 561-570.

[150] Chung, H. H, et al. XRCC1 R399Q polymorphism is associated with response to
       platinum-based neoadjuvant chemotherapy in bulky cervical cancer. Gynecol Oncol,
       (2006). , 1031-1037.

[151] Wang, Z. H, et al. Single nucleotide polymorphisms in XRCC1 and clinical response to
       platin-based chemotherapy in advanced non-small cell lung cancer]. Ai Zheng, (2004). ,
       865-868.

[152] Stoehlmacher, J, et al. A polymorphism of the XRCC1 gene predicts for response to
       platinum based treatment in advanced colorectal cancer. Anticancer Res, (2001). B): ,
       3075-3079.

[153] Yang, M, et al. Functional FEN1 polymorphisms are associated with DNA damage
       levels and lung cancer risk. Hum Mutat, (2009). , 1320-1328.

Chapter 5

New Potential Therapeutic Approaches by Targeting Rad51-Dependent Homologous Recombination

Axelle Renodon-Cornière, Pierre Weigel,
Magali Le Breton and Fabrice Fleury

Additional information is available at the end of the chapter

1. Introduction

Cellular DNA is constantly exposed to the effects of endogenous or environmental agents such as free radicals, radiation and chemicals. In higher organisms, these nucleic alterations are estimated at several thousands of lesions per cell [1] which can correspond to the loss of bases and also to the breaking of one or both strands of the DNA double helix. Among these DNA breaks, the double-strand break (DSB) is the most harmful because it is the most difficult to repair. A human cell can accumulate up to 50 DSBs per cell cycle [2]. Unrepaired DSBs can have serious consequences such as permanent cell cycle arrest or cell death by apoptosis. Imperfect repair can also lead to major syndromes such as genetic disorders, premature aging or malignant cell generation.

In response to DNA damage, the cell has developed a surveillance and DNA repair network. DSBs of DNA, which are the most severe nucleic acid alterations, are repaired mainly by either non-homologous end-joining (NHEJ) or homologous recombination (HR).

NHEJ repair leads to a direct rejoining of the separated DNA ends [3]. This pathway begins by the binding of the Ku 70/80 heterodimer to DNA ends (Figure 1) which recruits and induces the activation of the DNA-dependent protein kinase catalytic subunit (DNA-PKc). Kinase activity is required for NHEJ since it causes the recruitment of other proteins and promotes the bringing together of DNA ends. Finally, ligase VI and XRCC4/XLF co-factors are involved in the final step of ligation and the generation of DNA repair. This process involves mainly the G0-G1 and S phases of the cell cycle. Its disadvantage is the possible loss of genetic information due to deletions or insertions of nucleic acids during the ligation of DNA ends and thus NHEJ repair is considered error-prone.

DNA repair by HR is more complex and needs a homologous sequence, which can be present in the homologous chromosome or in a gene in multicopy [4]. HR predominates in the S and G2 phases, when the sister chromatids are present and can also be a model for DNA repair [5]. In eukaryotic cells, DNA repair is supported by several protein complexes. Protein ATM (Ataxia Telangiectasia Mutated) has a role in DSB signaling via its activation induced by the MRN protein complex (MRE11-Rad50-NBS1 complex). MRE11 is a 5'-3' exonuclease that leads to a 3' end of DNA which is required for the process [6]. This resection of single-stranded DNA is followed by the recruitment of many proteins such as RPA, BRCA1, BRCA2, Rad51, Rad52, and Rad54. Rad52 is one of the first to settle on the DSB. BRCA1 then recruits BRCA2, Rad54 and Rad51 to form the nucleoprotein filament with ssDNA, whose role is to move the blade to the homologous sequence required for HR. Rad51 protein is the main element involved in the HR process. This recombinase catalyzes the homology search and the strand exchange with a homologous sequence and thus ensures the accurate repair of the DSB. In eukaryotes, Rad51 recombinase (RecA homolog in *Escherichia coli*) catalyzes the essential steps of homologous recombination and interacts directly with protein suppressors of breast cancer (BRCA1, BRCA2) [8] and p53 [9] which also indicates the importance of Rad51 in apoptosis.

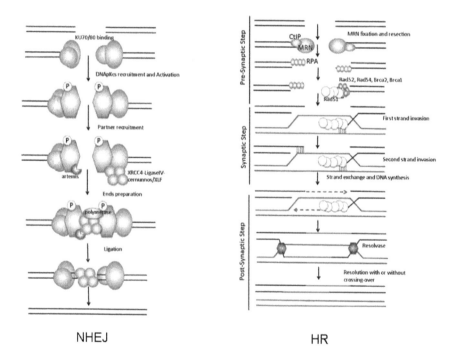

NHEJ   **HR**

Figure 1. Schematic representation of the mechanism of DNA DSB repair by NHEJ and HR (figures taken from [7]).

HR involves a large number of protein complexes and can be divided into three main stages (Figure 1):

Formation of the HsRad51/DNA nucleofilament or pre-synaptic step. DNA DSBs are resected by nuclease to generate 3′-protruding ends. The complex MRN (MRE11-RAD50-NSB1) contributes to DNA resection which is followed by formation of the replication protein A (RPA) complex. This ssDNA-binding factor removes secondary structures of ssDNA and is subsequently replaced by Rad51. Rad51 is recruited onto ssDNA to form the nucleofilament. Protein mediators such as Rad52 and Rad51 paralogs, Rad51B-C-D, BRCA1/2, facilitate the loading of Rad51 onto the ssDNA. The DNA binding sites of Rad51 are located in the N-terminal domain of each Rad51 monomer [8].

Homologous DNA pairing or synaptic phase. The nucleofilament of Rad51 is involved in the search for homologous DNA. Once a homologous sequence is located, the Rad51 filament invades the duplex DNA and generates a displacement of the homologous DNA strand to form a D-loop.

Exchange and resolution of the DNA intermediate structure or post-synaptic phase. The second 3′ ssDNA overhang anneals to the displaced DNA strand and serves as a model strand for DNA synthesis. Two Holliday Junctions (HJ) are then formed. Their resolution is the final step and generates two dsDNA. HJ can be either resolved or dissolved resulting in cross-over or non-crossover products.

It is clear that Rad51 plays an essential role at different levels of HR and several interactions are involved such as Rad51/ssDNA, Rad51/Rad51, Rad51/dsDNA, and Rad51/nucleotide. In addition, Rad51 interacts with its partners involved in HR (e.g. Rad52, Rad54).

Many cancer treatments using chemotherapy or radiotherapy target and disrupt the function of the DNA of tumor cells by inducing adducts or single- or double-strand breaks in DNA. However, these anticancer therapies are often faced with the emergence of radio- and chemoresistance, either induced or intrinsic to cancer cells. Since it was shown that some pathways of DNA repair can remove DNA damage induced by radio- or chemotherapy in cancer cells, these pathways have become potential therapeutic targets to sensitize tumors.

This observation is especially true for the Rad51 protein. A high level of HR induced by overexpression of Rad51 is frequently described in various types of cancer cell, including breast cancer, pancreatic, nonsmall-cell lung carcinoma and leukemia (AML and CML) [9-13]. In these cancer cells, this overexpression provides a degree of cancer resistance by promoting the repair of DSBs induced by cancer treatments [14]. Moreover, it has been shown that the survival of cancer patients expressing higher levels of Rad51 is shorter and that a reduced amount of Rad51, following antisense or ribozyme treatment, increases the effectiveness of cancer treatment by radiotherapy [15-17]. Modulation of HR to potentiate treatment is an option described in many publications [18]. It can be achieved by either acting directly on the recombinase activity of Rad51 or attempting to interfere with the interactions between Rad51 and some of its partners, which are not necessarily related directly to the repair of DNA.

The purpose of this chapter is to review all the chemical modulators that can act directly or indirectly on Rad51-mediated homologous recombination. First, the different steps of Rad51 activity which can be targeted will be detailed. Secondly, the chemical molecules that inhibit Rad51 activity or affect the expression level of Rad51 will be described. Finally, their applications in combination with anticancer treatments will be discussed in order to open up possibilities for counteracting chemo- and radioresistance.

2. Rad51 activity in homologous recombination

Human Rad51 (*Homo sapiens* Rad51 or HsRad51) is composed of 339 amino acids. It is the eukaryotic homolog of the RecA protein in prokaryotes. Homologs of the human recombinase are highly conserved between species: 98.8% similarity with *Mus musculus* Rad51 (MmRad51] and 81% with *Saccharomyces cerevisiae* Rad51 (ScRad51) [19,20].

To date, no complete structure of the HsRad51 protein has been determined. The crystalline structure of ScRad51 truncated at 79 N-terminal amino acids [21], one structure of *Pyrococcus furiosus* RadA [22], and five RadA of *Methanococcus voltae* [23,24] and *Sulfolobus solfataricus* [25] have been determined.

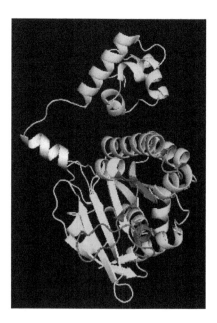

Figure 2. MvRadA subunit structure (PDB 1XU4). The N-terminal domain and the polymerization motif are colored in cyan and yellow, respectively. The putative DNA binding L1 and L2 loops are highlighted in orange and red. The Walker A and B motifs, corresponding to the ATP binding site, are labeled in blue and purple, respectively.

RadA and Rad51 are composed of two domains: a small N-terminal domain and a C-terminal domain entitled the core domain. The structures of the two HsRad51 domains were resolved by NMR [8] and crystallography [26], respectively.

The C-terminal domain of HsRad51 was crystallized in the form of a fusion protein comprising the BRC4 motif of BRCA2 protein (residues 1517-1551), a flexible linker and the central domain of HsRad51 (residues 97-339). The co-crystallization of HsRad51 with the BRC4 motif indicated the existence of a polymerization sequence located at the subunit-subunit interface of Rad51 [26]. The C-terminal domain contains an ATPase domain comprising units of Walker A (Hs: 127-134) and Walker B (Hs: 217-222) which are essential for ATP binding while loops L1 (Hs: 230-236) and L2 (Hs: 269-287) are involved in DNA binding.

The N-terminal domain of HsRad51 interacts with double-stranded DNA by a helix-hairpin-helix structure (residues 61-69). This type of protein-DNA interaction is conserved among many proteins interacting with DNA [27]. These sites of interaction, illustrated in Figure 2, are necessary for the formation of the nucleofilament, which is the key step of the recombinase activity of Rad51 (Figure 1). Nucleofilament formation is accompanied by a stretch modification of the DNA helix. The nucleofilament can adopt several conformations, only one of which is active for DNA strand exchange. The extended conformation is the functional form of the filament.

The conformation of the Rad51 filament depends on nucleotides: ATP promotes the extended conformation whereas ADP stabilizes the compressed form. Most of the structures have been solved in the presence of ATP analogs. Thus the conversion of an extended conformation to a compressed conformation accompanies the hydrolysis of ATP [28,29].

Several HsRad51 studies have also shown that ammonium sulfate [30], calcium [28,29] and AMPPNP [29] significantly increase the effectiveness of the strand exchange reaction *in vitro* by promoting the formation of an extended filament, which confirms that this structural form is the functional conformation of Rad51.

Since Rad51 protein is central to HR, all chemical molecules able to disrupt the interaction sites of Rad51 directly will also be able to modulate DNA repair by HR. Other ways of modulating HR via Rad51 are possible. Figure 3 presents the main ways and the catalytic steps of Rad51 being targeted to modulate HR.

3. Specific molecules targeting Rad51

The great majority of compounds identified as inhibitors of Rad51 have been selected by high-throughput screening from chemical libraries.

3.1. Modulators of Rad51 recombinase activity

This chemical group acts directly on the catalytic steps of Rad51 recombinase activity.

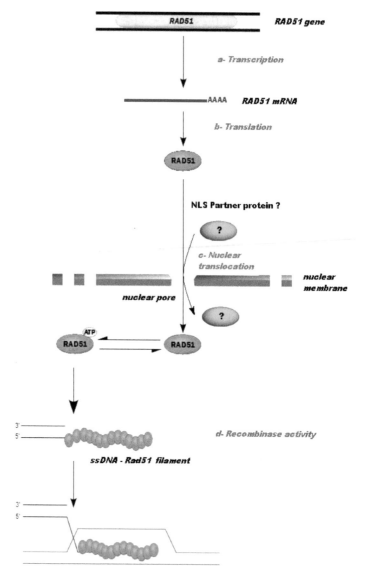

Figure 3. Intracellular pathways and catalytic steps of Rad51 as potential targets to modulate HR. (a) and (b) - The modulation of transcription and translation of Rad51 leads to changes in the recombinase activity and hence to the modification of HR repair. (c) - The intracellular localization and the delivery of Rad51 to the DNA damage sites are required for the HR process. Rad51 cellular distribution has an important regulatory role in HR and any element able to modulate the nuclear translocation of Rad51 is also able to modulate HR. (d) - Finally, molecules acting on the steps of Rad51 activity will target recombinase activity and HR-mediated DNA repair.

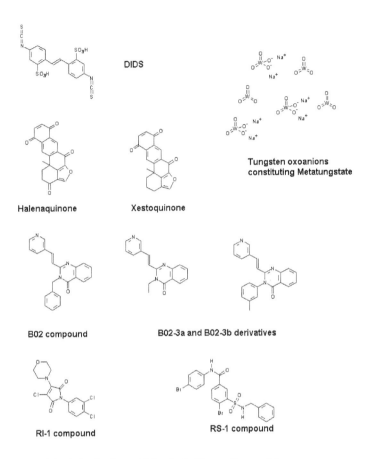

DIDS

Tungsten oxoanions constituting Metatungstate

Halenaquinone   **Xestoquinone**

B02 compound   **B02-3a and B02-3b derivatives**

RI-1 compound   **RS-1 compound**

Figure 4. Structure of chemical molecules modulating the Rad51 recombinase activity.

3.1.1. 4'-Diisothiocyanostilbene-2,2'-disulfonic acid or DIDS

DIDS (Figure 4) is a molecule known since the early 1970s for its ability to inhibit ionic channels and membrane transporters [31].

In the context of HR, Ishida T. et al. [32] have shown that DIDS can also inhibit *in vitro* the binding of Rad51 to DNA. In fact, strand exchange reactions are inhibited in the presence of DIDS. This process is dependent on salt concentration since the addition of 0.2 M KCl to the reaction medium changes the behavior of the inhibition. In the absence of KCl, increasing the concentration of DIDS gradually inhibits strand exchange with an IC50 (without KCl) close to 2 μM when the Rad51 concentration is 6 μM. In the presence of 0.2 M KCl, inhibition is considered to be shifted: the addition of DIDS has no effect up to a concentration of 1 μM. In this condition, the IC50 (with KCl) is close to 5 μM in the presence of Rad51 at 6 μM concentration. The interaction between Rad51 and DIDS probably involves electrostatic strength.

It has also been observed that the inhibition of the binding of Rad51 to ssDNA is not changed in the presence or absence of ATP. As previously described, HsRad51 contains a binding site for ATP and ATPase activity [33]. Analysis of the ATPase activity of Rad51 shows that ATP hydrolysis is greatly decreased when the protein is not bound to DNA. However, according to the results of Ishida and collaborators, Rad51 is able to hydrolyze ATP in the presence of DIDS and without DNA. The assumption is that this activation by DIDS of an asynchronous ATPase function results from the inhibition of the binding of Rad51 to DNA, which is ATP-dependent. However, Amunugama et al. [34] showed that the presence of ATP bound to Rad51 is required during strand exchange. In contrast, ATP hydrolysis does not appear to be essential for recombination. Hence, inactivation of the ATPase activity by DIDS cannot alone explain the mechanism of inhibition of Rad51.

DIDS interacts physically with Rad51 and dissociates it from ssDNA by competing with ssDNA for Rad51 binding.

3.1.2. Metatungstate

Metatungstate is a polyoxometalate (POM) consisting of 12 tungsten oxoanions (Figure 4) which dissociate into monotungstates in aqueous alkaline solution. This molecule is mainly used as a catalytic agent for chemical reactions of hydrocarbons.

Li and colleagues [35] have demonstrated that the metatungstate structure can bind *in vitro* to MvRadA protein, a homolog of Rad51 from Archaea *Methanococcus voltae* (MvRadA). The main contact zones established between the protein and metatungstate concern the L1 loop region with Arg_{218} and Arg_{230} and the L2 loop region with Arg_{224} of RadA. Both these L1 and L2 domains including Tyr_{232} and Phe_{203} are involved in DNA binding. It should be noted that the same pattern is found in HsRad51 [36]. In the MvRadA filament, these locations result in a distribution of molecules of tungstates on the longitudinal axis of rotation. It is shown that these tungsten clusters interact between the DNA-binding loops L1 and L2 stabilizing the inactive conformation of Rad51 [23,24,37].

Tungstate binding to MvRadA induces several effects on the functions of the recombinase protein activity. ATPase activity decreases by about 90% with equimolar amounts of MvRadA and metatungstate. By using gel electrophoresis *in vitro*, binding assays of MvRadA to ssDNA reveal that metatungstate inhibits ssDNA binding (IC50 = 0.13 μM for 1 μM MvRadA). The same *in vitro* assays using dsDNA also show an inhibition with a similar IC50 whereas the IC50 value of metatungstate for strand exchange activity is 0.5 μM in the presence of KCl. These observations indicate that, *in vitro*, metatungstate can inhibit the ssDNA and dsDNA binding of MvRadA, thus inactivating the functions essential for HR. As mentioned previously, the inactivation of ATPase activity does not alone explain the inhibition of Rad51 functions and probably those of RadA. It is therefore suggested that metatungstate acts as a competitive inhibitor of DNA binding by MvRadA.

Metatungstate is a potent inhibitor of ATPase and strand exchange activities of MvRadA and other experiments performed with HsRad51 have shown a significant increase in the IC50 of metatungstate for HsRad51 as compared with that for MvRadA (IC_{50}^{RadA} = 0.5 μM and IC_{50}^{Rad51} = 30 μM)[35].

3.1.3. Halenaquinone

Xestoquinone and halenaquinone molecules (Figure 4) are extracted from the marine sponge *Xestospongia exigua*. These molecules are similar except that xestoquinone does not contain the oxygen at the C-3 position in contrast to halenaquinone. Only halenaquinone presents inhibitory properties of phosphatidylinositol 3-kinase [38] and some anti-proliferative features [39].

Takaku et al. tested 160 crude extract fractions from marine sponge and used the D-loop formation assay to detect the homologous-pairing activity of Rad51. The authors reported that the halenaquinone inhibits HR at DNA pairing and D-loop formation stages but no inhibitory effect was observed with xestoquinone [40]. By Surface Plasmon Resonance (SPR) measurement, they showed that both halenaquinone and xestoquinone are able to bind to Rad51 but the affinity between halenaquinone and Rad51 is higher than between xestoquinone and Rad51. This result can explain the efficient inhibition of Rad51-mediated homologous pairing by halenaquinone.

Takaku and collaborators then examined whether both molecules affect ssDNA and dsDNA binding by Rad51. By an electrophoretic mobility shift approach, halenaquinone was found to inhibit Rad51-dsDNA binding specifically, but not Rad51-ssDNA binding. Interestingly, the authors showed that halenaquinone inhibits the secondary dsDNA binding by the Rad51-ssDNA complex. These results suggest that halenaquinone probably interacts near the dsDNA-binding site of Rad51. It can therefore inhibit the ternary complex formation containing ssDNA, dsDNA and Rad51 which promotes the DNA homologous pairing step during the HR process. In contrast, neither ssDNA binding nor dsDNA binding by Rad51 was affected by the presence of xestoquinone.

The authors then studied the intracellular effects of halenaquinone on the Ionizing Radiation (IR)-induced formation of Rad51 foci. When human cells were exposed to IR and treated with halenaquinone, Rad51 foci formation was significantly decreased. This result indicates that halenaquinone destabilizes the Rad51 foci, probably by inhibiting the ternary complex formation. Halenaquinone may be useful in medical research as a potential inhibitor of HR.

3.1.4. Compound B02 and derivatives

By high-throughput screening based on the quenching fluorescence method, Huang and colleagues have investigated the identification of specific inhibitors of the Rad51 strand exchange activity [18]. From 200,000 small molecules of the NIH repository, 174 compounds were positives and, after supplementary analyses and different controls, 13 molecules were identified as potential inhibitors of Rad51 with an inhibition higher than 30%. The IC50 values for the most potent inhibitors of Rad51-induced D-loop formation were determined.

Among these molecules, both compounds A04 and A10 were found to be inhibitors for Rad51 and RecA and their IC50 values were 5 μM and 26.6 μM, respectively. Another compound, the B02 molecule, was found to disrupt Rad51 binding to DNA and nucleoprotein filament formation. Although the B02 molecule presents an IC50 (27.4 μM) higher than A04 or A10, this molecule has a higher specificity for HsRad51.

Moreover, the study of B02 derivatives has revealed an efficient inactivation of Rad51 by both B02-3a and B02-3b, which contain an ethyl and an m-methylphenyl group, respectively (instead of the benzyl group located in the B02 molecule) (Figure 4). Modification of the pyridin radical of B02 suppresses the Rad51-induced D-loop inhibition, which demonstrates the importance of these chemical groups. The recent *in vivo* work of the same team has shown that B02 inhibits DSB-induced HR and increases cell sensitivity to the ICL agents, cisplatin and mitomycin C [41].

3.1.5. Compound RI-1 or 3-chloro-1-(3,4-dichlorophenyl)-4-(4-morpholinyl)-1H-pyrrole-2,5-dione

From a screening of 10,000 molecules of Chembridge DIVERSet™, the RI-1 compound was identified as an inhibitor of HsRad51 [42]. A first screening by fluorescence polarization (FP) enabled molecules that can bind to HsRad51 to be selected. A second screening based on the inhibition of homologous recombination in a cell line of human osteosarcoma (U2OS) was used and eight molecules were identified. A final test with the human embryonic kidney cell line (HEK293) identified RI-1, whose action is the specific inactivation of HsRad51.

RI-1 is composed of a chloromaleimide moiety (Figure 4) which promotes covalent binding to the thiol group of Rad51 cysteine 319 by a Michael addition mechanism. This binding potentiates the inhibition of the polymerization of HsRad51 during nucleofilament formation. It should be noted that the binding site is located on a surface which is highly conserved among mammalian homologs of Rad51. Experiments with *Saccharomyces cerevisiae* Rad51 (ScRad51) also show a fixation on the corresponding cysteine target (C377). However, this site is not present in RecA and inhibition was not found. RI-1 is potentially a specific inhibitor for mammalian homologs of Rad51. The binding site is located on the interface between two monomers of HsRad51 so it inhibits the polymerization of Rad51 onto ssDNA [21]. It is known that cysteine 319 is located in an ATP-binding loop [23], therefore the binding of RI-1 may disrupt the interaction of Rad51 with ATP. Moreover, this interaction area is also involved in the binding of other HR repair proteins such as Rad52 and Rad54 [43]. The IC50 of RI-1 is from 5 to 30 μM depending on the HsRad51 intracellular concentration. A synergistic anticancer effect is also observed for the association of RI-1 with mitomycin C (MMC) in U2OS, HeLa, MCF-7 and SH2038 cell lines.

3.1.6. Compound RS-1 (Rad51-Stimulatory-1) or 3-[(benzylamino)sulfonyl]-4-bromo-N-(4-bromophenyl)benzamide)

In contrast to inhibitors, few molecules stimulating HR have been reported. However, by screening 10,000 molecules (Chembridge DIVERSet™) using the FP method described above, Connell et al. identified a molecule, RS-1, which stimulates Rad51 binding onto ssDNA and increases the stability of the nucleofilament [44].

In the presence of RS-1 (Figure 4), the nucleofilament is in the active form characterized by the long length of the Rad51-ssDNA complex (100Å). The presence of nucleotide cofactors is also important since ATP is required for RS-1 to stimulate the formation of the active filament. However, RS-1 does not stimulate Rad51 by inhibiting its ATPase activity since it has no effect on the Rad51-dependent hydrolysis of ATP.

The RS-1-induced extension of the Rad51-ssDNA nucleofilament stimulates the exchange step of DNA strands, which can be evaluated by estimating D-loop formation. In the presence of non-hydrolyzable ATP or Ca2+, RS-1 increases D-loop formation by 5 to 11 times [28,44]. The stimulatory action of RS-1 is specific to HsRad51, since no effect with *E. coli* RecA or ScRad51 was found. This stimulation was then analyzed at the cellular level.

An analysis of cell survival (neonatal human fibroblasts) showed that the cells are more resistant to cisplatin treatment in the presence of 7.5 μM RS-1. This result is probably due to the ability of RS-1 to stimulate HR in response to DNA-damage agents like cisplatin.

3.2. Chemical modulators of Rad51 expression

3.2.1. Methotrexate drug

The structure of methotrexate (Figure 5) is similar to that of the folate metabolic precursor of coenzyme tetrahydrofolate (FH4) involved in the synthesis of nucleic bases. Methotrexate is a molecule used as an inhibitor of dihydrofolate reductase and acts in nucleic base synthesis occurring during the S phase of the cell cycle, as well as in non-restorative homologous recombination [45]. Therefore, methotrexate targets the S phase and the functions of HR by reducing the rate of repair of DNA damage, which can be shown by comet assay [46]. The study conducted by Du and colleagues [47] found that the inhibition of the formation of HsRad51 foci was effective in the presence of methotrexate in a human osteosarcoma cell line (HOS) after irradiation. This inhibition seems to be related to the Rad51 protein expression level, which is significantly decreased by the treatment. In addition, it was observed that the expression levels of BRCA2 and Rad52 were therefore not affected by methotrexate. It induces a specific downregulation of HsRad51. It should be noted that the treatment of HOS cells with 0.1 μM methotrexate causes a decrease in the transcription of 70% and > 95% of HsRad51 after 12 and 24 hours, respectively. However, these studies cannot determine the interactions involved in decreasing the mRNA levels of HsRad51.

3.2.2. Phenylhydroxamic acid or PCI-24781

Phenylhydroxamic acid (Figure 5) belongs to the inhibitory molecules of histone deacetylases (HDACs) used for their antitumor activities [48]. In particular, PCI-24781 inhibits HDAC2, which is one of the HDAC family involved in the regulation of the factors of HR. According to the work of Adimoolam et al. [49], inhibition of HDAC by PCI-24781 reduces the expression of Rad51 in the cell line derived from human colorectal carcinoma (HTC116) and thus reduces the HR response to DSBs induced by therapy. The first observation by immunofluorescence showed a complete inhibition of the formation of HsRad51 foci in the presence of PCI-24781 after irradiation and an apoptosis rate of 7% with 0.2 μM. After 24 hours, the level of protein synthesis in the absence or presence of 0.2 mM PCI-24781 showed a decrease in transcription of 60% BRCA1 and 80% for both BRCA2 and HsRad51. BRCA2 protein is also involved in the DNA repair protein complex with Rad51 [50]. The HR inhibition observed results in an additive effect of the reductions in expression levels of Rad51 and BRCA2. Moreover, the fall in BRCA2 in the nuclear compartment probably removes the inhibition of caspase-3 protease, which is able to cleave Rad51 and thus to inactivate HR DNA repair [51]. Therefore, PCI-24781 may indirectly activate the cleavage of Rad51 in addition to inducing a decrease in its synthesis. Interestingly, PCI-24781 induces the expression of the gene of the GADD45y protein which is a factor of cellular growth arrest [52]. This third effect of PCI-24781 can limit the growth of tumor cells.

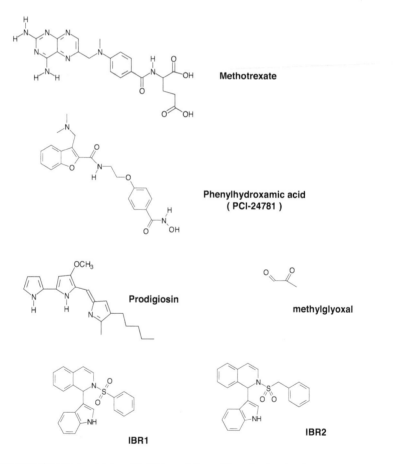

Figure 5. Structure of chemical modulators of Rad51 transcription and its nuclear translocation.

3.2.3. Prodigiosin

Prodigiosin is a tripyrrole red pigment (Figure 5) from bacteria *Serratia marcescens*. Recently, immunosuppressive and anticancer properties have been identified for this molecule [53-55]. The first action of prodigiosin is an increase in DNA DSBs, which probably results from the inhibition of both topoisomerase I and II [56]. Lu and collaborators described a significant reduction in the level of HsRad51 protein and mRNA in breast tumor cell lines (MCF-7, MDA-MB-231, T- 47D, A549, HCT116) in the presence of 50 nM prodigiosin [57]. The downregulation of Rad51 is mainly induced by lowering mRNA expression and not by proteasome-mediated Rad51 degradation. Although the tumor suppressor protein p53 is known as a repressor of Rad51 [58], prodigiosin downregulates Rad51 in a p53-independent manner. This result is an advantage for the prodigiosin-mediated therapy of cancer in which p53 is deficient compared to molecules such as flavopiridole or roscovitine, which need an activation of this protein. On the other hand, prodigiosin activates JNK and p38-MAPK signaling pathways, which are known to mediate the pro-apoptotic effect of numerous anticancer drugs [59]. By using specific inhibitors of both these signaling pathways, Lu and collaborators have shown that the level of Rad51 mRNA is restored. This result confirms the involvement of JNK and p38-MAPK signaling pathways in the prodigiosin-induced Rad51 downregulation. However, this is in contradiction with the work of Chuang et al. [60], which showed that the activation of p53-MAPK could increase the level of Rad51 protein, improving its stability and not significantly altering the level of HsRad51 mRNA. Similarly, Ko and colleagues [61] have shown a decrease in mRNA levels of HsRad51 with curcumin treatment and an inactive ERK/p38-MAPK signaling pathway. Although the Rad51 downregulation mechanism is not fully understood, prodigiosin seems to be a potent suppressor of Rad51 which may be used to overcome HR-mediated drug resistance in cancer.

4. Modulation of Rad51 by inactivation of nuclear translocation

4.1. Nuclear Localization Signal

Eukaryotic proteins are expressed in the cytoplasm. If their functions are carried out in the nucleus, they have to pass through the nuclear membrane. In contrast to small biomolecules, numerous proteins larger than 20 Kda require active transport via a signal peptide of recognition: a Nuclear Localization Signal (NLS) [62]. This signal peptide may be recognized by karyopherins [63] to form a nuclear protein complex (NPC) [64,65]. Proteins involved in HR are not exempt from this obligation for nuclear translocation. Thus, it is useful to analyze the possibility of blocking the passage of Rad51 through the nuclear membrane and thereby inhibit HR. Rad51 protein does not contain an NLS so its nuclear translocation requires an association with another protein. Interestingly, among the Rad51 paralogs, Rad51C has an NLS [66] as do Rad54 [67] and Rad52 [68] proteins. Other proteins related to DNA repair such as BRCA2 have been reported as being involved in Rad51 transport [69]. The mechanism of Rad51 transport is not clearly understood although Rad51C seems to be an interesting candidate. In particular, Gildemeister and colleagues have found that Rad51C deficiency significantly reduces the amount of Rad51 in the nucleus before and after DNA damage [69]. Another option is protein kinase B or AKT-1 protein kinase which is involved in the cytoplasmic sequestration of Rad51 [70]. The modulation of Rad51 transport offers an excellent tool to potentiate anticancer therapy through inhibition of Rad51 nuclear translocation.

4.2. AKT-1 kinase and BRCA1 proteins

Activation of AKT-1 promotes cell proliferation and the activated form is regularly found in cancer cells. In addition, to reducing malignant cell division, the inhibition of the AKT-1 signaling pathway has been investigated for the purpose of co-therapeutic approaches [71]. AKT activation occurs through a series of successive phosphorylation steps at thr-450, thr-308 and ser-473 by JNK kinases, phosphoinositide-dependent kinase 1 and by several kinases (PKD2 and others), respectively [72]. Plo and collaborators have demonstrated another aspect of the activation of AKT-1 in HR DNA repair of chemotherapy-induced DSBs [70]. This group studied the level of HsRad51 and BRCA1 in cell lines MCF7 and MDA-MB-231 and observed a decreased level in the nucleus while both these proteins accumulated in the cytoplasm. Although the HsRad51 and BRCA1 features are not modified, AKT-1 activation induces a retention signal of these proteins in the cytoplasm. Thus, their absence in the nucleus confers a deficiency of recombinase activity. The retention mechanism is still unknown, but it seems to be related to AKT-1-mediated BRCA1 NLS phosphorylation. In fact, it has been observed that AKT-1 phosphorylates BRCA1 on two sites located in the region of the NLS [73] and some mutations of these sites show a suppression of nuclear translocation of Rad51 and BRCA1, irrespective of the activated AKT-1. In this context, an activator of AKT-1 phosphorylation, such as methylglyoxal (Figure 5) [74,75] may promote the cytoplasmic sequestration of Rad51.

4.3. Modulation of the interaction between Rad51 and BRCA2

Human BRCA2 protein is constituted of 3418 amino acids (384 kDa) and contains several interaction domains. There is an interaction site with N-terminal RPA and in the central region of BRCA2 there are 8 repeated motifs called BRC motifs [76]. BRC1, BRC4, BRC7 and BRC8 motifs are able to interact with Rad51 with different affinities [77]. Pellegrini et al. [26] have shown that the BRC4 motif interacts with HsRad51 by mimicking the motif of Rad51 which is responsible for its polymerization. These BRC motifs can bind monomeric or oligomeric forms of Rad51 in a cell cycle-dependent manner and in response to DNA damage. HsRad51 regulation is also mediated by serine 3291 of the BRCA2 C-terminal domain. In the absence of DNA damage, this serine is phosphorylated by CDK1 whereas it is in a dephosphorylated form with inactivated CDK1. The ser-3291 can bind only to the oligomeric form in the nucleoprotein filament. This binding plays a role in stabilizing the Rad51-DNA complex since the phosphorylation of ser-3291 inhibits oligomerization in the absence of DSB and then synchronizes the repair mechanism [78]. It has been proposed that the BRCA2 protein is directly involved in the nuclear transport of Rad51 [50]. The pancreatic adenocarcinoma cell line CAPAN-1 is known to be defective in BRCA2 [79]. It has a deletion of the BRCA2 domains for DNA repair and the nuclear localization signals [80]. Rad51 exhibits impaired nuclear translocation in CAPAN-1. Therefore, it has been proposed that Rad51 requires BRCA2 for its nuclear translocation and that C-terminally trunkated BRCA2 retains Rad51 in the cytoplasm. BRCA2-Rad51 interaction is also essential in the HR process and many works have described those derivative peptides of BRCA2 that are able to mimic and bind to this interaction site [81-83]. Small molecules have been proposed to disrupt the interaction and two patents have been deposited [84]. By using the two-hybrid system in yeast, Lee and Chen suggested several molecules from a drug screening. Two hydrophobic molecules (phenylsulfonyl indolyl isoquinoline derivatives) IBR1 and IBR2 (Figure 5) were found to be able to disrupt the interaction and can thus potentiate anticancer treatments. The authors suggest that the benzene ring of IBR2 interacts in the hydrophobic pocket of Rad51 which is involved in the subunit-subunit interaction during filament formation and also in the interaction with the BRC4 motif of BRCA2. This phenyl moiety of IBR2 may be a competitor with the Rad51 F86 or BRC4 F1524 [85].

The authors also analyzed the effect of IBR2 at cellular level. After irradiation, breast cancer cells (MCF-7) presented a lower number of Rad51 foci than when these cells were pre-treated with IBR2. Another result was the fast degradation of Rad51 in the treated cells where the HR was impaired. This work, which is ongoing, has led to the development and synthesis of other IBR2 analogs [85,86].

5. Conclusion

DNA repair by homologous recombination is now a potential target in cancer therapy. The induction of DNA damage is one of the means of action against uncontrolled cell proliferation systems, while repairs are causes of resistance to radio- and chemotherapy. DNA repair is frequently found to be deregulated in tumor cells. Rad51 is the central protein of HR and its expression level is correlated with resistance to chemotherapeutic drugs. This observation suggests that targeted inhibition of Rad51 through small chemical molecules may improve the response to drug treatment by reducing HR.

Among antitumoral strategies, several studies have proposed numerous molecules that inhibit the recombinase activity; these are described in Figure 6.

DIDS and metatungstate are molecules that deregulate the ATPase activity of Rad51. DIDS thus causes a random hydrolysis of ATP without ssDNA-bound Rad51, while metatungstate inhibits the ATPase activity. The ATPase center is located at the Rad51 subunit-subunit interface which binds and hydrolyzes ATP and regulates the conformation of the DNA binding site. Although both molecules act differently, they induce an inhibition of the binding of Rad51 onto DNA.

Inhibition of the Rad51 polymerization is also interesting since it directly affects filament formation. The compound RI-1 can bind to the thiol group of cys-319 which inhibits the interaction between monomers of Rad51.

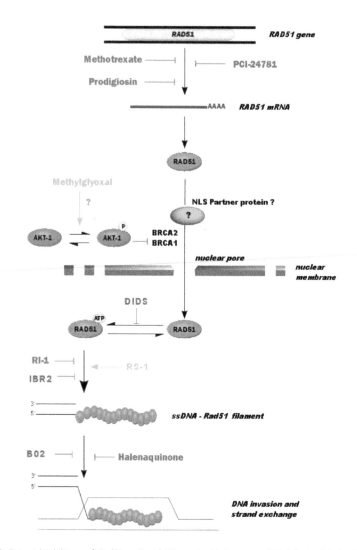

Figure 6. Potential inhibitors of Rad51-mediated HR repair: Methotrexate (100nM), Prodigiosin (100nM) and PCI-24781 (200nM) treatments reduce the levels of Rad51 mRNA between 20% and 50% in cancer cells. The decrease of Rad51 transcript level also induces an inhibition of Rad51 foci formation [47,49,57]. A 10μM concentration of DIDS significantly inhibits the binding of Rad51 to DNA, leading to the strand exchange inhibition [32]. RI-1 and B02 (20μM) decreases 50% of DNA binding by Rad51 and disrupts the formation of Rad51 foci after DNA damage in cells [18,42]. IBR2 (20μM) inhibits the Rad51 oligomerization in vitro by binding to Rad51 hydrophobic pocket [85]. 20μM of RS-1 promotes the binding of Rad51 to DNA [44]. Halenaquinone (30μM) inhibits the step of DNA homologous pairing mediated by Rad51 in vitro and the Rad51 foci formation after irradiation of cells [40]. It is noteworthy that in vitro IC50 values depend on the protein concentration and technical conditions, which makes difficult to compare them each other and to values obtained in cells.

Halenaquinone modulates the recombinase activity by inhibiting the binding of the Rad51-ssDNA complex with dsDNA. The hypothesis is that the presence of halenaquinone destabilizes the Rad51-ssDNA binding or the interaction between Rad51 subunits. This instability results in a disassembly of the complex, before the recognition of the homologous sequence. Compound B02 is also capable of modulating the function of Rad51 by disrupting Rad51 binding to DNA and formation of the nucleoprotein filament. Moreover, this compound increases cell sensitivity to DNA damaging agents and to PARP1 inhibitors. Thus, small molecules acting directly on the recombinase activity steps offer a potential development for new anticancer treatment associated with chemo- or radiotherapy. Another approach is to decrease the expression of the *RAD51* gene. For these purposes, several studies have demonstrated that methotrexate and molecule PCI-24781 significantly reduce the synthesis of Rad51 mRNA. Note that only the effects on mRNA levels were observed but the mechanism of transcription control remains unclear.

Prodigiosin also decreases the level of Rad51 mRNA which seems to be related to the activation of the JNK and p58-MAPK signaling pathways. Therefore, chemical compounds that up- and downregulate Rad51 production and/or activity may be useful for the suppression of tumor progression but the therapeutic applications of this strategy are currently inconceivable and unlikely. The last mode of action focuses on the transport of Rad51 from cytoplasm to nucleus. Rad51 is a protein whose nuclear functions require partners to facilitate its entry into the nucleus. It has been noted that the activation of the anti-apoptotic protein AKT-1 inactivates the nuclear translocation of Rad51 and BRCA1. The mechanism is still poorly understood but retention in the cytoplasm causes a significant decrease in HR. Several studies also suggest that BRCA2 is involved in this transport. Molecules such as IBR, which interfere with the Rad51-BRCA2 interaction, may induce cytoplasmic sequestration of Rad51 which decreases HR. These molecules capable of inhibiting the transport of Rad51 appear attractive candidates.

Most of these Rad51 inhibitors have been identified from screening libraries. These small molecules were tested in vitro and in cellular models but it will be necessary to quantify their efficacy, identify their toxicities and their potentially additional pharmacokinetic and pharmacological properties by animal assays. An understanding of toxicities, adverse effects, and special dosing considerations of existing anticancer compounds is important to the design of effective drug combinations and to the interpretation of the toxicological profile of new chemical entities.

Afterwards, a major challenge is to design new analogues to these molecules that will be more selective for Rad51 so that their efficacy will be improved and their potential toxicities will be decreased. The process of identifying and selecting these analogues has undergone a sea change in the recent decades with the development of solid-state and combinatorial chemistry and computer modeling of drug–target interactions.

By sensitizing cells to DNA damage, Rad51 inhibitors open up new perspectives in the search for agents capable of suppressing homologous recombination and thereby potentiating chemo- and radiotherapy treatments for cancer. Moreover, these molecules may be not only instrumental in the development of combination anticancer therapies but also excellent tools to analyze Rad51 activities and cellular functions.

Acknowledgements

This work was supported by grants from the Ligue contre le Cancer Comité de Loire Atlantique et de Vendée and the Région pays de la Loire (CIMath project). We thank G. Levillain for his help.

Author details

Axelle Renodon-Cornière, Pierre Weigel, Magali Le Breton and Fabrice Fleury*

*Address all correspondence to: fleury-f@univ-nantes.fr

Unité UFIP, CNRS FRE 3478, University of Nantes, France

References

[1] Jackson, S. P., and Bartek, J. (2009) *Nature* 461, 1071-1078

[2] Vilenchik, M. M., and Knudson, A. G. (2003) *Proc Natl Acad Sci U S A* 100, 12871-12876

[3] Krejci, L., Chen, L., Van Komen, S., Sung, P., and Tomkinson, A. (2003) *Prog Nucleic Acid Res Mol Biol* 74, 159-201

[4] Shinohara, A., and Ogawa, T. (1995) *Trends Biochem Sci* 20, 387-391

[5] Takata, M., Sasaki, M. S., Sonoda, E., Morrison, C., Hashimoto, M., Utsumi, H., Yamaguchi-Iwai, Y., Shinohara, A., and Takeda, S. (1998) *EMBO J* 17, 5497-5508

[6] Uziel, T., Lerenthal, Y., Moyal, L., Andegeko, Y., Mittelman, L., and Shiloh, Y. (2003) *EMBO J* 22, 5612-5621

[7] Rass, E., Grabarz, A., Bertrand, P., and Lopez, B. S. (2012) *Cancer Radiother* 16, 1-10

[8] Aihara, H., Ito, Y., Kurumizaka, H., Yokoyama, S., and Shibata, T. (1999) *J Mol Biol* 290, 495-504

[9] Bearss, D. J., Lee, R. J., Troyer, D. A., Pestell, R. G., and Windle, J. J. (2002) *Cancer Res* 62, 2077-2084

[10] Raderschall, E., Stout, K., Freier, S., Suckow, V., Schweiger, S., and Haaf, T. (2002) *Cancer Res* 62, 219-225

[11] Slupianek, A., Hoser, G., Majsterek, I., Bronisz, A., Malecki, M., Blasiak, J., Fishel, R., and Skorski, T. (2002) *Mol Cell Biol* 22, 4189-4201

[12] Maacke, H., Opitz, S., Jost, K., Hamdorf, W., Henning, W., Kruger, S., Feller, A. C., Lopens, A., Diedrich, K., Schwinger, E., and Sturzbecher, H. W. (2000) *Int J Cancer* 88, 907-913

[13] Maacke, H., Jost, K., Opitz, S., Miska, S., Yuan, Y., Hasselbach, L., Luttges, J., Kalthoff, H., and Sturzbecher, H. W. (2000) *Oncogene* 19, 2791-2795

[14] Vispe, S., Cazaux, C., Lesca, C., and Defais, M. (1998) *Nucleic Acids Res* 26, 2859-2864

[15] Ohnishi, T., Taki, T., Hiraga, S., Arita, N., and Morita, T. (1998) *Biochem Biophys Res Commun* 245, 319-324

[16] Christodoulopoulos, G., Malapetsa, A., Schipper, H., Golub, E., Radding, C., and Panasci, L. C. (1999) *Clin Cancer Res* 5, 2178-2184

[17] Collis, S. J., Tighe, A., Scott, S. D., Roberts, S. A., Hendry, J. H., and Margison, G. P. (2001) *Nucleic Acids Res* 29, 1534-1538

[18] Huang, F., Motlekar, N. A., Burgwin, C. M., Napper, A. D., Diamond, S. L., and Mazin, A. V. (2011) *ACS Chem Biol* 6, 628-635

[19] Barlow, A. L., Benson, F. E., West, S. C., and Hulten, M. A. (1997) *EMBO J* 16, 5207-5215

[20] Haaf, T., Golub, E. I., Reddy, G., Radding, C. M., and Ward, D. C. (1995) *Proc Natl Acad Sci U S A* 92, 2298-2302

[21] Conway, A. B., Lynch, T. W., Zhang, Y., Fortin, G. S., Fung, C. W., Symington, L. S., and Rice, P. A. (2004) *Nat Struct Mol Biol* 11, 791-796

[22] Shin, D. S., Pellegrini, L., Daniels, D. S., Yelent, B., Craig, L., Bates, D., Yu, D. S., Shivji, M. K., Hitomi, C., Arvai, A. S., Volkmann, N., Tsuruta, H., Blundell, T. L., Venkitaraman, A. R., and Tainer, J. A. (2003) *EMBO J* 22, 4566-4576

[23] Wu, Y., Qian, X., He, Y., Moya, I. A., and Luo, Y. (2005) *J Biol Chem* 280, 722-728

[24] Wu, Y., He, Y., Moya, I. A., Qian, X., and Luo, Y. (2004) *Mol Cell* 15, 423-435

[25] Ariza, A., Richard, D. J., White, M. F., and Bond, C. S. (2005) *Nucleic Acids Res* 33, 1465-1473

[26] Pellegrini, L., Yu, D. S., Lo, T., Anand, S., Lee, M., Blundell, T. L., and Venkitaraman, A. R. (2002) *Nature* 420, 287-293

[27] Doherty, A. J., Serpell, L. C., and Ponting, C. P. (1996) *Nucleic Acids Res* 24, 2488-2497

[28] Bugreev, D. V., and Mazin, A. V. (2004) *Proc Natl Acad Sci U S A* 101, 9988-9993

[29] Ristic, D., Modesti, M., van der Heijden, T., van Noort, J., Dekker, C., Kanaar, R., and Wyman, C. (2005) *Nucleic Acids Res* 33, 3292-3302

[30] Sigurdsson, S., Trujillo, K., Song, B., Stratton, S., and Sung, P. (2001) *J Biol Chem* 276, 8798-8806

[31] Makara, G. B., Stark, E., Karteszi, M., Palkovits, M., and Rappay, G. (1981) *Am J Physiol* 240, E441-446

[32] Ishida, T., Takizawa, Y., Kainuma, T., Inoue, J., Mikawa, T., Shibata, T., Suzuki, H., Tashiro, S., and Kurumizaka, H. (2009) *Nucleic Acids Res* 37, 3367-3376

[33] Stark, J. M., Hu, P., Pierce, A. J., Moynahan, M. E., Ellis, N., and Jasin, M. (2002) *J Biol Chem* 277, 20185-20194

[34] Amunugama, R., He, Y., Willcox, S., Forties, R. A., Shim, K. S., Bundschuh, R., Luo, Y., Griffith, J., and Fishel, R. (2011) *J Biol Chem* 287, 8724-8736

[35] Li, Y., He, Y., and Luo, Y. (2009) *Biochemistry* 48, 6805-6810

[36] Matsuo, Y., Sakane, I., Takizawa, Y., Takahashi, M., and Kurumizaka, H. (2006) *FEBS J* 273, 3148-3159

[37] Qian, X., Wu, Y., He, Y., and Luo, Y. (2005) *Biochemistry* 44, 13753-13761

[38] Fujiwara, H., Matsunaga, K., Saito, M., Hagiya, S., Furukawa, K., Nakamura, H., and Ohizumi, Y. (2001) *Eur J Pharmacol* 413, 37-45

[39] Schmitz, F. J., and Bloor, S. (1988) *J.Org.Chem* 53, 3922-3925

[40] Takaku, M., Kainuma, T., Ishida-Takaku, T., Ishigami, S., Suzuki, H., Tashiro, S., van Soest, R. W., Nakao, Y., and Kurumizaka, H. (2011) *Genes Cells* 16, 427-436

[41] Huang, F., Mazina, O. M., Zentner, I. J., Cocklin, S., and Mazin, A. V. (2012) *J Med Chem* 55, 3011-3020

[42] Budke, B., Logan, H. L., Kalin, J. H., Zelivianskaia, A. S., Cameron McGuire, W., Miller, L. L., Stark, J. M., Kozikowski, A. P., Bishop, D. K., and Connell, P. P. (2012) *Nucleic Acids Res*

[43] Krejci, L., Damborsky, J., Thomsen, B., Duno, M., and Bendixen, C. (2001) *Mol Cell Biol* 21, 966-976

[44] Jayathilaka, K., Sheridan, S. D., Bold, T. D., Bochenska, K., Logan, H. L., Weichselbaum, R. R., Bishop, D. K., and Connell, P. P. (2008) *Proc Natl Acad Sci U S A* 105, 15848-15853

[45] Mariani, B. D., Slate, D. L., and Schimke, R. T. (1981) *Proc Natl Acad Sci U S A* 78, 4985-4989

[46] Ostling, O., and Johanson, K. J. (1984) *Biochem Biophys Res Commun* 123, 291-298

[47] Du, L. Q., Du, X. Q., Bai, J. Q., Wang, Y., Yang, Q. S., Wang, X. C., Zhao, P., Wang, H., Liu, Q., and Fan, F. Y. (2012) *J Cancer Res Clin Oncol* 138, 811-818

[48] Buggy, J. J., Cao, Z. A., Bass, K. E., Verner, E., Balasubramanian, S., Liu, L., Schultz, B. E., Young, P. R., and Dalrymple, S. A. (2006) *Mol Cancer Ther* 5, 1309-1317

[49] Adimoolam, S., Sirisawad, M., Chen, J., Thiemann, P., Ford, J. M., and Buggy, J. J. (2007) *Proc Natl Acad Sci U S A* 104, 19482-19487

[50] Davies, A. A., Masson, J. Y., McIlwraith, M. J., Stasiak, A. Z., Stasiak, A., Venkitaraman, A. R., and West, S. C. (2001) *Mol Cell* 7, 273-282

[51] Brown, E. T., Robinson-Benion, C., and Holt, J. T. (2008) *Radiat Res* 169, 595-601

[52] Zhang, X., Sun, H., Danila, D. C., Johnson, S. R., Zhou, Y., Swearingen, B., and Klibanski, A. (2002) *J Clin Endocrinol Metab* 87, 1262-1267

[53] Chang, C. C., Chen, W. C., Ho, T. F., Wu, H. S., and Wei, Y. H. (2011) *J Biosci Bioeng* 111, 501-511

[54] Pandey, R., Chander, R., and Sainis, K. B. (2009) *Curr Pharm Des* 15, 732-741

[55] Perez-Tomas, R., Montaner, B., Llagostera, E., and Soto-Cerrato, V. (2003) *Biochem Pharmacol* 66, 1447-1452

[56] Montaner, B., Castillo-Avila, W., Martinell, M., Ollinger, R., Aymami, J., Giralt, E., and Perez-Tomas, R. (2005) *Toxicol Sci* 85, 870-879

[57] Lu, C. H., Lin, S. C., Yang, S. Y., Pan, M. Y., Lin, Y. W., Hsu, C. Y., Wei, Y. H., Chang, J. S., and Chang, C. C. (2012) *Toxicol Lett* 212, 83-89

[58] Lazaro-Trueba, I., Arias, C., and Silva, A. (2006) *Cell Cycle* 5, 1062-1065

[59] Fan, M., and Chambers, T. C. (2001) *Drug Resist Updat* 4, 253-267

[60] Chuang, S. M., Wang, L. H., Hong, J. H., and Lin, Y. W. (2008) *Toxicol Appl Pharmacol* 230, 290-297

[61] Ko, J. C., Tsai, M. S., Weng, S. H., Kuo, Y. H., Chiu, Y. F., and Lin, Y. W. (2011) *Toxicol Appl Pharmacol* 255, 327-338

[62] Adam, S. A., Marr, R. S., and Gerace, L. (1990) *J Cell Biol* 111, 807-816

[63] Radu, A., Blobel, G., and Moore, M. S. (1995) *Proc Natl Acad Sci U S A* 92, 1769-1773

[64] Ryan, K. J., and Wente, S. R. (2000) *Curr Opin Cell Biol* 12, 361-371

[65] Gorlich, D., and Kutay, U. (1999) *Annu Rev Cell Dev Biol* 15, 607-660

[66] French, C. A., Tambini, C. E., and Thacker, J. (2003) *J Biol Chem* 278, 45445-45450

[67] Golub, E. I., Kovalenko, O. V., Gupta, R. C., Ward, D. C., and Radding, C. M. (1997) *Nucleic Acids Res* 25, 4106-4110

[68] Shen, Z., Cloud, K. G., Chen, D. J., and Park, M. S. (1996) *J Biol Chem* 271, 148-152

[69] Gildemeister, O. S., Sage, J. M., and Knight, K. L. (2009) *J Biol Chem* 284, 31945-31952

[70] Plo, I., Laulier, C., Gauthier, L., Lebrun, F., Calvo, F., and Lopez, B. S. (2008) *Cancer Res* 68, 9404-9412

[71] Martelli, A. M., Evangelisti, C., Chiarini, F., Grimaldi, C., Manzoli, L., and McCubrey, J. A. (2009) *Expert Opin Investig Drugs* 18, 1333-1349

[72] Xiao, L., Gong, L. L., Yuan, D., Deng, M., Zeng, X. M., Chen, L. L., Zhang, L., Yan, Q., Liu, J. P., Hu, X. H., Sun, S. M., Liu, J., Ma, H. L., Zheng, C. B., Fu, H., Chen, P. C., Zhao, J. Q., Xie, S. S., Zou, L. J., Xiao, Y. M., Liu, W. B., Zhang, J., Liu, Y., and Li, D. W. (2010) *Cell Death Differ* 17, 1448-1462

[73] Altiok, S., Batt, D., Altiok, N., Papautsky, A., Downward, J., Roberts, T. M., and Avraham, H. (1999) *J Biol Chem* 274, 32274-32278

[74] Chang, T., Wang, R., Olson, D. J., Mousseau, D. D., Ross, A. R., and Wu, L. (2011) *FASEB J* 25, 1746-1757

[75] Jia, X., Chang, T., Wilson, T. W., and Wu, L. (2012) *PLoS One* 7, e36610

[76] Bork, P., Blomberg, N., and Nilges, M. (1996) *Nat Genet* 13, 22-23

[77] Wong, A. K., Pero, R., Ormonde, P. A., Tavtigian, S. V., and Bartel, P. L. (1997) *J Biol Chem* 272, 31941-31944

[78] Davies, O. R., and Pellegrini, L. (2007) *Nat Struct Mol Biol* 14, 475-483

[79] Jasin, M. (2002) *Oncogene* 21, 8981-8993

[80] Holt, J. T., Toole, W. P., Patel, V. R., Hwang, H., and Brown, E. T. (2008) *Cancer Genet Cytogenet* 186, 85-94

[81] Chen, C. F., Chen, P. L., Zhong, Q., Sharp, Z. D., and Lee, W. H. (1999) *J Biol Chem* 274, 32931-32935

[82] Nomme, J., Renodon-Corniere, A., Asanomi, Y., Sakaguchi, K., Stasiak, A. Z., Stasiak, A., Norden, B., Tran, V., and Takahashi, M. (2011) *J Med Chem* 53, 5782-5791

[83] Nomme, J., Takizawa, Y., Martinez, S. F., Renodon-Corniere, A., Fleury, F., Weigel, P., Yamamoto, K., Kurumizaka, H., and Takahashi, M. (2008) *Genes Cells* 13, 471-481

[84] Lee, W.-H. C. L. (2006) Compositions and methods for disruption of BRCA2-Rad51 interaction. USA

[85] Lee, W.-H., Chen, P.-L., Zhou, L., and Zhu, J. (2009) Compositions and methods related to Rad51 inactivation in the treatment of neoplastic diseases and especially CML. the regents of the university of California, CA, US, USA

[86] Qiu, X. L., Zhu, J., Wu, G., Lee, W. H., and Chamberlin, A. R. (2009) *J Org Chem* 74, 2018-2027

Chapter 6

Nucleotide Excision Repair Inhibitors: Still a Long Way to Go

K. Barakat and J. Tuszynski

Additional information is available at the end of the chapter

1. Introduction

The last few decades have witnessed a new astounding trend emerge in cancer research. The new strategy materialized as a glimmer of hope to improve current standard cancer treatments that target DNA. These DNA-damaging agents induce lesions into the genome, which are aimed at preventing cancer cells from proliferating and invading the surrounding tissue. However, as was shown by many experiments, in response to that cancer cells mobilize DNA repair pathways that tend to remove the induced damage. As a consequence, they exhibit increased resistance towards what would otherwise be an efficacious treatment [1]. These findings have validated DNA repair enzymes as new molecular targets in the context of the battle against cancer [2]. Fortunately, the proof of the concept of targeting DNA repair as a cancer-therapeutic-strategy has been provided by several convincing studies, many of which are advancing through pre-clinical and clinical trials [3]. A particular example of a novel target in such pathways is the nucleotide excision repair (NER) mechanism, which correlates with the induced resistance to platinum treatments [4].

In normal cells, NER removes a broad range of DNA lesions, protecting cell integrity [5]. In cancer cells exposed to DNA damaging agents that distort the DNA helix or form bulky injuries to the genome, NER comes into play and removes the damage, in order to prevent cancer cells from lethal consequences of this damage [5, 6]. A striking example of this mechanism is represented by the use of platinum compounds such as cisplatin, the principal component of many treatments involving solid tumors including testicular, bladder, ovarian, head and neck, cervical, lung and colorectal cancer [7]. It has been demonstrated that NER is the major DNA repair mechanism that removes cisplatin-induced DNA damage, and that resistance to platinum-based therapy correlates with high expression of ERCC1, a major enzymatic element of the NER machinery. In this context, a reasonable way to increase the efficacy of platinum-based therapy and decrease drug resistance would be to regulate NER by inhibiting the activity of ERCC1 and interacting proteins using yet to be discovered therapeutic compounds [8-11].

The protein ERCC1 forms a heterodimer with XPF. The resulting complex is an endonuclease enzyme that cleaves the 5` end of the damaged DNA strand whereas XPG cleaves it in the 3′ position [6]. ERCC1-XPF is recruited to the damage site through a direct interaction between ERCC1 and XPA, an indispensible element of the NER pathway. No cellular function beyond NER has been observed for XPA and competitive inhibition of the XPA interaction with peptide fragments is considered effective at disrupting NER. Furthermore, based on clinical data, cancer patients that have been shown to have low expression levels of either XPA or ERCC1 demonstrate a correlation with a higher sensitivity to cisplatin treatments [12, 13].

This chapter reviews the state-of-the-art efforts that have been made to date to identify inhibitors of the NER pathway. These efforts have been mainly focused on targeting either the ERCC1-XPA or the ERCC1-XPF interactions. We discuss the various methods that were used toward this aim and illustrate the mode of action of the identified inhibitors. We hope that the compiled knowledge in this chapter will help researchers and clinicians in their efforts to develop new drug candidates that can improve the efficacy of and reduce resistance against platinum treatments and other DNA damaging agents as a way to arrest tumor progression.

2. Nucleotide excision repair pathway

The nucleotide excision repair process, shown in Figure 1, occurs as a stepwise mechanism and involves more than 30 different proteins. It is a "cut-and-paste" mechanism that replaces a ~30 nucleotide DNA strand that contains the lesion with a correct base pair sequence. This pathway has been extensively studied so that all the genes that are involved in it have been cloned and expressed as recombinant proteins. The main players within NER include the seven Xeroderma Pigmentosum (XP) complementation groups, XPA to XPG proteins; the Excision Repair Cross Complementing group 1 protein (ERCC1); the human Homolog of yeast RAD23 (hHR23B), the Replication Protein A (RPA), the subunits of Transcription Factor that possess Helicase activity (TFIIH), and the Cockayne Syndrome proteins A and B (CSA and CSB) [14]. Depending on the location of the DNA damage within the genome, one can recognize two NER sub-pathways. First is the transcription-coupled repair (TCR-NER), if the DNA damage is located within the actively transcribed genes of the genome. The second is the global genome repair (GGR-NER), if the damage is located within the whole genome. The two types are thought to be identical except for the initial damage recognition step. The two mechanisms involve five sequential steps [15] described below.

The foremost step is the detection of the damage. As mentioned above, the recognition step is the only difference between TCR and GGR. In the GGR subpathway, the XPC-hHR23B-XPE complex continuously scans the genome for bulky DNA damage until it recognizes a lesion and, consequently, initiates the rest of the NER sequence. On the other hand, a stalled RNAPII and Cockayne syndrome proteins, CSA and CSB, recognize the damage and activate the TCR-NER pathway. Once the damage is recognized the second step starts by recruiting the TFIIH complex in order to unwind the DNA helix surrounding the lesion. TFIIH is composed of two major sub-complexes. The core is formed from the association of a large number of proteins including XPB, XPD, p62, p52, p44, p34 and p8. The rest of TFIIH is the cdk-activating kinase sub-complex, which contains cdk7, cyclin H and MAT1. Interestingly, TFIIH possesses both 3'-5' and 5'-3' helicase activities through the two ATP-dependent helicases XPB and XPD, respectively [16]. It opens the DNA structure forming a ~30 base pair bubble around the lesion. The two proteins RPA and XPA stabilize the opened DNA structure and recruit the two endonucleases that are necessary for the subsequent incision step. The interaction of XPA with the 34-kDa subunit of RPA (RPA34) activates XPA to recruit the other components of NER.

The Damaged strand-incision is the rate-limiting step for the whole pathway. The two endonucleases XBG and XPF-ERCC1 cut the two ends of the strand that contains the damage. The correct location of XPA is crucial for the recruitment of the XPF-ERCC1 heterodimer endonuclease. XPG cuts the 3' end of the damage, while XPF-ERCC1 cuts the 5' end [17]. The damaged strand is then released. DNA polymerases fill the single strand gap using the complementary intact strand as a template and DNA ligase I closes the 3' nick as a final step [15].

3. ERCC1 over-expression correlates with cisplatin resistance

ERCC1 is a 33-kDa protein that forms a tight heterodimer endonuclease complex with XPF. As described above, this endonuclease cleaves the DNA strand at the phosphodiester bonds on the 5' side of the damage. It is important to mention that the ERCC1-XPF complex has additional functions in other DNA repair pathways including inter-strand cross-link repair, double-strand break repair, and homologous recombination. Many studies have shown considerable correlation between resistance to cisplatin and the over-expression of ERCC1 [19]. This profoundly significant conclusion has been reached from several independent clinical trial investigations on ovarian [20], colorectal [21], and non–small cell lung cancer [22]. For example, a study on ~750 patients who suffer from late stages of lung cancer revealed that patients with low levels of ERCC1 and who received platinum therapy had better survival rates than those with the same levels of the protein but did not receive the platinum treatment [23]. A more recent study on 444 patients who experienced non-small lung cancer concluded that non-platinum-containing chemotherapy is more effective than platinum-based therapy on patients with high ERCC1 levels [24]. Very recently, Stefanie and coworkers [25] performed a retrospective study investigating the correlation of ERCC1 expression with patients' survival in ovarian cancer after platinum-based treatment. Their work revealed that patients with ERCC1-negative ovarian cancer had significantly better survival rates than those with ERCC1-positive ovarian cancer. They concluded that ERCC1 protein over-expression is a marker for poor survival of high-grade ovarian cancer even in patients operated on who had residual disease. All of these investigations lead to the conclusion that ERCC1 is not only a gene that is usually activated in patients subjected to platinum-based therapy but it may also act as a predictive criterion for identifying those patients who could benefit from platinum treatments [26, 27]. This latter role of ERCC1 as a biomarker is important because it can guide clinicians in their therapeutic decision-making and select the best treatment approach for a particular group of patients.

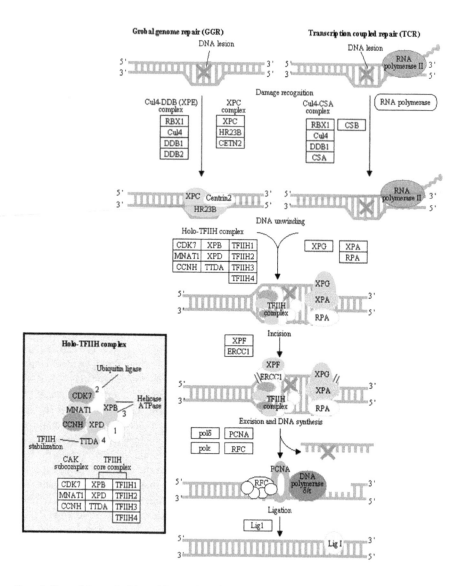

Figure 1. Steps of the nucleotide excision repair pathway. See text for details (adopted from the KEGG database [18]).

4. The ERCC1-XPA interaction is essential for a functional NER pathway

Regardless of the type of NER that is initiated, the XPA protein is equally essential to complete both pathways [28]. It plays a vital role in DNA lesion recognition and in the attraction of many other NER repair proteins. For example, prior to the incision step in NER, the ERCC1-XPF endonuclease is recruited to the damaged DNA site through a secondary interaction between ERCC1 and XPA [29, 30, 31]. Therefore, this protein-protein interaction is necessary for a functional NER mechanism. The NMR crystal structure was resolved by Tsodikov's group [13] and the critical residue-residue interactions were determined [4] through our binding energy predictions (see Figure 2). A 14-residue peptide from XPA that includes three essential consecutive glycines (residues 72–74) is buried within a hydrophobic cleft within the central domain of ERCC1. This peptide has two critical characteristics. First, it is necessary and sufficient for binding to ERCC1. Second, and more importantly, it can compete with the full-length XPA protein in binding to ERCC1 and disrupting NER *in vitro*.

In a recent study, Barbara et al. [32] reported mutations in the central domain of ERCC1 that had a significant impact on NER activity *in vitro* and *in vivo*. These mutations occur at the XPA binding site within ERCC1, preventing the interaction between the two proteins. Due to these mutations, the ERCC1-XPF nuclease was not recruited to the damaged DNA sites after exposing cells to ultra violet (UV) radiation. Consequently, the last incision step that is performed by ERCC1-XPF was never completed leading to a dysfunctional NER mechanism in these cells and, hence, a hypersensitivity to UV radiation. These results are consistent with previous findings on the importance of XPA in NER, where no cellular function beyond NER has been observed for XPA [12]. Interestingly, these mutations did not affect the activity of ERCC1-XPF in other DNA repair pathways leading to two distinctive conclusions. First, the XPA-ERCC1 interaction is only necessary for NER but not for other DNA repair pathways in which ERCC1-XPF is important for their activity. Second, the involvement and recruitment of ERCC1-XPF to the different DNA repair pathways is coordinated through different and not overlapping protein-protein interactions mediated by ERCC1. Based on these findings, one can selectively disrupt the activity of ERCC1-XPF within these DNA repair pathways by inhibiting its interactions with the recruitment factors to the damaged sites. These observations, coupled with the available crystal structure of this interaction make ERCC1 and XPA an extremely attractive target for computationally assisted development of small molecule inhibitors targeted for use in combination therapies involving cisplatin.

5. ERCC1 interacts massively with XPF

As shown in Figure 3, ERCC1 in engaged in a tight interaction with XPF in which almost every residue from XPF is either interacting or being affected by an interaction with ERCC1 residues. The main interaction sites are located within the C-terminal domains of the two proteins. The most tightly interacting regions in XPF include residues 828 to 835, 859 to 862, 878 to 882 and 892 to 905. These exhibit almost no flexibility in the bound structure, demonstrating a contribution to binding with ERCC1. The two proteins form the heterodimer enzyme that is responsible for the cleavage of one side of the damaged nucleotides chain.

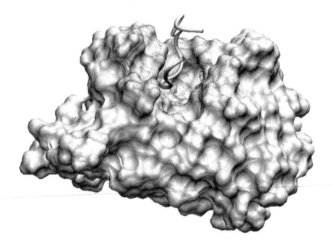

Figure 2. XPA-ERCC1 protein-protein interaction. The binding between ERCC1 (whight) and XPA (yellow) is predicted [4] to be primarily mediated by 5 residues from XPA peptide, namely; G72, G73, G74, F75 and I76. On the other hand, the contribution from the ERCC1 binding site is distributed among the following 10 residues: R106, Q107, G109, N110, P111, F140, L141, S142, Y145 and Y152.

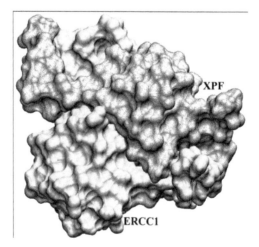

Figure 3. The ERCC1-XPF complex. The C-terminal domain of each protein interacts massively with its counterpart from the other protein forming the heterodimer endonuclease enzyme.

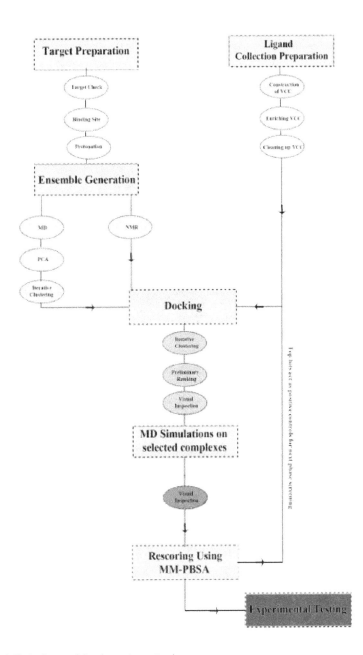

Figure 4. The implemented virtual screening protocol.

6. Earlier efforts to identify NER inhibitors

Although the NER pathway has been recognized as one of the most important factors that increase the resistance against platinum-based therapy, little work has been done so far on regulating its activity. Here, we wish to point to three major studies that identified inhibitors for the NER mechanism. First is the work done by Barret et al. [33] and their discovery of F11782. Second are the findings of Jiang and Yang [34] on the effects of the cell cycle checkpoint abrogator UCN-01 (7-hydroxystaurosporine) on NER. Finally, the work on the DNA damaging agent Et743 is a landmark result [35]. We briefly describe these outcomes below.

6.1. F11782

Using the 3D (DNA Damaged Detection) assay first proposed by Wood *et al.* [36] and later modified by Salles's team [37], Barret et al. [33] screened for NER inhibitors and identified F11782. The compound was already known as an inhibitor of both the topoisomerases II and I [38]. Moreover, F11782 did not show any activity toward other enzymes such as DNase I or T4 polynucleotide kinase, indicating that the compound targets one of the proteins that are involved in NER. Further investigations on F11782 limited its NER inhibitory activity to one of the earlier steps of the pathway, specifically either the helicase or the incision steps, with more preference given to the incision step [33].

6.2. UCN-01

Jiang and Yang [34] analyzed the effects of UCN-01, which is a well-known protein kinase C inhibitor and cell cycle checkpoint abrogator [39], on the NER pathway. These findings showed that UCN-01 inhibited the repair of cisplatin-induced DNA damage both *in vitro* and *in vivo* and indicated that UCN-01 has a dramatic inhibitory effect on the interaction of NER proteins. The drug enhanced the activity of cisplatin only in NER-proficient cells, but not in the deficient ones. However, no direct binding of UCN-01 to any of these proteins has been reported and it has been speculated that the observed inhibitory activity may result from UCN-01-mediated regulation of the signaling pathway that involves post-translational modifications of repair proteins. Although Jiang and Yang [34] attributed the loss of NER activity to an attenuation in the ERCC1-XPA protein-protein interaction, their careful and detailed binding analysis of the compound to the two proteins revealed that UCN-01 did not interact directly with either of them. However, in this work we used UCN-01 as a positive control, assuming it can bind to the XPA binding site within ERCC1, particularly because the drug can fit within the binding pocket despite its limited interactions with the protein.

6.3. Ecteinascidin 743

A final compound that has been shown [35] to interfere with NER is Ecteinascidin 743 (Et743). At the time of writing this article, Et743 is in phase II/III clinical development and its main mode of action is as a DNA damaging agent. The drug seems to specifically obstruct the TCR-NER sub-pathway, however, it does not act as an inhibitor of any of the proteins that are involved in the NER mechanism. A model proposed by Gregory et al. [35] suggests that the DNA adducts formed by Et743 are more efficient than those of cisplatin in dealing with NER. These authors suggest that the Et743-guanine adducts trap the TCR-NER pathway at the incision or ligation steps, preventing the pathway from being completed.

7. Recent attempts to discover novel NER inhibitors

As mentioned above, most of the earlier NER inhibitors listed above were not discovered to be potent or specific NER inhibitors. In other words, they were found mainly by chance to partially inhibit the NER pathway. Given the impact of regulating the NER pathway on improving many of the chemotherapeutic drug cocktails currently in clinical use, it is very important to directly target elements of NER pathway itself. Following this path, our group has been focusing on this problem in hope of implementing a novel strategy that would reverse resistance and potentiate the efficacy of cisplatin and other similar chemotherapeutic agents. The foremost endeavor is to specifically and separately target the two protein-protein interactions described above, namely the XPA-ERCC1 [4, 40] and XPF-ERCC1 [41] interactions. These efforts have already resulted in two successful examples where inhibitors identified by us via virtual screening were able to sensitize cells to ultra violet radiation (UV) and potentiate the efficacy of cisplatin in cancer cells. Here, we briefly describe the methods used and their outcomes. The studies described below primarily utilized computational tools to develop inhibitors that disturb these interactions. This was then followed by experimental validation of the predicted effects of these inhibitors on cancer cells.

7.1. The method

In the following studies, virtual screening identified small molecules that bind to and fit within the binding site within the interacting proteins in order to disturb its binding to the other protein in the complex. The virtual screening (VS) protocol that was used is shown in Figure 4. It is an improved version of the relaxed complex scheme (RCS) technique reported by McCammon and his team [42]. In the original RCS approach, all-atom MD simulations (e.g., 2-5 ns simulation) are applied to explore the conformational space of the target, while docking is subsequently used for the fast screening of drug libraries against an ensemble of receptor conformations. This ensemble is extracted at predetermined time intervals (e.g., every 10 ps) from the simulation, resulting in hundreds of thousands of protein conformations. Each conformation is then used as a target for an independent docking experiment.

The RCS methodology has been successfully applied to a number of cases. An excellent example is that of an HIV inhibitor, raltegravir which became the first FDA approved drug targeting HIV integrase [43, 44]. Other successful examples include the identification of novel inhibitors of the acetylcholine binding protein [45], RNA-editing ligase 1 [46], the influenza protein neuraminidase [47] and *Trypanosoma brucei* uridine diphosphate galactose 4'-epimerase [48]. These applications employed alternative ways to solve two main problems with the method, namely, reducing the number of extracted target conformations and deciding on how to select the final set of hits after carrying out the screening process. For the first problem, a number of studies suggested extracting the structures at larger intervals of the MD simulation, e.g. every 5ns or so [45], condensing the structural ensemble generated from MD simulations using QR factorization [46], or clustering the MD trajectory using root-mean-square-deviation (RMSD) conformational clustering [47, 48]. On the other hand, to rank the screened compounds and suggest a final set of top hits, some studies used only docking predictions [45-47], while others suggested using a more accurate scoring method (e.g. MM/PBSA (Molecular Mechanics/ Poisson Boltzmann Surface Area)) to refine the final selected hits [42]. All of these approaches, similar to the work presented here, were aimed at keeping the balance between significantly reducing the number of target structures and retaining their capacity to describe the conformational space of the target. Figure 4 describes the approach that was used to put together and improve the RCS to target the strong protein-protein interactions described above.

Our implementation follows the same guidelines as in the RCS method. We first use MD simulations and generate large enough trajectories that can progress through the phase space of the binding site. The length of the MD simulations (usually on the order of 100 ns) is determined by applying metrics that employ principal component analysis (PCA). Once the trajectory reaches an adequate sampling of target conformations, clustering analysis extracts representative structures that describe the dominant dynamics of the binding site. The extracted structures are then used as rigid targets to screen the whole library of compounds and suggest models for the most preferred ligand-protein complexes, hence, utilizing the "conformational sampling" model. These bound structures are then solvated and used to run all-atom MD simulations to relax the two molecules and generate new trajectories that represent their "induced fit" models. The MM-PBSA method finally ranks the newly generated structures and suggests a set of top hits for experimental testing.

7.2. XPA-ERCC1 inhibitors

Our earliest challenge was to directly disturb the interaction between the ERCC1 and XPA proteins. Two subsequent screening experiments were used. The initial study screened two compound databases for inhibitors of the ERCC1-XPA interaction and constructed a pharmacophore model demonstrating the crucial features necessary for their inhibition. The databases used included the National Cancer Institute Diversity Set (NCIDS) and DrugBank compounds.

The NCIDS is a collection of approximately 2,000 compounds that are structurally representative as scaffolds of a wide range of molecules, representing almost 140,000 compounds that are available for testing at the NCI. A number of its ligands contain rare earth elements and cannot be properly parameterized for docking experiments, leaving us with 1,883 compounds that can be actually used. This work exploits a cleaned 3D version of the NCIDS formatted for use in AutoDock and it was prepared by the AutoDock Scripps team. What makes the NCIDS so valuable and extensively screened by many groups (even in HTS) is that its individual molecules have distinctive structures and are the cluster representatives of their parent families. Once screened and a number of its molecules rank high in the hit list, one can return back and screen the whole family of the representative structure, instead of screening the actual NCI set of compounds. On the other hand, the DrugBank database is not only a set of molecules representing FDA-approved drugs, but it also represents a unique bioinformatics and chem-informatics resource. It relates each drug to its target(s). It includes details about the different pathways, structural information and chemical characteristics of these targets and the way they take part in inducing a particular disease. This information is stored in a freely available website that is linked to other databases (KEGG, PubChem, ChEBI, PDB, Swiss-Prot and Gen-Bank) and a range of structure displaying applets. The DrugBank collection includes ~4,800 drug structures including >1,350 FDA-approved small molecule drugs, 123 FDA-approved biotech (protein/peptide) drugs, 71 nutraceuticals and >3,243 experimental drugs. Once a hit is identified from this library, it simply represents a drug. This means many barriers of pre-clinical and clinical tests can be readily overcome and the molecule can be tested directly for its novel biological activity. Moreover, a hit from this collection may explain a mysterious side effect that would not be discovered before its identification as a regulator for the examined target.

This initial study utilized a minimized model of the XPA binding site within ERCC1 to employ flexible residue docking as implemented in AutoDock 4.0. This was then followed by RCS docking, where MD simulations and RMSD conformational clustering were used to generate a set of forty-four representative conformations of the binding site within ERCC1. AutoDock was then used to screen against a set of seven target conformations, composed of the six most dominant cluster-representative structures along with an equilibrated folded conformation for the binding site produced by employing principal component analysis on the ERCC1 trajec-tory. Top hits were rescored by docking them to the whole set of cluster-representative struc-tures and ranked by their weighted average binding energy. The non-redundant hits from these screens were then used to identify a dynamic binding-site pharmacophore that target the ERCC1-XPA interaction. The pharmacophore model was then compared to docking results for the weak inhibitor of NER, UCN-01 (7-hydroxystaurosporine). A number of selected hits from this study are shown in Figure 5.

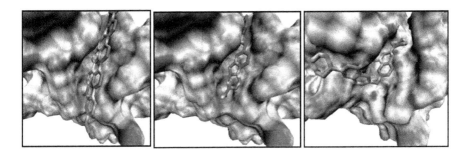

Figure 5. Three selected hits within their preferred binding site conformations. Adopted from [4].

Comparing the methodology that was used here to the workflow discussed in the above, one can make three observations. First, the virtual screening methodology depended mainly on docking scoring to rank the compounds. Second, the clustering analysis that was used to extract dominant conformations of the target was not iterative, it used a cut off RMSD value that is commonly employed in the literature. Finally, no post-docking refinements were performed on the final set of compounds. These shortcomings were properly adjusted in the subsequent study [49].

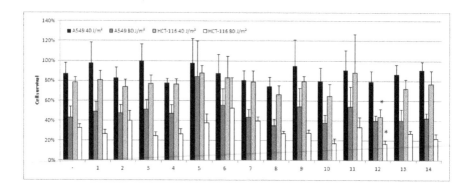

Figure 6. Sensitivity of cancer cells to UVC irradiation alone or in combination with potential inhibitors of the interaction between ERCC1 and XPA. IC50 values (J/m²). Compound 12 showed promising effects on cancer cells and was henceforth termed NERI01.

The new study used the CN chemical library for virtual screening. The CN chemical library (~50,000 compounds) is a repository of all synthetic, natural compounds and natural extracts in the existing French public laboratories. The whole database is divided into two main categories. The first part includes information about all synthetic products, while the second contains the natural compounds and extracts. In this work, we used the whole CN database in our screening. In contrast to the previously mentioned databases, compounds in this library are represented by 2D SDF structures with no hydrogen atoms attached. This required a number of cleaning and preparation steps before using them in VS simulations.

The second ERCC1-XPA study exactly followed the screening protocol described above. The hit rate of the new study was higher than that of the one described here, indicating the importance of utilizing more accurate scoring, performing iterative clustering and refining the docked structures using MD simulations. A promising hit, shown in Figure 6 as compound 12, was discovered and validated on a UV radiation sensitivity cell-based assay [40]. The validated hit was termed NER inhibitor 01 (NERI01) has been shown to be effective in sensitizing colon cancer cells to UV radiation, which induces the same type of damage as cisplatin and its lesions are removed by ENR.

7.3. XPF-ERCC1 inhibitors

The final study focused on the more challenging problem of interfering with the ERCC1-XPF interaction. As shown in Figure 4, the two proteins have a very close interaction with each other. A comparison of atomic fluctuations (as revealed by the corresponding B-factor values) between the unbound-XPF and the bound-XPF structures is shown in Figure 7. Almost all XPF residues are rigid in the bound case compared to the free structure. This indicates a massive interaction between ERCC1 and XPF in which almost every residue from XPF is either interacting or being affected by an interaction with ERCC1 residues. The most flexible regions in XPF include residues 828 to 835, 859 to 862, 878 to 882 and 892 to 905. These have almost no flexibility in the bound structure, demonstrating a contribution to binding with ERCC1.

The enthalpic contribution, as calculated by the MM-PBSA analysis, to the binding energy between the two proteins is exceptionally large (-123 kcal/mol). While the solvation energy contributed passively to the interaction (298 kcal/mol), compensation from the electrostatic and van der Waals interactions dominated the overall interaction (-238 kcal/mol and -184 kcal/mol, respectively). From these analyses we showed that ERCC1-residues shared ~50% of the total energy with PHE293 being the residue that contributes the most to the ERCC1-XPF interaction (-11 kcal/mol). On the XPF side, PHE894 has been found to contribute -7.7 kcal/mol to the binding energy. With the exception of ASP839 from XPF which disfavored the interaction by ~1 kcal/mol, the indicated residues favored the binding between ERCC1 and XPF. This allowed us to identify a binding site on the XPF surface that was used to identify putative inhibitors of this protein-protein interaction.

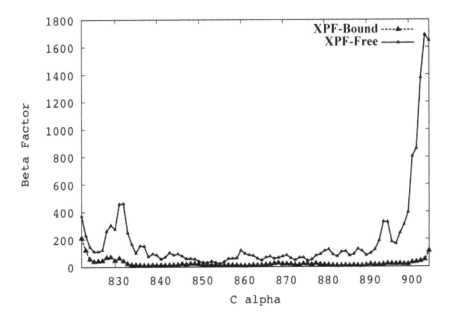

Figure 7. Flexibility of the XPF residues. Atomic fluctuations for the free and bound XPF proteins are shown here. Binding of ERCC1 to XPF considerably stabilized the protein, indicating a wide range of protein-protein interaction.

The screening methodology adopted the VS protocol shown in Figure 4 and used to screen the CN chemical library, NCI diversity set and DrugBank compounds for inhibitors of this interaction. A number of promising hits were experimentally validated and were very effective in disrupting the NER pathway and potentiating cisplatin efficacy. The most promising compounds with binding modes are shown in Figure 8.

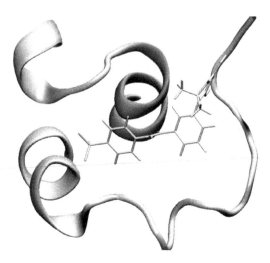

Figure 8. Binding mode of most promising XPF-ERCC1 inhibitor.

8. Conclusions

DNA damaging agents induce lesions into the genome aiming at preventing cancer cells from proliferating and invading the surrounding tissue. However, DNA repair pathways remove the induced damage and, hence, increase resistance to an otherwise efficacious treatment [1]. This approach has validated DNA repair enzymes as new molecular targets in the context of the battle against cancer. Nucleotide excision repair (NER) is a major DNA repair mechanism that removes mainly DNA lesions that distort the DNA helix or form bulky injuries to the genome. Among the most affected drugs with NER activity are platinum compounds such as cisplatin, the backbone for many treatments of solid tumors including testicular, bladder, ovarian, head and neck, cervical, lung and colorectal cancer. It has been demonstrated that NER is the major DNA repair mechanism that removes cisplatin-induced DNA damage, and that resistance to platinum-based therapy correlates with high expression of ERCC1, a major element of the NER machinery. Therefore, one way to improve such drugs and reduce their acquired resistance is by developing inhibitors that would regulate the NER machinery.

This chapter reviewed the state-of-the-art efforts that were made to identify inhibitors of the NER pathway. We discussed the various methods that were used toward this aim and illustrated the mode of action of the identified inhibitors. The earlier efforts were not focused on NER as a target. However, the first identified NER-inhibitors were discovered unintentionally. These efforts include the examples of finding the three drugs F11782 [33], UCN-01 (7-hydrox-ystaurosporine) [34] and Et743 as weak inhibitors of the NER activity. Recent studies exploited the fact that ERCC1 and its associated proteins XPA and XPF have a considerable correlation between resistance to cisplatin and their over-expression in cancer cells [19]. The latter studies were aimed at discovering specific inhibitors that target these interactions in the hope of dis-turbing their binding and hence reducing the NER activity. In this regard, we described the development of two different classes of NER inhibitors. The first class, represented by the lead compound NERI01 target the ERCC1-XPA interaction, while the second class represented by NERI02 targets the ERCC1-XPF interaction. Future directions of this research include the de-velopment of derivative structures for the identified hits and their optimization for improved drug-like properties and higher specificity to target their representative protein interactions. While great efforts have been done both *in silico* and *in vitro* to identify and validate novel inhibitors for the two mentioned NER targets, no *in vivo* studies have been performed on them yet. This is mainly due to the fact that the two proteins have been very recently recognized as druggable targets and no one in the past thought of regulating NER as a way to improve cancer therapy. However, we think that the studies presented here offer a proof-of-concept that in-hibiting the interaction of ERCC1 with either XPA or XPF has a considerable impact on the NER mechanism and, therefore, enhances the efficacy of chemotherapeutic treatments that are associated with acquired resistance due to over expression of the NER elements. We hope that this chapter will be found of value to the researchers and clinicians interested in developing new drug candidates that can improve the efficacy of and reduce resistance against platinum treatments and other DNA damaging agents as a way to arrest tumor progression.

Acknowledgements

This research was supported by funding from NSERC, Alberta Cancer Foundation, Alberta Cancer Research Institute, Canadian Breast Cancer Foundation, Alberta Advanced Education and Technology and the Allard Foundation.

Author details

K. Barakat[1,2] and J. Tuszynski[3,4]

1 Department of Physics, University of Alberta, Canada

2 Department of Engineering Mathematics and Physics, Fayoum University, Fayoum, Egypt

3 Department of Physics, University of Alberta, Canada

4 Department of Oncology, University of Alberta, Canada

References

[1] Harper JW, Elledge SJ. The DNA damage response: ten years after. Mol Cell. 2007;28:739-45.

[2] Basu B, Yap TA, Molife LR, de Bono JS. Targeting the DNA damage response in oncology: past, present and future perspectives. Curr Opin Oncol. 2012;24:316-24.

[3] Barakat KH, Gajewski MM, Tuszynski JA. DNA polymerase beta (pol beta) inhibitors: A comprehensive overview. Drug Discov Today. 2012;17:913-20.

[4] Barakat KH, Torin Huzil J, Luchko T, Jordheim L, Dumontet C, Tuszynski J. Characterization of an inhibitory dynamic pharmacophore for the ERCC1-XPA interaction using a combined molecular dynamics and virtual screening approach. J Mol Graph Model. 2009;28:113-30.

[5] Rouillon C, White MF. The evolution and mechanisms of nucleotide excision repair proteins. Res Microbiol. 2011;162:19-26.

[6] Nouspikel T. DNA repair in mammalian cells : Nucleotide excision repair: variations on versatility. Cell Mol Life Sci. 2009;66:994-1009.

[7] Koberle B, Tomicic MT, Usanova S, Kaina B. Cisplatin resistance: preclinical findings and clinical implications. Biochim Biophys Acta. 2010;1806:172-82.

[8] Metzger R, Leichman CG, Danenberg KD, Danenberg PV, Lenz HJ, Hayashi K, et al. ERCC1 mRNA levels complement thymidylate synthase mRNA levels in predicting response and survival for gastric cancer patients receiving combination cisplatin and fluorouracil chemotherapy. J Clin Oncol. 1998;16:309-16.

[9] Handra-Luca A, Hernandez J, Mountzios G, Taranchon E, Lacau-St-Guily J, Soria JC, et al. Excision repair cross complementation group 1 immunohistochemical expression predicts objective response and cancer-specific survival in patients treated by Cisplatin-based induction chemotherapy for locally advanced head and neck squamous cell carcinoma. Clin Cancer Res. 2007;13:3855-9.

[10] Bellmunt J, Paz-Ares L, Cuello M, Cecere FL, Albiol S, Guillem V, et al. Gene expression of ERCC1 as a novel prognostic marker in advanced bladder cancer patients receiving cisplatin-based chemotherapy. Ann Oncol. 2007;18:522-8.

[11] Jun HJ, Ahn MJ, Kim HS, Yi SY, Han J, Lee SK, et al. ERCC1 expression as a predictive marker of squamous cell carcinoma of the head and neck treated with cisplatin-based concurrent chemoradiation. Br J Cancer. 2008;99:167-72.

[12] Rosenberg E, Taher MM, Kuemmerle NB, Farnsworth J, Valerie K. A truncated human xeroderma pigmentosum complementation group A protein expressed from an adenovirus sensitizes human tumor cells to ultraviolet light and cisplatin. Cancer Res. 2001;61:764-70.

[13] Tsodikov OV, Ivanov D, Orelli B, Staresincic L, Shoshani I, Oberman R, et al. Structural basis for the recruitment of ERCC1-XPF to nucleotide excision repair complexes by XPA. EMBO J. 2007;26:4768-76.

[14] Wood RD. DNA damage recognition during nucleotide excision repair in mammalian cells. Biochimie. 1999;81:39-44.

[15] de Laat WL, Jaspers NG, Hoeijmakers JH. Molecular mechanism of nucleotide excision repair. Genes Dev. 1999;13:768-85.

[16] Sung P, Bailly V, Weber C, Thompson LH, Prakash L, Prakash S. Human xeroderma pigmentosum group D gene encodes a DNA helicase. Nature. 1993;365:852-5.

[17] Sijbers AM, de Laat WL, Ariza RR, Biggerstaff M, Wei YF, Moggs JG, et al. Xeroderma pigmentosum group F caused by a defect in a structure-specific DNA repair endonuclease. Cell. 1996;86:811-22.

[18] Yano JK, Wester MR, Schoch GA, Griffin KJ, Stout CD, Johnson EF. The structure of human microsomal cytochrome P450 3A4 determined by X-ray crystallography to 2.05-A resolution. J Biol Chem. 2004;279:38091-4.

[19] Martin LP, Hamilton TC, Schilder RJ. Platinum resistance: the role of DNA repair pathways. Clin Cancer Res. 2008;14:1291-5.

[20] Kang S, Ju W, Kim JW, Park NH, Song YS, Kim SC, et al. Association between excision repair cross-complementation group 1 polymorphism and clinical outcome of platinum-based chemotherapy in patients with epithelial ovarian cancer. Exp Mol Med. 2006;38:320-4.

[21] Shirota Y, Stoehlmacher J, Brabender J, Xiong YP, Uetake H, Danenberg KD, et al. ERCC1 and thymidylate synthase mRNA levels predict survival for colorectal cancer patients receiving combination oxaliplatin and fluorouracil chemotherapy. J Clin Oncol. 2001;19:4298-304.

[22] Lord RV, Brabender J, Gandara D, Alberola V, Camps C, Domine M, et al. Low ERCC1 expression correlates with prolonged survival after cisplatin plus gemcitabine chemotherapy in non-small cell lung cancer. Clin Cancer Res. 2002;8:2286-91.

[23] Olaussen KA, Dunant A, Fouret P, Brambilla E, Andre F, Haddad V, et al. DNA repair by ERCC1 in non-small-cell lung cancer and cisplatin-based adjuvant chemotherapy. N Engl J Med. 2006;355:983-91.

[24] Cobo M, Isla D, Massuti B, Montes A, Sanchez JM, Provencio M, et al. Customizing cisplatin based on quantitative excision repair cross-complementing 1 mRNA expression: a phase III trial in non-small-cell lung cancer. J Clin Oncol. 2007;25:2747-54.

[25] Scheil-Bertram S, Tylus-Schaaf P, du Bois A, Harter P, Oppitz M, Ewald-Riegler N, et al. Excision repair cross-complementation group 1 protein overexpression as a predictor of poor survival for high-grade serous ovarian adenocarcinoma. Gynecol Oncol. 2010;119:325-31.

[26] Altaha R, Liang X, Yu JJ, Reed E. Excision repair cross complementing-group 1: gene expression and platinum resistance. Int J Mol Med. 2004;14:959-70.

[27] Ozkan M, Akbudak IH, Deniz K, Dikilitas M, Dogu GG, Berk V, et al. Prognostic value of excision repair cross-complementing gene 1 expression for cisplatin-based chemotherapy in advanced gastric cancer. Asian Pac J Cancer Prev. 2010;11:181-5.

[28] Sugasawa K, Ng JM, Masutani C, Iwai S, van der Spek PJ, Eker AP, et al. Xeroderma pigmentosum group C protein complex is the initiator of global genome nucleotide excision repair. Mol Cell. 1998;2:223-32.

[29] Li L, Elledge SJ, Peterson CA, Bales ES, Legerski RJ. Specific association between the human DNA repair proteins XPA and ERCC1. Proc Natl Acad Sci U S A. 1994;91:5012-6.

[30] Buchko GW, Isern NG, Spicer LD, Kennedy MA. Human nucleotide excision repair protein XPA: NMR spectroscopic studies of an XPA fragment containing the ERCC1-binding region and the minimal DNA-binding domain (M59-F219). Mutat Res. 2001;486:1-10.

[31] Saijo M, Kuraoka I, Masutani C, Hanaoka F, Tanaka K. Sequential binding of DNA repair proteins RPA and ERCC1 to XPA in vitro. Nucleic Acids Res. 1996;24:4719-24.

[32] Orelli B, McClendon TB, Tsodikov OV, Ellenberger T, Niedernhofer LJ, Scharer OD. The XPA-binding domain of ERCC1 is required for nucleotide excision repair but not other DNA repair pathways. J Biol Chem. 2010;285:3705-12.

[33] Barret JM, Cadou M, Hill BT. Inhibition of nucleotide excision repair and sensitisation of cells to DNA cross-linking anticancer drugs by F 11782, a novel fluorinated epipodophylloid. Biochem Pharmacol. 2002;63:251-8.

[34] Jiang H, Yang LY. Cell cycle checkpoint abrogator UCN-01 inhibits DNA repair: association with attenuation of the interaction of XPA and ERCC1 nucleotide excision repair proteins. Cancer Res. 1999;59:4529-34.

[35] Aune GJ, Furuta T, Pommier Y. Ecteinascidin 743: a novel anticancer drug with a unique mechanism of action. Anticancer Drugs. 2002;13:545-55.

[36] Wood RD, Robins P, Lindahl T. Complementation of the xeroderma pigmentosum DNA repair defect in cell-free extracts. Cell. 1988;53:97-106.

[37] Salles B, Rodrigo G, Li RY, Calsou P. DNA damage excision repair in microplate wells with chemiluminescence detection: development and perspectives. Biochimie. 1999;81:53-8.

[38] Perrin D, van Hille B, Barret JM, Kruczynski A, Etievant C, Imbert T, et al. F 11782, a novel epipodophylloid non-intercalating dual catalytic inhibitor of topoisomerases I and II with an original mechanism of action. Biochem Pharmacol. 2000;59:807-19.

[39] Wang Q, Fan S, Eastman A, Worland PJ, Sausville EA, O'Connor PM. UCN-01: a potent abrogator of G2 checkpoint function in cancer cells with disrupted p53. J Natl Cancer Inst. 1996;88:956-65.

[40] Barakat KH, Jordheim LP, Dumonte C, Tuszynski J. Virtual screening and biological evaluation of inhibitors targeting the XPA-ERCC1 interaction. Accepted in PLoS ONE. 2012.

[41] Jordheim LP, Barakat KH, Heinrich-Balard L, Matera E-L, Cros-Perrial E, Bouledrak K, et al. Small molecule inhibitors of ERCC1-XPF protein-protein interaction synergize alkylating agents in cancer cells. Submitted to Molecular Pharmacology. 2012.

[42] Lin JH, Perryman AL, Schames JR, McCammon JA. The relaxed complex method: Accommodating receptor flexibility for drug design with an improved scoring scheme. Biopolymers. 2003;68:47-62.

[43] Schames JR, Henchman RH, Siegel JS, Sotriffer CA, Ni H, McCammon JA. Discovery of a novel binding trench in HIV integrase. J Med Chem. 2004;47:1879-81.

[44] Markowitz MN, B.Y. Gotuzzo, F.; Mendo, F.; Ratanasuwan, W.; Kovacs, C.; Zhao, J.; Gilde, L.; Isaacs, R.; Teppler, H. Potent antiviral effect of MK-0518, novel HIV-1 integrase inhibitor, as part of combination ART in treatment-naive HIV-1 infected patients. 16th International AIDS Conference, Toronto, Canada. 2006.

[45] Babakhani A, Talley TT, Taylor P, McCammon JA. A virtual screening study of the acetylcholine binding protein using a relaxed-complex approach. Comput Biol Chem. 2009;33:160-70.

[46] Amaro RE, Schnaufer A, Interthal H, Hol W, Stuart KD, McCammon JA. Discovery of drug-like inhibitors of an essential RNA-editing ligase in Trypanosoma brucei. Proc Natl Acad Sci U S A. 2008;105:17278-83.

[47] Durrant JD, McCammon JA. Potential drug-like inhibitors of Group 1 influenza neuraminidase identified through computer-aided drug design. Comput Biol Chem. 2010;34:97-105.

[48] Durrant JD, Urbaniak MD, Ferguson MA, McCammon JA. Computer-aided identification of Trypanosoma brucei uridine diphosphate galactose 4'-epimerase inhibitors: toward the development of novel therapies for African sleeping sickness. J Med Chem. 2010;53:5025-32.

[49] K. Barakat, L. Jordheim, C. Dumontet, Tuszynski. J. Virtual screening and biological evaluation of inhibitors targeting the XPA-ERCC1 interaction. Molecular Cancer Therapeutics. Submitted Jan 2012

Chapter 7

Aspects of DNA Damage from Internal Radionuclides

Christopher Busby

Additional information is available at the end of the chapter

1. Introduction

In this chapter, there is insufficient space to exhaustively review the research which has been carried out on internal radionuclide effects. I hope only to highlight evidence which shows that internal radionuclides cannot be assessed by the current radiation risk model, and to suggest some research directions that may enable a new model to be developed, one which more accurately quantifies the real effects of such exposures. The biological effects of exposure to ionizing radiation have been studied extensively in the last 70 years and yet very little effort has gone into examining the health effects of exposure to internal incorporated radionuclides. This is curious, since the biosphere has been increasingly contaminated with novel man-made radioactive versions of naturally occurring elements which living creatures have adapted to over evolutionary timescales, and intuition might suggest that these substances could represent a significant hazard to health, one not easily or accurately modelled by analogy with external photon radiation (X-rays and gamma rays).

The question of the health effects of internal radionuclide exposures began to be asked in the early 1950s when there was widespread fallout contamination of food and milk from atmospheric nuclear tests. It quickly became the subject of disagreements between two committees of the newly formed International Commission on Radiological Protection (ICRP)[1]. The questions of the equivalence of internal and external radiation exposure, which were the basis of these disagreements, have still not been resolved. In the West, up to very recently, the whole spectrum of health effects from internal incorporated radionuclides has focused on animal studies of Radium, Plutonium and Strontium-90 and human retrospective studies of those individuals exposed to Radium-226 and Thorium-232 in the contrast medium "Thorotrast". These studies suffer from a number of problems which will be discussed.

Soviet scientists were more interested in internal radiation effects from fission-product radio-nuclides, but unfortunately their valuable studies have been difficult to access since they are published in Russian. In 1977 Gracheva and Korolev published a book summarising work in this area which was translated in India in 1980 as *Genetic Effects of the Decay of Radio-nuclides in* Cells [2]. This presented a wealth of interesting data relating to beta emitter ge-netic effects in various systems and drew attention to the distinction that must be made between external and internal radiation. This is important since the whole assessment of ra-diation in terms of health has been through the quantity "absorbed dose" and what can be called the bag-of-water model.

In this bag of water model, illustrated in Fig 1, the total energy transferred by the radiation to living tissue is diluted into a large mass, greater than a kilogram, as if the effects were uniform throughout the tissue being considered. In Fig 1 the tissue mass A represents an ex-ternal irradiation by X-rays or gamma rays and here the effects are uniform across the tissue. But in the case B, for internal irradiation, it is clear that it is possible, for certain kinds of exposure, for tissue local to the source to receive very large amounts of radiation energy at the same overall energy transfer to the tissue mass.

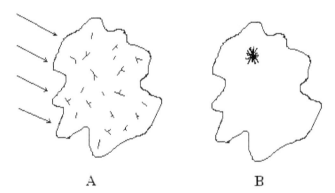

A   B

Figure 1. Comparing external and internal irradiation: the ICRP/ ICRU bag of water model. In case A, external radia-tion (X-rays or gamma rays) there are 20 events uniformly spaced throughout the tissue and the "absorbed dose" (see text) at any microscopic point is evenly distributed. In case B, for internal irradiation (here from a radioactive particle) there is a very large transfer of energy to a small tissue volume and the concept of "absorbed dose" does not apply.

Thus, in the historic and also the current system of radiation protection, those experts who assess radiation risk, who are termed *Health Physicists,* calculate the cumulative ab-sorbed dose in Grays, i.e. in terms of the total energy in Joules imparted by the beta elec-tron or alpha particle decays of the internal radionuclide contamination to one kilogram of tissue. For this calculation, the tissue is modelled as water. For example, those whose body contains 100 Bq of Strontium-90 are assessed, for the purposes of radiation protec-tion, as having received a cumulative absorbed dose of 100 x w where w is the "cumulative (absorbed) dose coefficient", obtained from measurements of the biological half life of the Strontium in the body and the decay energy of each decay in Joules. This number **w** is to be found in a Table published by the ICRP. In the case of the Strontium-90 contaminated individual, if the person weighed 50 kg, then the mean activity concentration would be 2 Bq/kg. The resulting absorbed dose would then be 2 x 2.8 x 10^{-8} (this is the ICRP 72 dose coefficient [3]). In other words, the committed dose is 5.6 x 10^{-8} Sv (0.056 µSv). But can this be safely compared with a dose from a chest X-ray (40 µSv) or from natural background radiation (2500 µSv) or from a high dose acute exposure to gamma rays from an atomic bomb linearly scaled to zero dose (the current way of modelling radiation effects)? This chapter explores this question. It is one which has become increasingly necessary as serious health effects, including cancer and leukemia, have been reported in those exposed to internal radioactivity in areas contaminated by radionuclides released from nuclear sites, weapons testing fallout and accidents like Chernobyl and Fukushima, at very low conventionally calculated "absorbed doses".

The matter has been discussed in some detail since 1998 by the independent European Committee on Radiation Risk (ECRR) whose reports [4, 5] provide a methodology for assessing health effects through a system of weighting factors based on available data. As more and more evidence emerged after 1995 that something was very wrong with the ICRP absorbed dose approach to internal radiation, the UK government set up a Committee Examining Radiation Risk from Internal Emitters (CERRIE). Since there were (and are) political dimensions to the issue, the committee was composed of scientists and experts from the nuclear industry and the official radiation protection organisations in the UK. Unfortunately the 4-years process ended in acrimony, legal threats to member of the committee, and failure to agree a final report. Two reports were issued [6, 7]. However, there was agreement that there were reasonable concerns about the safety of employing "absorbed dose" for certain internal radionuclide situations, and similar concerns about the safety of the ICRP model were made in 2005 by the French IRSN [8]. The error factor that these discussions led to was believed by different ends of the CERRIE process to be between 10-fold and 1000-fold. More recently, the value put on this error factor by the retired Scientific Secretary of the ICRP at a meeting in Stockholm in 2009 was "two orders of magnitude". What this means, in our Strontium-90 case above, is that the dose from 100Bq contamination to the whole body is no longer 0.056µSv but may now be between 0.56µSv and 56µSv and the risk of fatal cancer is proportionately increased. To put this in perspective, the mean Sr-90 dose over the period 1959-1963 to individuals in the northern hemisphere was given as about 1 mSv [9]. The ICRP risk model gives a 0.45% per Sievert excess lifetime cancer risk. Epidemiological studies suggest that the cancer "epidemic" which began in the 1980s in areas of high rainfall and fallout is a consequence of the earlier fallout exposures [10]. The weighting of dose necessary to explain this is greater than 300 if calculated from the ICRP absolute risk factor of 0.05/Sv [5, 11]. Many other instances of anomalous health effects from exposure to internal radionuclides require hazard weighting factors of between 100-fold and more than 1000-fold, and these are consequences of mechanisms which will be presented.

2. Fundamental principles

Ionising radiation, however it is delivered, creates harmful effects by causing mutations in genetic material both at the somatic level (cellular DNA) and germ cell level (heritable mutations). The mutations are caused by alterations in the cellular DNA in the nucleus and in mitochondria. These are brought about by three mechanisms:

a. Direct ionisation of the DNA and subsequent chemical alteration of the bases to molecules which are not recognised as a coding signal.

b. Indirect ionisation of the DNA by reactive species produced by ionisation of water (called Reactive Oxygen Species ROS).

c. A mechanism termed "Genomic Instability" which is an inducible cell-cell signal consequence of the production of ROS in the cytoplasm (non-DNA region) of an irradiated cell. This process is communicable between cells in some way and even between individuals and has been termed the "bystander effect".

These mechanisms are well described in the literature and in textbooks, and the processes described here can be found in the reports of radiation protection agencies e.g. [12].

Ionising radiation always transfers its energy to matter in the form of structured tracks of charged particles. Photon radiation (gamma and X-radiation) is absorbed by matter mainly through Compton Effect, Photoelectron, and Pair-production. All these cause the creation of tracks of energetic electrons which carry the energy of the original photon and collide with molecules in the absorbing medium causing ionisation. The ionised fragments (in the case of living tissue mainly of water) then recombine or react with local molecular entities causing chemical changes in the molecular structure. Various chemical reactions take place e.g.

$$H_2O \text{ (radiation)} \rightarrow H_2O^+ + e^-$$

$$H_2O^+ + H_2O \rightarrow OH^* + H_3O^+$$

The free radical OH^* has an unpaired electron and is highly reactive; it will combine with local species including DNA if that is close to the track. If it reacts again with water species the result is a range of highly reactive fragments which are collectively described as Reactive Oxygen Species. The process can be written:

$$H_2O \text{ (radiation)} \rightarrow e_{aq}, H^*, H_2O_2, H_2, OH^*.$$

The relative concentrations of the main ROS are [12]:

e_{aq} (hydrated electron) 45%

OH^* (hydroxyl radical) 45%

H^* (hydrogen radical) 10%

These reactive species attack molecules in the cell and cause damage; because it is an oxidising agent the OH^* radical is likely to be the most effective DNA damaging agent, abstracting a hydrogen atom from the deoxyribose moiety of DNA yielding a highly reactive DNA radical. This will then rearrange or react with local molecules to produce a new molecule in the DNA coding sequence, the gene, a molecule which is unrecognizable to the coding transfer process and alters the message of the gene.

It seems that evolution has recognised the dangers of high levels of cellular ROS and has developed a process to deal with the threat to the species or to the organism. At the organism level the process involves firstly the existence of double strands of DNA which permit repair of ionisation damage to a base located on one strand by copying from the opposite strand. This type of lesion, termed a "point mutation" is a more likely result for chemical mutagenesis or random attack by ROS species present in the cell at some background concentration (as a by product of other chemical processes in the cell). In some cells, the result of DNA damage is programmed cell suicide, termed apoptosis. But at the organism level, one response is the induction of genomic instability, whereby a signal is switched on in the DNA resulting in increased levels of random mutagenesis built into cell replication of the damaged cell and also bystander cells. The exact purpose of this process, which is well documented, is uncertain [13]. If the damage is more extensive, involving locally multiply damaged sites (LMDS) or both strands, it becomes more difficult to accurately repair the material and either a fixed mutation or cell death results.

Internal exposure results from the radioactive decay of radionuclides incorporated into tissue through inhalation or ingestion. There are three principle types of decay which represent the majority of all internal exposures. Gamma decay, which produces fast electron tracks, β decay which also produces fast electron tracks, and alpha decay. In addition there are also short range electron tracks from Auger decays. The main internal nuclides of environmental and radiobiological importance are listed in Table 1.

Apart from effects at the nuclide (recoil, transmutation) β decay is indistinguishable from the fast (photoelectron) electron tracks produced from gamma and X-ray interactions. With β-decay, unstable elements change into elements with one greater atomic number Z and emit an electron in the process; they may also emit a gamma ray. Sometimes the daughter nuclide is also unstable and may further decay. An example is Strontium-90 which emits a β-particle of endpoint energy 546 keV (kiloelectron volts) and transmutes into Yttrium-90 which further emits a β-particle of endpoint energy 2280 keV and transmutes into stable Zirconium-90. There are several series decay sequences in which ten or more unstable nuclides are formed, one from another. An example is the natural α-emitter Uranium-238 which decays through twelve sequential unstable radionuclides until the sequence stops at stable Lead-206. Transmutation involving α-decay involves the change of the chemical element to one with Atomic Number Z four places lower on the Periodic Table. Thus U-238 emits an α-particle and decays to Thorium-234.

There is strong evidence that damage to DNA is the cause of the effects of ionising radiation. For example, experiments have been carried out with nuclides which have short range electron emissions (Auger emissions) or Tritium chemically incorporated into DNA precursors so that these elements become covalently bonded to the DNA. The measured harmful effects are up to 100-times greater than would be predicted from the "absorbed dose" showing that it is the ionisation in the DNA that is key to the destruction of the cell [14, 15]. Another argument is based on the effects of the weak β-emitter Tritium, as tritiated water HTO. The measured effects of Tritium exposure are not too different from that expected on the basis of the absorbed dose (although it may be higher, see below). But clearly the Tritium will be evenly distributed throughout the cell. Since the beta energy of Tritium is only 6 keV the electron track range will be less than 0.5 μ and the ionisations will occur in clusters, uniformly distributed in the cell but with no overlap. It is clear that only those clusters which are close to the DNA will have an effect on the DNA, and the great majority of the energy will be "wasted" in the cytoplasm. Thus for a Tritium dose modelled by ICRP as 1 mSv, only a very small fraction of the Tritium decays will contribute to the effect.

The main target DNA, in the cell nucleus, represents a very small fraction of the total material in the cell. In a 10 μ diameter cell (mass 520 pg) there is 6 pg of DNA made up of 2.4 pg bases, 2.3 pg deoxyribose, 1.2 pg phosphate. In addition, associated with this macromolecule are 3.1 pg of bound water and 4.2 pg of inner hydration water [16]. Since absorbed dose is given as Joules per kilogram, if it were possible to accurately target the DNA complex alone, a dose to the cell (mass 520 pg) of 1 milliJoule per kilogram (one milliGray, one milliSievert) would, if absorbed only by the DNA complex (6 pg), represent a dose of 520/6 = 87 mSv to the DNA. It is possible to imagine the DNA as an organ of the body, like the thyroid gland or the breast. If this is done, then there should be a weighting factor for its radiobiological sensitivity of 87 which would be based on spatial distribution of dose alone. Of course, for external photon irradiation, to a first approximation, tracks are generated at random in tissue. Therefore only a small proportion of these tracks will intercept the DNA but the interception will be mainly uniform, and the health effects from such external exposure may be assumed to be described by the averaging approach of "absorbed dose". This is not the case for internal exposures from radionuclide decays in a number of quite specific circumstances which will be described below (see [5]).

The calculations of "absorbed dose" also assume that the medium irradiated has uniform isotropic qualities with an absorption coefficient roughly equivalent to that of water. However the absorption of gamma radiation is proportional to the 4th power of the atomic number Z. It follows that the probability of absorption of gamma radiation will be location specific, and this is highly relevant to a number of high Z elements, either biologically necessary (Iodine, Z=53) or as contaminants (Uranium Z=92) [17].

Radionuclides are primarily chemical elements with the affinities and reactivities of the non-radioactive forms of these elements. They will therefore have quite specific biochemical pathways in the body and may accumulate at positions in cells as a result of their chemical group, valency, ionic volumes, charge etc. This will result in high local doses at sites where they accumulate. In addition, the decay of a nuclide attached to some cell structure or macromolecule will result in the alteration of the radionuclide into a different element with a different charge, with resultant recoil energy. This will always break the chemical bond and result in ionisation. Thus there will be local ionisation and this may be on some critical macromolecule like DNA. These localisation and transmutation effects were studied in the 1960s but no attempt has been made to incorporate them for radiation protection purposes.

The decay of a radionuclide attached chemically to the DNA is illustrated schematically in Fig 2.

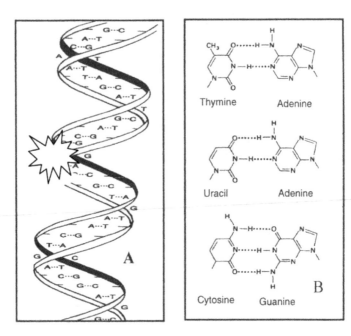

Figure 2. Certain radionuclides (Sr-90, Uranium) bind to DNA and when they decay cause (a) transmutation ionisation and (b) local electron emission ionisation (Auger, β particle) on or close to the critical target for radiation effects.

Cells have two phases of activity during their lifespan. They are mostly in a quiescent phase where DNA is not localised spatially. For a short period at the end of their lifespan, when they replicate, they are in a cell cycle phase. In this phase they are much more sensitive to irradiation. Therefore this repair replication phase represents a critical window for mutagenesis if it can be engineered. The radiation sensitivity of the repair replication phase has been studied extensively and it was suggested [18, 19] that two irradiation events separated by about 10-12 hours could represent an enhanced hazard since the first pushed quiescent cells into repair replication and the second damaged them during the sensitive 12 hour period. The idea is termed the Second Event theory. There is some evidence for it from work with split doses of X-rays. It will be discussed below.

3. Concerns about internal radionuclides

To summarise, the position is that the current assessment of harm from radiation exposure is based on a quantity which does not assume any structure in the tissue being irradiated. It does not distinguish between different radionuclides on the basis of their chemical properties except at the organ level (Iodine/ thyroid) and it does not include any weighting for chemical affinities for DNA, nor for transmutation effects. It does not consider the fractionation of doses within cell cycle repair times. Risk factors are based almost entirely on acute external gamma ray exposures. The main concerns are for radionuclides which are significant environmental contaminants and which are listed in Table 3.

3.1. Proximity effects on local doses

Since the genotoxic effects of radiation are mediated by ionisation and the local concentration of reactive oxygen species, it is firstly this local ionization density that is the proper measure of the effectiveness of a radiation exposure. The current risk model acknowledges this by weighting the highly ionizing α particle tracks by an arbitrary factor of 20. But of concern is the overlap of such tracks, and of electron tracks from β–decays or Auger electron showers with active DNA in the nucleus, and especially at the time when this is in some critical state, as in cell repair/division. For externally delivered photon radiation, it can be assumed (in the absence of high-Z photoelectron effects) that ionization is uniform across tissue. Under these circumstances it is only a matter of probability whether a cell is intercepted by a track or not. It has been calculated [20] that, at normal Natural Background levels, each cell in the body will, on the basis of probability, receive one hit per year (a hit being the traversal of the cell by an electron track). Of these hits, some small proportion will involve a track that intercepts the DNA and may cause damage. This damage, if it results in a point mutation, will be repaired before cell division. The ionisation density in a photoelectron track is assumed to be low. Therefore, for external exposure, a dose of 1 mSv to the whole body can be assumed to provide a dose of 1 mSv to the cell on average. At the cell level, this is not the case. A cell can be intercepted by a track or not. If not, then the cell dose is zero and there is no ionisation. If so, then the cell dose can be greater than 1 mSv. The dose to the DNA from such processes will again be either zero or some dose greater than 1 mSv.

For internal exposures, the probability of interception of the track is clearly a function of the distance of the nuclide from the DNA. In addition, internal exposures may be to α tracks which carry significantly more ionisation density. The range of most α tracks (which carry about of 5 MeV energy) is about 4 cell diameters and so, theoretically, the track dose to the cell from one decay is in the region of 500 mSv. The matter becomes serious when the nuclide is an alpha emitter but also has a high chemical affinity for DNA. This is the case for Uranium. Anomalous effects from internal nuclides have been known for a long time. Early studies of cell doses from Tritium were carried out by Apelgot [21] and Robertson and Hughes [22] and reviews of Tritium and of S-35 and P-32 studies are found in ref [2]. In order to emphasise the profound effects which can be identified in internal exposures, the case of Carbon-14 will be examined in greater detail below.

The cell dose from any decay is fairly simple to approximate on the basis of a continuous slowing down approximation and the assumption that the energy delivered along a track is a constant function of the track length. Electron track lengths in tissue for a range of energies are given in Table 1. These apply also to photoelectrons which have energies almost equal to the gamma photons that produced them. Assuming a cell diameter of 10 μ, the energy deposited in the cell is merely the decay energy divided by the track length in the cell. This is then converted into Joules (1 keV = 1.6 x 10^{-19} J) and divided by the mass of the cell in kg. For a 10 μ diameter cell this is 5.2 x 10^{-13} kg. For the Strontium-90 example, a single decay track will deposit approximately 1 mSv in each cell traversed by the track.

Energy (keV)	Range (cms)	*Linear energy transfer keV/μ	Examples (maximum β–energy, keV)
5	1.2 E-4	4	Tritium (5.7)
15	5.2 E-4	2.9	
20	8.6 E-4	2.3	
150	2.8 E-2	0.53	Sulfur-35 (167);Carbon-14 (155)
500	1.78 E-1	0.28	Strontium-90 (546) Caesium-137 (514) Iodine-131 (607) Caesium-134 (658) Barium-140 (168)
1000	4.42 E-1	0.22	Iodine-132 (1610,1210,1040) Barium-140 (1020,1010)
2000	9.92 E-1	0.201	Yttrium-90 (2280) Iodine-132 (2160)

* this is simply the loss of energy of the particle over unit distance

Table 1. Continuous slowing down range in muscle tissue for electrons in g cm^{-2} (values very similar to range in water) (from ICRU Report 35 Table 2.5 [23])

3.2. Calculating the spatial effect enhancement

The spatial effect enhancement is the probability of an ionisation track from an internal nuclide intercepting the DNA at some given level of ionisation density, compared with the probability of this happening from external radiation.

Thus we take the mutagenic event of interest to be associated with absorbed dose (energy per unit mass) in a volume element of a track which is coincident with active DNA in space and time. For nuclides with chemical affinity for DNA this ratio is clearly very large. In the limiting case of covalent binding it can be assumed that approximately half of the decays of the bound nuclide traverse the DNA, and in addition the *transmutation* of the nuclide causes a point ionisation at its position. In the limit this probability will be 1; for example, C-14 which is incorporated into one of the DNA bases will decay and change into Nitrogen. This will immediately destroy the purine or pyrimidine base which it is part of and will introduce a mutation which may or may not be repaired.

The probability of the interception of a charged particle track intercepting the DNA depends on the distance of the point source and the dimensions of the DNA target employed. The cross sectional diameter of one strand is about 0.3 nm but, in mitosis, various much larger condensed targets exist. The principle is the same, however: the probability of intercepting the target falls off rapidly with distance. The result for a condensed DNA target of cross section $0.1 \times 1\ \mu$ is given in Fig 3. The calculation is given in Appendix A. The result confirms what is intuitively obvious: the effect of radionuclide decay in the cytoplasm is much less harmful than for nuclides bound to DNA. This is particularly significant for the α-emitters which have chemical affinity for DNA, Uranium (as UO_2^{++}) and, possibly, Plutonium.

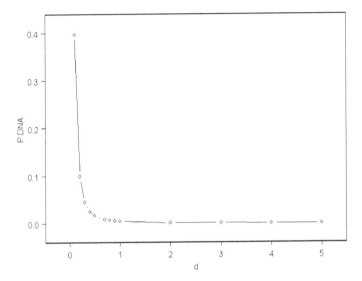

Figure 3. Approximate probability of a track interception of a DNA target modelled as a strip of $0.1 \times 1\ \mu$ by distance in μ from target. In this model, the maximum probability is 0.5 for a nuclide located on the surface of a flat strip.

One other simple way to illustrate this spatial effect is merely to consider the tissue as two compartments, an organ A which may be called "DNA" and one B which may be called "everything else". The current ICRP risk model calculates the absorbed dose of any internal exposure by dividing the total decay energy by the mass. This would not distinguish between compartments A and B; both would receive the same dose. But as far as cancer is concerned (or other consequences of genetic damage) all the ionisation in compartment B is wasted. It has no effect. Therefore it is the dose to compartment A that is the cause of the effect. This would suggest that the spatial enhancement is at minimum the ratio MassB/MassA or about 90-fold. This assumes that *all* the DNA in the cell is a critical target which is unlikely to be the case. If the critical DNA represented even $1/10^{th}$ of the total cellular DNA, the spatial enhancement from track interception alone would be 900-fold.

3.3. Double strand breaks

At natural background radiation levels, where there is one "hit" per cell per year, the Poisson probability of multiple tracks across the DNA strand is low. Most of the "hits" are repairable and the biological response is proportional to the dose. But it is believed on the basis of good evidence that genetic mutations result from multiply damaged sites [12]. If two adjacent DNA strands are broken, then repair is not possible since there is no template from which to copy the correct sequence. Ward et al. (1988) [16] compare DNA damage necessary to inactivate exposed cells between radiation and chemical mutagens. Table 2 lists some of the results:

Agent	DNA lesion	No of lesions per cell
Ionizing radiation	SSB	1000
	DSB	40
	Total LMDS	440
Benzo[a]pyrene 4,5 oxide	Carcinogenic adduct	100,000
methylnitrosourea	7-methylguanine	800,000
Aflatoxin	Carcinogenic adduct	10,000

Table 2. Yields of DNA damage necessary to kill 63% of exposed cells [16]

From Table 2 it is clear that ionizing radiation is more effective than the most powerful chemical carcinogens in causing genetic lesions to the DNA, but it is the double strand breaks (DSBs) and LMDS which are the most efficient processes. From simple kinetic theory it is clear that the probability of inducing double strand breaks or LMDS will increase as the number of tracks per unit time increases. At low background external doses this is very unlikely. But as the dose rate increases, so the likelihood of multiple tracks increases (for a discussion see [18]). This is not true for a number of internal exposure situations where multiple tracks can occur at very low doses, conventionally assessed. The first is exposure to particulates.

Radioactivity from releases from nuclear explosions, e.g. accidents like Chernobyl, or from weapons tests or Uranium weapons is partly in the form of sub-micron particulates which are respirable and can be translocated from the lung. Tissue near such particles will receive multiple tracks even though the dose, as assessed as energy per unit mass may be very low. Similar multiple track effects can occur close to high Z element particles whether they are intrinsically radioactive or not, e.g. platinum (catalysers), bismuth, gold (prostheses), due to secondary photoelectron conversion from natural background gamma radiation [7, 17]. The second is where a relatively immobilised nuclide has a sequential decay pathway and so there is more than one decay from approximately the same position. This situation is more genotoxic when the decays occur within the repair replication cycle; the Second Event [7] and this situation will be discussed separately.

Radionuclide	Half life (Decay product)	Decay	Reasons for concern	Other remarks
Tritium H3	12.32 y Helium-3	Low energy β	Ubiquitous; Discharged in large amounts by all nuclear sites and weapons tests; present as tritiated water and easily incorporated into body. Can be present as organically bound tritium which may accumulate in the body.	Evidence of serious genetic effects in invertebrate development at very low doses; short range of β decay causes high ionisation density.
Carbon-14	5730 y Nitrogen	β emitter	Discharged by nuclear sites, particularly reprocessing sites (Sellafield) and weapons tests. Incorporated into the carbon of the body	Doses by ingestion mainly of vegetables, milk, fish. Both Carbon and hydrogen (Tritium) make up the structure of living systems. Transmutes to a gas, nitrogen.
Sulphur-35	87 days Chlorine	β emitter	Discharged from nuclear sites. Concentrates in foods.	Sulphur also a part of internal macromolecules in living systems. Transmutes to a reactive gas, Chlorine
Strontium-90	28.9 y Yttrium-90	β emitter	Globally Widespread. Atmospheric test fallout, nuclear sites, accidents (Chernobyl, Fukushima); Group 2 affinity for DNA	Second event nuclide with daughter Y-90 of concern since it binds to DNA
Krypton-85	10.7 y Rubidium-85	β emitter	Very large amount routinely released from nuclear sites is building up in atmosphere.	Very soluble in fats and therefore can build up in body fat (beast tissue, lymphatic tissue) following inhalation
Barium-140	12 d Lanthanum-140	β emitter	Large quantities from nuclear weapons tests; Group 2 affinity for DNA	Second event emitter binds to DNA. Of concern in assessing effects of nuclear atmospheric tests and accidents
Iodine-131	8 days Xenon-131m	βγ emitter	Large amounts from accidents, licensed releases. Affinity for Thyroid and Thyroxine in circulating blood	Second event emitter with daughter Xe-131m short half life. Transmutes to a gas.
Tellurium-132	3.25 d Iodine-132	βγ emitter	Released in large amounts from accidents; daughter is Iodine 132	Second event series
Caesium-134	2 y Barium-134	βγ emitter	Released from nuclear explosions, accidents	Binds to muscle
Caesium-137	30 y Barium-137m	βγ emitter	Released from nuclear explosions, accidents, nuclear sites under licence	Binds to muscle; concerns over effects on heart in Chernobyl contaminated areas.

Radionuclide	Half life (Decay product)	Decay	Reasons for concern	Other remarks
Radium-226	1599 y Radon-222	α emitter	NORM Contamination near oil and gas processing sites; widely studied but problems with the studies (see text). Decays to Radon gas.	Group 2 Calcium seeker. Binds to DNA. Evidence of non-cancer reduction in lifespan in human studies.
Polonium-210	139d Lead-204	α emitter	Releases from nuclear sites; daughter of Lead 201 which can build up in environment as a result of contamination from NORM	
Uranium-238	4.5 x 10⁹y Series	α emitter	Releases from nuclear sites; contamination from mining and processing; from weapons fallout and accidents; from battlefield weapons usage and testing. Widespread in the environment but generally not measured near nuclear sites	High Z photoelectron effects; binds to DNA; considerable evidence for its anomalous genotoxicity
Plutonium-239	2.4 x 10⁴ y Uranium-235	α emitter	Releases from nuclear sites, weapons test fallout, widespread environmental contaminant	Binds to DNA (?) evidence for anomalous genotoxicity

Table 3. Internal radionuclides of concern

Third, if an alpha emitting nuclide is either randomly positioned near or chemically attracted to the DNA, there is a significant probability that the highly ionising track will traverse the two strands of the DNA and damage multiple sites. This is the origin of the high efficiency of alpha emitters which resulted in their being weighted by ICRP. Fourth, there are situations where dose is delivered by very low energy beta emitters; the best example is Tritium. Because dose is assessed as energy per unit mass, the very low decay energy of Tritium means that there is a large number of decays from different atoms of Tritium (90 tracks) to deliver the same dose as one 500 keV β decay from Caesium-137 or from the traversal of a cell by a 500 keV photoelectron track. This would suggest a mechanism backing the evidence (see below) that Tritium represents a greater mutagenic hazard than is calculated on the basis of its absorbed dose.

3.4. Summary of enhancement mechanisms; caveats over high dose studies

The target for radiation effects is the DNA, the nuclear DNA and the sensitivity to radiation varies depending on whether the cell is in quiescent phase or in repair replication. Within the 12 hour repair replication period there are other sensitive windows. The end point for radiation damage to the DNA can be genetic mutation leading to heritable damage (in germ cells) or cancer, but if the ionisation density is too great, or the sequential hits to close together then the cell will die. The interesting thing then is that this will *decrease the fixed muta-tion rate* and therefore *will decrease the cancer rate.* Thus we would not expect studies of high dose and high dose rate to elicit information which informs on low dose and low dose rate.

The dose/ dose rate response in cancer studies will inevitably have a complex character for this reason. This is clear from the results of retrospective studies of Radium and Thorotrast contamination, studies which have been influential in supporting the current radiation risk model, an issue with will be discussed further below. The key point is that, for certain inter-nal exposure regimes, the ionisation density at the DNA and the damage to the DNA can be extremely high even though the absorbed dose, as calculated by the current methodology, may be extremely low.

The regions of internal and external dose, are illustrated in Fig 4.

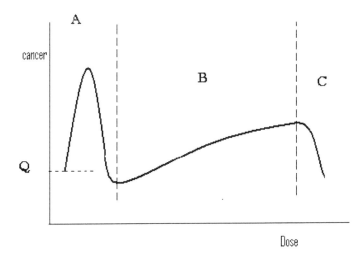

Figure 4. Regions of interest in a theoretically predicted dose response relation (see text and ECRR2010). Exactly this dose response is seen in infant leukemia rates after Chernobyl in Greece, Germany (3 dose regions) Wales, Scotland and Belarus (see [25]).

The analysis from ECRR 2010 [5] is described in Fig 4, the end point is assumed to be cancer rate. Q is the background rate. There are three regions. In the first region A, sensitive cells in repair replication are first mutated (positive slope) and then overwhelmed (negative slope). Next, in region B the cells in quiescent phase are mutated and eventually overwhelmed in C. This is also the organism response since at high doses C the organism suffers from non can-cer causes of death which affect the cancer rates, reducing them as the dose increases. These responses are seen in many epidemiological and animal studies but are generally misinter-preted. Burlakova has made a special study of dose response relationships and has shown the type AB response for a wide range of objective markers of DNA damage and also whole organism end-points [24].The dose response is seen in, and most easily explained in, infant leukemia after internal radiation exposures. As the exposure increases, foetal death ensues at some point, and the leukemia rate in the infant falls [25]. If the dose response is assumed to be linear, and the low dose data points assumed to be data scatter, a line drawn between the background cancer rate Q and the peak in region B cuts the response line in such a way as to suggest that radiation is actually reducing cancer rate, the so-called hormesis theory. The analysis in ECRR2010 points out that this is a misinterpretation of the data.

From what has been discussed, it is possible to summarise the mechanisms that may lead to increased risk of damage to DNA, and indeed to decreased risk in the case of high local doses which will kill rather than fix mutations. The mechanisms are listed in Table 4 where enhancements from alpha emitters with affinity for DNA may deliver such high local doses as to inactivate the cell.

Mechanism	Range	Examples
Spatial location		
DNA affinity	0.1-100	Uranium, Strontium, Barium, Radium Plutonium?
Membrane affinity	?	Caesium, Potassium, Rubidium, Chlorine, Sodium
DNA incorporation	Very high	Tritium, Carbon-14
Particulates	10-1000	Uranium. Plutonium
Protein incorporation	?	Sulphur-35, Tritium, Carbon-14
Transmutation	5-100	All covalently bound internal nuclides e.g. Sulphur-35, Tritium, Carbon-14
Temporal location		
Critical cell lifespan phase interception by immobilised source	0.01-100	Strontium-90 , Tellurium-132, Tritium, Radium-226, particulates
Critical repair replication window interception	0.01-1000	Strontium-90 , Tellurium-132, Tritium, Radium-226, particulates
Fat soluble noble gases	?	e.g. Kr-85
High Atomic Number photoelectron amplification	U-238 100-1000	Uranium, Platinum, Gold, Bismuth, potentially all elements with Z"/>53

Table 4. Main mechanisms of enhancement of genetic hazard from internal irradiation (see ECRR2010).

4. Specific concerns and new research directions

4.1. Location enhancement and chemical affinity

Concern has been shown since the 1950s that radionuclides of Group 2 in the Periodic Table, notably Strontium-90 and Barium-140, may have high affinity for DNA. These elements exist in solution as dipositive ions which are known to concentrate in organs (bones, teeth) which have high phosphate concentrations. Calcium and Magnesium are also known to bind electrostatically to the DNA Phosphate backbone and to stabilise its conformation. It is therefore likely that Strontium, Barium and Radium also have such affinity. The concentration of the radiation risk establishment on Radium epidemiology has been based on an end-point of bone cancer because the nuclide concentrates in bone. The affinity for DNA has been overlooked.

In the 1960s, for the reason that it was believed that Strontium would bind to DNA, and because some experiments showed that this was the case, there was significant concern about Strontium-90 contamination of milk. Mouse experiments demonstrated effects on intrauterine foetal death [26], and studies on rats showed development effects from Sr-90 [27]. There were effects at very low doses from Sr-90 [28], and by 1970 the director of the UK Medical Research Council suggested that further interest be taken in research on Sr-90 [29]. However nothing was done. In 2004, the CERRIE committee unanimously called for there to be further research into the effects of exposure to Sr-90 [6]. Also classified with these Group 2 is Uranium which exists in solution as the dipositive ion UO_2^{++} the Uranyl ion. This has very high affinity for DNAP [30] which led to its introduction as a chromosome stain for electron microscopy as early as 1960 [31].

The most necessary research is to measure the affinity of Strontium, Radium and Uranyl ion for chromosomes *in vivo*. Owing to the high opacity of Uranium there are certainly potential electron microscope methods for examining its location in cells *in vivo*. It might be possible to employ autoradiography to measure the affinity constants in vivo for Ra-226, Sr-90 and Ba-140. Affinity constants for DNAP can be easily measured *in vitro* for Strontium, Barium and Radium but this does not appear to have been done.

Animal studies of Radium and Uranium have assumed that the end point must be bone cancer or leukemia, and that only high doses will cause cancer. Effects at low doses have been assumed to be random scatter. It is suggested that low dose animal studies be undertaken with lifespan observation of all possible conditions to resolve this issue.

1. *There is the question of membrane affinity. If certain ions congregate at certain* membranes, the local ionisation density from radioactive decay will be higher than if these were uniformly distributed in cytoplasm. Experiments with the nuclide Sodium-22 by Petkau showed a supralinear dose response and effects at very low doses as calculated by using the total solution volume as a denominator [32]. If such effects occur *in vivo* there are a number of critical membranes which might be destroyed from internal radionuclide ions. Experiments *in vitro* might involve K-40, and Cs-137.

2. *DNA is made from Carbon, Oxygen, Hydrogen and Nitrogen. Carbon-14 and* Tritium can both therefore become covalently bonded into the molecule, and Tritium can easily exchange with labile hydrogen atoms on -SH, -OH and –NH moieties. The resultant decay will result in the total internal rearrangement or local reaction resulting in permanent alteration of the molecule. This will produce a point mutation with 100% efficiency. The electrons from the decay or reactive species created during the trans-

mutations through abstraction of protons from water may damage other local DNA leading to LMDS or DSBs. In the case of C-14, the transmutation to N-14 will totally destroy the molecule since the two elements have different valency, outer electron structure, and reactivity. Owing to the long half-life of C-14, experiments on its genetic effects have been difficult to carry out. Nevertheless, some studies have been published which show that these transmutation effects dominate the hazards of C-14 and Tritium incorporation (see below).

4.2. Particulates

The problem of the anisotropy of dose from internal "hot particles" was raised by Tamplin in the 1980s [33]. It was discussed by CERRIE and was the subject of a review by Charles et al in 2003 [34]. Since the 1950s, there has been a new class of internal radionuclide exposure which has not existed throughout evolution. This is the sub-micron or nanometre diameter radioactive particle. Particles below 1μ diameter can be inhaled and translocated from the lung to the lymphatic system. They are created in nuclear explosions, from power station accidents, from nuclear site releases and from Uranium weapons on battlefields. Depending on their nuclide composition they can produce very high local doses to tissue in which they become immobilised, but may also, depending on their diameter and composition, produce lower doses. Two concerns are Uranium and Plutonium oxide particles. Both contaminated large areas of land in Europe after Chernobyl. Both are resuspended from coastal sediments where contamination exists e.g. the Baltic Sea and the Irish Sea and plutonium from this latter source has been measured in coastal autopsy specimens [35], sheep faeces, and childrens teeth [36]. The well known Seascale child leukemia cluster [37] was discussed by the authorities [40] who dismissed the idea that the leukemia was caused by inhalation of plutonium and uranium on the basis that the doses to the lymphatic system were below natural background [38, 39]. However, the methodology employed diluted the particulate energy into a lymphatic system modelled as several kilograms of tissue [38] rather than the tracheobronchial lymph nodes which weigh about then grams and which are known to be the origin of leukemia in some animals.

The problem with the hot particle issue is that there will be a range of local energies (local dose) which will have either little effect (A), a genetic effect (B) or a killing effect (C). This was pointed out in 1986 following Chernobyl [41] and the idea is illustrated in Fig 5. Regions A to C will have dimensions resulting (a) from the activity and composition of the particle and (b) from its diameter. A particularly interesting case is that of a weakly radioactive particle like U-238 produced from battlefield use of Uranium weapons, so called depleted Uranium. Such a particle may be more carcinogenic than the much more radioactive plutonium particles found in the Irish Sea and epidemiology seems to bear this out. Of interest also is the photoelectron amplification of natural background radiation by internal high atomic number particles like Uranium-238, but also other elements (see below). It is not sufficient to dispute the hazards from particulates by pointing out that they will have such high activities that cells will be killed rather than mutated e.g. [34, 6].

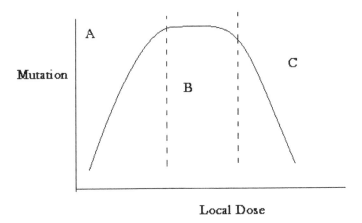

Figure 5. Effects in cells local to "hot particles" (see text).

4.3. Protein and DNA incorporation and transmutation

The inactivation of key enzymes or DNA by incorporation through biosynthesis of radionuclide substituted precursors is a matter that seems to have been entirely overlooked in radioprotection. The environmental contamination nuclides which will inactivate biological molecules are those from which they are constituted, namely Carbon (C-14), Hydrogen (Tritium), and Sulfur (S-35). Some results for C-14 and Tritium will be briefly presented. There is an important experiment which shows the contribution of transmutation to the lethal effects of C-14. Apelgot and Latarjet [42, 43] incorporated C-14 into the cells of the bacteria *e.coli* by culturing in a medium containing 2-14C-thymidine. The samples were stored at -196 C, The specific activity of the 2-^{14}C thymidine was 166 MBq/milliMol. The experiments continued for a year. To evaluate the role of the β-radiation, a control non-radioactive bacteria sample was stored in the presence of 2-^{14}C thymidine in such a way that the radioactivity per cm^3 of this suspension was the same as the study sample. From a comparison of the results, the authors concluded that the predominant lethal effect was from transmutation with an efficiency of 160-times that which would be obtained from the β-radiation. Similar results have been obtained from studies of C-14 by Anderson and Person [44, 45] who put the hazard coefficient relative to β-radiation at 10-fold. These authors studied the mutagenic effect of C-14 and compared transmutation with external X-ray doses. Pluchennik [46, 47] studied the mutagenic effect of C-14 decay in *Chlorella* grown in a medium containing a single carbon source with different fractions of C-14. The number of mutants from the C-14 rose rapidly at low fractions and quickly saturated due to killing effects; the data showed that the mutagenic effect considerably exceeded that due to external radiation. Other research carried out in the 1960s has largely confirmed this generalisation [2].

The genetic effects of incorporation of C-14 are of concern since the atmospheric nuclear tests in the 1950s and 1960s greatly increased the C-14 in the atmosphere. The genetic hazard to man was first pointed out by Totter et al [48] in 1958 and also by Pauling [49]. A number of studies were carried out on different systems. These include onion bulbs [50, 51] grown in an atmosphere of $^{14}CO_2$ resulting in chromosome aberrations, micronuclei and elongated cells. Onion bulbs were also studied by Friedkin and Atchison [52] who compared chromosome aberration in the roots between labelled thymidine (incorporated in DNA) and thymine (not incorporated). The frequency of aberrations was 3.95% for the thymidine but only 0.43% for the thymine, showing that the effect of transmutation was 9-times that of the β –radiation. A study of the effect of the C-14 position in the thymidine [53] showed quite clearly that it was transmutation that was the cause of the effects.

Kuzin et al [54, 55] compared the transmutational component of C-14 incorporation with external γ radiation in the broad bean. The amount of chromosomal aberration in 2 days was found to be 25-times per rad for the transmutational component. Other studies on drosophila [56, 57] give results which suggest that the mutagenic efficiency of C-14 is about three times that of chronic external γ radiation. Valuable reviews of effects from Tritium and Sulphur-35 are presented in [2].

Tritium has been increasing in the biosphere since the nuclear atmospheric testing. The main form in which it exists is tritiated water (HTO) but the nuclide also is incorporated into carbon compounds e.g. CH_3T and this is termed organically bound Tritium. Tritium is also employed for radioactively labelling compounds in chemical, medical and biochemical research. Tritium has a half-life of 12.6 years and radiates low energy β-particles (0-18 keV) and when incorporated in a molecule it transmutes to Helium with molecular restructuring and ionisation and realises a recoil energy of 0-3 keV. These events convey a high probability of destruction or inactivation of the parent molecule. If this is a macromolecule, local restructuring may alter the tertiary folding structure and inactivate the entire molecule. Thus the effects of Tritium are amplified in the ratio of the molecular mass to the Tritium mass, which may be by orders of magnitude. The question of whether these results show enhancement of effect relative to externally calculated absorbed dose does not seem to have been addressed either for lethality or mutation. Experiments with very low dose exposures of Tritium to invertebrates have identified significant developmental effects [58]. Tritium is also of interest as a pseudo-second event nuclide (see below) owing to the fact that the number of events associated with unit dose is far greater than the mean event number associated with background gamma radiation.

4.4. Temporal location: The second event theory

It is well known that dividing cells are more sensitive to radiation than quiescent cells. Once cells are committed to division, they enter the active part of the cell cycle, during which DNA repair takes place followed by cell division. It is therefore clear that any damage or signal which moves cells from quiescence into the repair replication sequence puts the cells into a condition where a second damaging event will carry an enhanced risk of mutation or lethality. This is the basis of the Second Event Theory [18, 19]. This postulates that split doses to the cell DNA, separated by 10-12 hours, will represent an enhancement of hazard. The sequence is vanishingly unlikely for external natural background irradiation but exceedingly likely for a number of specific internal sequential emitters. These include exposure regimes involving Sr-90/Y-90, Te-132/I-132 and various others. They include hot particles (since there are continuous releases of tracks from these) and also Tritium which, due to its very low decay energy, produces many more tracks per unit dose than natural background radiation.

The probabilities of second event processes occurring can be calculated but depend on basic assumptions. A paper by Cox and Edwards of the UK National Radiological Protection Board [59] concluded that the cell dose enhancements were finite but low. However it was pointed out that there were major faults in the cell dimensions employed in this study [60]. Clearly, the enhancement is a function of the location of the Second Event nuclide, the factor increasing sharply as the critical volume is reduced. For location on the DNA the potential enhancement becomes enormous. Table 5 shows results for Sr90/Y90. A number of studies have indeed shown anomalous genetic hazard from Sr-90/Y-90 [7, 18]. However, since Strontium also binds to DNA it carries enhancement from other mechanisms. An interesting experiment which suggests that there are 2nd event effects from Sr-90/Y90 was a comparison of the genetic damage effectiveness of Sr-90 and the singly decaying Sr-89 on yeast suspensions at the same doses. The results showed that the Sr90 was four-times as genetically damaging as the Sr89 for the same dose [61]. Further support comes from cell culture experiments with split doses of X-rays which show an enhancement of effect for split dose regimes during the repair replication period [18, 62, 63]. In view of the important implications this has for medical X-ray and radiology the question should be examined by further research. Such research might include (a) split dose research on living animals, e.g drosophila, zebra fish, (b) comparison of sequential decay effects from indentical elements with different decay sequences e.g. Sr-90/Y-90 vs. Sr-89.

External dose comparison	2nd Event enhancement probability [19]	Cox and Edwards (2000)[59] Cox Edwards and Simmonds (2004) [6]
1 mGy	30	1.3
0.1 mGy	200	8.6
0.01 mGy	1900	82
0.001 mGy	9400	407
1 atom per g of tissue	5×10^9	

Table 5. Second Event Enhancements for Sr-90/Y-90 (From Busby 1998 [19])

4.5. Secondary photoelectron effects

The quantity employed in radiation protection, *absorbed dose*, is defined as $D = \Delta E / \Delta M$. Hitherto, the mass into which the energy has been diluted is that of living tissue; ICRU provide tables of absorption coefficients for different living tissue, adipose, bone, muscle etc. which can be employed for calculations involving doses, but generally all these denominator quantities have the absorption characteristics of water (H_2O) (ICRU35 1984). The absorption of electromagnetic (photon) radiation is due to a number of processes, the main three being pair-production, Compton scattering and photoelectron production. For elements of atomic number greater than about 30, and for photon energies of less than about 500 keV, the photoelectric effect predominates. Even for the low atomic number elements that make up living systems, there is fairly quantitative conversion of incident photon radiation below 200 keV (and induced photon radiation from second order and third order processes) into photoelectrons. These are fast electrons which are indistinguishable from beta radiation and have the energy of the incident photon minus their binding energy (which is generally far less than the incident photon energy and can be ignored). The absorption of photon radiation by elements is proportional to the fourth or fifth power of the atomic number Z. Thus the predominant absorber in water is the Oxygen atom Z=8 and it is reasonable to give the effective atomic number of water as 7.5. Of course, there are elements in tissue with higher atomic numbers, but interestingly, apart from Iodine (Z=53) few elements with Z>26 (Iron, Fe).The incorporation of high Z elements into living systems would generally be harmful since it would increase the radiation dose, and therefore such developments have been lost though evolutionary selection. Iodine is an exception, but it should be noted that the main sites for radiation damage in terms of sensitivity are the main sites for Iodine concentration, the thyroid gland and the blood. It has been suggested that the metabolic and cell repair status controls exercised by the thyroid gland are the reason why Iodine has been incorporated into living systems and is employed as a kind of radiation-repair control mechanism [17].

A problem in radiation protection arises when high Z elements are incorporated into living tissue, since the enormously greater absorption of photon radiation by such material will result in enhanced doses to tissues adjacent to the high Z material. The problem was first addressed in 1947 in relation to X-rays of bone [64] and has been studied in the past in relation to prostheses. More recently, interest has shifted to the use of high Z material to enhance photon radiotherapies for tumour destruction where it has been shown to be effective. Gold nanoparticles have been successfully employed (and patented) for radiotherapy enhancement [65].

Despite this knowledge, the enhancement of photon radiation by high Z contaminants has not been addressed in radiation protection. The situation may have arisen out of the fact that prosthetic materials are not intrinsically radioactive and contamination from high Z elements like Lead (Z=82) are considered under the heading of chemical toxicity. The issue was raised in 2005 [66, 67]. It was pointed out that there are two circumstances where the Secondary Photoelectron Effect (SPE) would have significant radiological implications. These are (a) for high Z elements that bind to DNA and (b) for internal high Z particulates. In the latter case, the effect will increase as the particle size is reduced, since for massive high Z contamination e.g. prostheses, most of the photoelectrons are wasted inside the bulk material. The emergence of the photoelectrons into tissue is a function of the mean electron path in the material, and the absorbed dose in local tissue is a function of the electron range and thus its energy.

The radiological implications of the idea emerged in considering the anomalous health effects of Depleted Uranium weapons and were presented to the CERRIE Committee in 2003 and the UK Ministry of Defence in 2004 although nothing was done. More recently there have been attempts to quantify the effects for particles through Monte Carlo modelling [68, 69], but these have not generally been very credible treatments or able to cope with the small volumes of complex media involved, and the results have been far removed from the few experimental data published [70, 65].

The particular concern is for the element Uranium, since this has been employed since 1991 as a weapon; the Depleted Uranium (DU) penetrators, used from the 1991 Persian Gulf War onward, produce a fallout comprising sub-micron Uranium Oxide particles which are environmentally mobile and respirable. Uranium has another quality which makes it of interest in SPE; as the uranyl ion UO_2^{++} it has a very high affinity for DNA phosphate: some 10^{10} M^{-1} [30]. This affinity has been known since the 1960s when it was first employed as an electron microscope stain for imaging chromosomes [31].

The SPE is therefore likely here to cause enhanced photoelectron ionization at the DNA due to enhanced absorption of natural background radiation (or medical X-rays). A similar process occurs with the Platinum chemotherapeutic agent cisplatin which binds to the DNA and acts as an antenna for background radiation and radiotherapy beams.

For SPE phantom radioactivity in other elements of high atomic number, the tissue doses are enhancements of the incident photon dose at the point of the atom or particle being considered. Due to the complex interactions these local doses must be determined by experiment. However, these experiments are straightforward and involve X-irradiation of high Z element contaminated tissue at different doses. In principle, this development suggests that the internalization of any high Z particle which is biologically long-lived will cause continuous irradiation of local tissue cell populations, which would represent a carcinogenic hazard. This has implications for those employing prosthetic materials and also for the dispersion of high Z particles (Tungsten, Platinum, Bismuth, Lead) in the environment. It also suggests that it may be of interest to examine tumours for the presence of high Z particles at their centre. Table 6 lists a number of potentially hazardous SPE elements.

Finally it should be pointed out that physical modelling through Monte Carlo codes is unlikely to establish useful data and certainly should not be employed as an attempt to dismiss the importance of the proposed mechanism.

Nevertheless, a FLUKA Monte Carlo model of the absorption by nanoparticles of Gold and Uranium carried out by [71] Elsaesser *et al* 2007 graphically confirmed the effect. The results for photoelectron track production following absorption of 100 keV photons is shown in Fig 6 below. Enhancement factor in this calculation for the 10nm Uranium particle relative to water was approximately 8000.

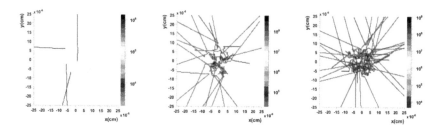

Figure 6. Photoelectron tracks emerging from (left to right) 10 nm particles of water (Z=7.5), Gold (Au; Z =79) and Uranium (U;Z=92) after irradiation with 100keV photons. Monte Carlo (FLUKA code) analysis. Track numbers are in proportion to the 4th power Z law (tracks are shown as projections on a flat plane). Note that the model uses 1000 incident photons for Au and U but 10,000 for water [71]

Material	Z	Z⁴/tissue	Source	Note
U	92	22642	Weapons particles, nuclear fuel cycle, atomic and thermonuclear bomb tests	Binds to DNA; known to cause cancer in animals and genomic damage at very low concentrations
Th	90	20736	Incandescent mantles Contrast media	Highly insoluble
Bi	83	14999	General contaminant	Insoluble
Pb	82	14289	General contaminant	Toxic; SH binding
Hg	80	12945	General contaminant	Toxic; enzyme binding
Au	79	12310	Prostheses; colloid used for rheumatism	Friction particles may travel in body; inert and insoluble
Pt	78	11698	Vehicle catalysers, general contaminant	Inert and insoluble
W	74	9477	Weapons; general particle contaminant	Associated with child leukemia cluster Fallon Nevada; known to cause genomic damage and cancer in animals.
Ta	73	8975	Capacitors	
I	53	2493	Thyroid, blood plasma	Radiation sensitivity

Table 6. Biologically significant environmental contaminants and materials exhibiting phantom radioactivity through the Secondary Photoelectron Enhancement (SPE) of natural background and medical X-rays

4.6. Fat soluble radioactive noble gases

The nuclide Krypton-85 has been released to the biosphere continuously since 1945 and increasingly from nuclear energy processes. With a half-life of 10.7 y and a β decay of 672 keV the concentration in the atmosphere has been building up to the extent that liquid air is now significantly radioactive. The assessment of harm from Kr-85 has generally been associated with skin doses from β decays in air. However Krypton (and Radon) are far more soluble in fats than in water and this water/oil partition driven equilibrium might cause build up of these nuclides in lymphatic tissue as a result of equilibria in the lung.

5. Conclusions and recommendations.

5.1. Epidemiology: Uranium effects

The current radiation risk position, that of the ICRP and its associated organisations, has been adequately reviewed by Harrison and Day [72]. With regard to the questions raised in the present overview, the only useful discussion in this paper, as in the CERRIE majority report [6], is the belief that the application of external risk models to internal exposures is supported by epidemiological studies of Thorotrast and Radium. It is therefore worth briefly looking more closely at these.

5.2. Radium and thorotrast studies: Re-examining the data

The increasing pressure brought to bear on the ICRP risk model focuses intensely on the arguments about internal and external radiation exposure rehearsed in the previous section. The ICRP and the radiation protection agencies have to concede much of the science, but fall back on the epidemiology. The problem is, very little human epidemiologic research has been done on internal radionuclide exposures. There are, however, two sets of studies which are said to broadly support the arguments that the current risk model is correct. These are the studies of individuals medically treated with Radium and Thorotrast. The studies originally were carried out because of doubt over the use of the external based risk model to deal with internal radionuclide exposures at a time when internal exposures from alpha emitters like plutonium were increasing in proportion to the development of the A-Bombs and H-Bombs. All of these studies were of roughly the same type. A group of individuals was formalised and then records were traced, or the individuals themselves were traced to see what the number of cancers were. The end point was always cancer, since the project was to see if the ICRP cancer risk model was accurate for these internal exposures. The medical and other (e.g. laboratory) exposures to Radium had been largely before 1960; e.g radium clock dial painters, and there were many of these who had survived from the period when they were employed. In addition there were individuals who had been exposed to Ra-224 as a treatment for various illnesses. There had been a fashion to treat syphilis, hypertension, gout, infectious polyarthritis, "muscular rheumatism", anaemia, epilepsy and multiple sclerosis [29] with radium. Then there were many individuals who had been injected with the substance Thorotrast, an X-ray contrast medium based on the nuclide Th-232, the daughter of which is Ra-228. So these are all internal radium exposures. What was reported in studies was that the cancer yields, mainly of liver cancer, bone cancer, and leukaemia could be roughly related to the exposures and that the yield was not too far away from the yield predicted by the

ICRP external type of risk model, i.e. the A-Bomb survivors. These studies are the last remaining defence that the current risk agencies can mobilise. There are a number of fatal problems with all the radium studies:

• The study groups were assembled long after the exposures and so not all those who had been exposed were in the study group: only the survivors. Many were dead. This biased the samples.

• A number of published studies give sufficient data to show that there was a high rate of death in the early period before the groups were assembled.

• The doses were not isotropic; for Thorotrast, the material was stored in depots in parts of the body where cells were quite resistant to radiation.

In addition, the doses were very large, so these studies were not of low dose chronic exposure but were in fact high dose internal chronic exposure.

Some of these problems were raised in 1970 in relation to the pioneering work by Robley Evans. Evans was a physicist and was concerned with the question of physical dosimetry of small quantities of internal emitters. Writing in the *British Journal of Cancer* in 1970, JF Loutit [29] took issue with the methodology of the Radium studies and pointed out that the massive bone marrow damage resulting from Radium exposure (which had been reported by many authors before Evans) would result in a very large excess death rate from a range of diseases. Loutit wrote that the limiting hazard from internally retained radium acquired occupationally being bone cancer needed to be reconsidered. He pointed out that evidence already existed in the 1930s from the work of Martland that those with substantial body burdens of radium had considerable life shortening and that the associated pathology had not been clarified. Loutit re-examined the radium dial case reports and found that internal radium had a profound effect on the bone marrow, best described as leukopenic anemia. This identifies one source of increased risk from non cancer illness and death which would have removed individuals from Radium and Thorotrast study groups. Indeed, the problem with all these studies is that they exclude about half of the exposed population who may have been lost to the researchers but are very likely to have died of cancer or a range of non cancer illnesses. In the better reported studies, where more data is made available, it is possible to see that this is indeed the case. An example is Wick et al. (1983) who examined cancer in Ra-224 patients. I have reduced the data from a diagram in this paper to produce the graph in Fig 7 which shows the percentage dead in the age group at exposure by the period between exposure and death. It is clear from the trend that for all the groups, the most deaths will have occurred in the first five years in individuals that were not in the study group.

This Ra-224 study by Wick et al. [73] is of the exposure group of German patients who were treated between 1948 and 1975 with Ra-224 for ankylosing spondylitis. There were 1501 total patients, among them 69 were missing and 433 were dead. What did they die of? We don't hear. But 3 of them developed bone cancer, 5 developed leukemia, and 6 bone marrow failure (cf Loutit above). This tiny cancer yield may approximate to the range predicted by the ICRP model (assuming that the dose could be accurately described) but what about the missing people? What about the 433 who died? If they died of conditions caused by the stress on their immune system (bone marrow failures and silent bone marrow problems) then the cancer yield is not a proper representation of the effects of the radium exposures on this group. And the cancer yield to produce an approximation to the ICRP risk predictions for leukemia is lower than in the control group. Addition of a handful of cases from the missing individuals or a handful of pre-leukemic immune-compromised individuals from the 633 dead would have a profound effect on the outcome of the study.

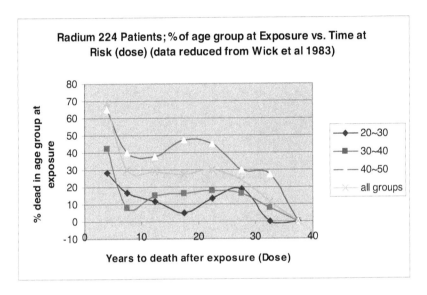

Figure 7. Percentage of each age group at exposure plotted against years to death from exposures in the Ra-224 study of Wick et al 1983. [73]

A similar picture is found in the thorotrast studies, where it is possible to see enough data. For example, in the paper by Mori et al 1983, 282 Japanese war wounded ex-servicemen thorotrast cases are followed up [74]. There were deaths from liver cancer, cirrhosis of the liver and also blood diseases. But in 170 deaths in the group, 42% were from cancer and 37% from other causes. There was no dose response for the cancers and the cancer yield was about 20-times greater than expected from ICRP. But the most interesting aspect is that from analysis of this group, the death rate was very high and the age at death very low compared with all Japanese populations. This is missed in the report since the method employed was to choose sick pathology controls from a hospital pathology records sample. I have compared their age specific death rates with all-Japan. Plots of the survival curves in the females in this group show that 100% were dead by age 75 compared with 65% for the equivalent all-Japan population. Results are given in Fig 8.

Of course, about 40% of these study group women died of cancer: the effects of the thoro-trast. But note that the others died from something else; they didn't live to a ripe old age nor did they live as long as the all Japan population. This is clear from the survival curves in Figure 8 which show almost a 20 years age effect in the women. For men, the shift was about 9 years (my unpublished results, not shown).

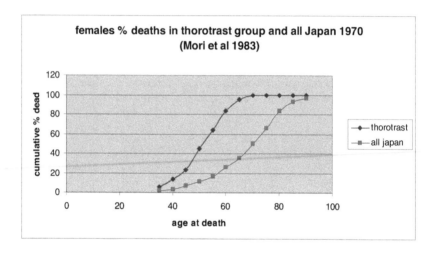

Figure 8. Survival curves for female thorotrast patients studied by Mori et al 1983 compared with all Japan. Data reduced from tables in Mori et al 1983 [74] and Japanese government publications.

The conclusions of this brief account of the re-examinations of the radium and thorotrast studies show that they cannot be used as indicators for low dose chronic risk to internal radionuclides. Apart from the fact that the doses were (like the A-Bomb doses) very large, the main fatal flaw was and is that confounding causes of death make the cancer yield conclusions unsafe. Loutit 1970 makes the point that the damage to the bone marrow would be likely to occur in the case of the weapons-fallout component Strotium-90, and he urged the research community to concentrate on examining risk from that nuclide, an exhortation which the research community entirely failed to take notice of. Loutit was a Medical Research Council MRC (Harwell) director.

5.3. Uranium

The anomalous health effects of exposure to Uranium, especially in the form of particulates, have been increasingly clear in the last 10 years. The radiobiological evidence is reviewed in ECRR2009 [75] and there is insufficient space here to do more than note that the current risk external radiation based model cannot begin to explain or predict what is found empirically. Despite the massive evidence including studies by nuclear industry and military scientists, the agencies ICRP, UNSCEAR, BEIR et al persist in their assertions that the observed effects cannot be due to Uranium. Most recently there have been studies of French Uranium workers showing leukemia and lymphoma excess, lung cancer excess and heart disease at doses which are too low by some 2000-times to explain them on the basis of current risk models [76, 77, 78]. There is an urgent need to carry out research into this issue. The effects of photoelectron amplification can easily be examined by studies involving varying external X-ray doses at different concentrations of Uranium particulates and molecular Uranyl ion in cell culture and animal studies. There is no routine measurement of Uranium in the vicinity of nuclear sites. This should also be remedied.

5.4. other epidemiological evidence

5.4.1. Childhood cancer near nuclear installations

There have been reports in peer reviewed journals of increased risk of childhood leukemia and non Hodgkin lymphoma near many nuclear sites in Europe. A list and discussion may be found in ECRR2010. Child leukemia excesses are found near nearly all the sites that have been examined [5]. e.g the reprocessing sites at Sellafield [37] Dounreay UK [79] and La Hague (France) [80] near the Atomic Weapons Establishment Aldermaston (UK) [81], the Atomic Energy Research Establishment Harwell (UK) [82], near Hinkley Point nuclear power station (UK) [83] and recently near all the combined nuclear sites in Germany (KiKK study) [84] and near all the combined nuclear sites in France [85], GB, and Switzerland.

The radiation risk community [86, 87] basing calculations on the ICRP risk model have worked out the dose ranges and say they cannot be more than a few microSieverts, well below Natural Background. The ICRP risk model predicts an excess risk of 0.05 cancers per Sievert. 100 microSieverts is 1/10,000 th (10^{-4} of a Sievert). An Excess Absolute Risk of 0.05/Sv is Excess Relative Risk (ERR) of 5E-8 per μSv. This, divided by the spontaneous risk of 3E-4 for 0-4 y old children, is 1E-3 per 6 microSv. But there are twice as many child leukemias as are expected: a doubling of risk: the ERR observed in the KiKK study was ~ERR=1. So ICRP predicts a 1000-fold lower risk than found in the KiKK study.)

The ICRP does not give a risk factor for childhood leukaemia but to define a difference between external and internal exposure we can employ the Excess Relative Risk based on the obstetric X-ray studies analysed by Wakeford and Little [88]. This gives an Excess Relative Risk of 50/Sv and based on the 40/Sv Obstetric X-rays results of Alice Stewart.

Stewart found a 40% excess risk after an X-ray dose of 10 mSv [88]. That would suggest a 4% increase after 1 mSv, 0.4% after 100 μSv. But we are seeing a 100% increase at this level. The error is now 100/0.4 = 250-fold.

5.4.2. Infant leukemia after Chernobyl

Five different groups [89-93] reported a statistically significant increase in infant leukemia in 5 different countries of Europe in those children who were in the womb at the time of the Chernobyl Caesium-137 fallout as measured by whole body monitoring. The effect was also reported from the USA [94]. Thus the Chernobyl exposure is the only explanation for the increase. This occurred and was reported from Greece, Germany, Scotland, Wales, Belarus, USA and the error this shows in the ICRP model was the subject of two peer reviewed papers in 2000 [92] and 2009 [25]. Using the Alice Stewart relation between dose and leukemia above, the error is about 400-fold (depending on the country) [25]. Using the ICRP model it is upwards of 1000-fold. This analysis is most relevant since it unequivocally supports the causal relation revealed by the nuclear site child leukemias yet in this case fission product internal radiation can be the only cause.

5.4.3. Cancer following Chernobyl in Northern Sweden

The study by Martin Tondel found a 11% increase in cancer for every 100 kBq/sq metre of Cs-137 from Chernobyl [95]. It is possible to calculate that 100 kBq/m^2 Cs-137 including a further 100kBq/ m^2 of Cs-134 if reduced exponentially due to rain washout to rivers and lakes with half life of 6 months would give a committed effective dose of about 1 mSv. The ICRP model [96] predicts an Excess Relative Risk of 0.45 per Sv, so the ICRP expected excess relative risk, including a Dose Rate Reduction Factor of 2 (as used by ICRP) is 0.0225%. The error in ICRP model defined by Tondel's result is thus 490-fold.

5.4.4. Human sex ratio at birth perturbed by low doses of internal fission-product ionising radiation

Studies by Hagen Scherb and Kristina Voigt [97] show clear and highly statistically significant alterations in the human sex ratio at birth (the number of boys born to girls) after (a) atmospheric bomb testing, (b) Chernobyl and (c) near nuclear facilities. Effects are shown to be local, European (several countries were studied) and global, supporting earlier evidence of increases in infant mortality during the period of atmospheric weapons testing [98, 99]. Sex ratio has been accepted as a measure of genetic damage with the preferential killing of one or other sex depending on the type of exposure (mothers or fathers). According to Scherb and Voigt, millions of babies were killed *in utero* by these effects [100]. A recent reanalysis of the sex ratio effect in Hiroshima reveals the effect in those populations also [101], evidence which was overlooked by the USA researchers through poor epidemiology and questionable decisions. This evidence objectively confirms the serious genotoxic effect of internal ionising radiation on germ cells and the exquisite sensitivity of humans and other living creatures to releases from Uranium fission. The ICRP does not consider such effects nor are they included in any assessment of harm.

5.4.5. Cancer and genotoxic effects in Iraq following DU exposure

A series of studies of the population of Fallujah Iraq shown [102- 104] to have been exposed to Uranium following the 2003-2004 battles have revealed extremely high rates of congenital malformations at birth and cancer and leukemia/lymphoma in adults. The studies also draw attention to significant sex ratio effects at birth beginning after 2004. These results, and the increases in genotoxic effects in the offspring of Gulf veterans support and are supported by the other sets of observations reviewed above which show that inhaled Uranium nanoparticles represent a very serious hazard which is entirely overlooked by ICRP.

5.4.6. Chernobyl effects as reported in the Russian peer-reviewed literature

The effects of the Chernobyl accident exposures have been reported in the Russian language peer review literature since 1996. These results have been reviewed by Busby and Yablokov 2006 [105] Yablokov et al 2010 [106] and Busby et al 2011 [107] but have been largely ignored by ICRP. They constitute a very large body of peer reviewed work which show that the effects of the Chernobyl accident exposures are massive and extremely serious [108]. They range from cancer and leukemia to heart disease especially in children together with a range of illnesses which can be best described by the term premature ageing [108]. They include congenital transgenerational diseases and are reported in animals and plants which cannot be affected by the kind of psychological processes (radiophobia) which have been employed by the radiation risk establishment to account for the early reports coming out of the affected territories. In addition, there are objective measurements of serious biological harm to humans and other living creatures affected by the exposures. The germline mutations found by minisatellite tests [109] in humans were also associated with real morphological effects and fitness loss in birds [110] and were shown to have caused significant sex ratio changes in the birds and also population loss [111] which is in agreement with the findings of Scherb and Voigt and the infant mortality findings [98, 99]. The implications for the understanding of the historic effects of the nuclear project on human health are alarming.

5.5. Summary and conclusions

The current radiation risk model is insecure for internal radionuclide effects. Massive evidence exists from epidemiology and also published studies of the effects of internal radionuclide exposures that the effects of location, chemical binding or affinity, temporal decay patterns and transmutation of internal radionuclides can have much greater genetic or lethal effects on cells than are predicted by the absorbed dose model. These data have been published since the 1950s but ignored for the purpose of radioprotection. Many critical research issues should have been pursued but have not been. It is recommended that those issues and research studies highlighted in this contribution are seen as a priority.

Appendix

Calculating the probability of a track interception with DNA as a function of distance of the point source

The model is given in Fig 9 and Fig 10. It locates the source at the centre of a sphere S radius r distance d from the DNA which is modelled as a cylinder of length 2R. We put r<D. Any decay which intercepts an infinitesimal strip of area A on the DNA cross section can be mapped onto a small area B on the surface of the sphere S. The required probability assumes that the decay can be in any direction. It is thus equal to the area B / total area of the sphere.

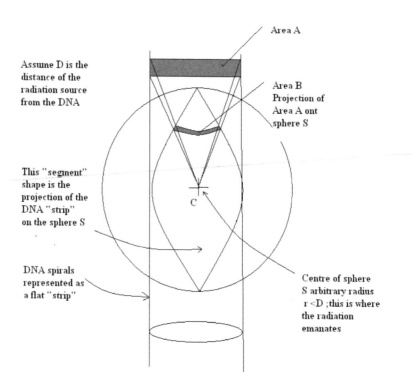

Figure 9. Model

From Fig 2, tan θ = R/d; θ = arctan (R/d)

Length of arc A = 2rθ = 2arctan (R/d)

d' = d/cos θ; θ = arctan (R/d') = arctan ((Rcos α)/d)

Area B (Fig 1) = 2rθ. rdα = 2r^2 arctan((Rcos α)/d)

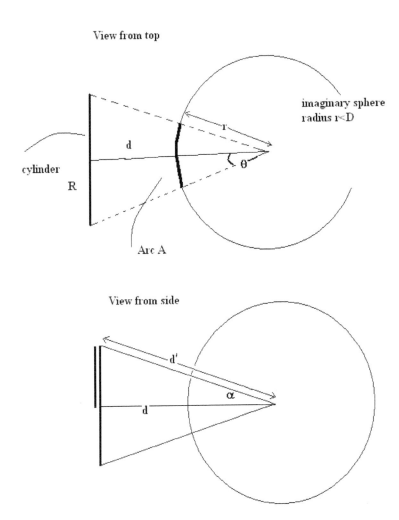

Figure 10. Model from top and side

Area B (Fig 1) = 2rθ. rdα = 2r^2 arctan((Rcos α)d)

Whole area of segment =

$$4r^2 \int_0^{\pi/2} \arctan((R\cos\alpha)/d)d\alpha$$

And the required probability is this divided by the surface area of the sphere 4πr^2

$$P\,(DNA) \;=\; \frac{1}{\pi}\int_0^{\pi/2} \arctan((R\cos\alpha)/d)d\alpha$$

Author details

Christopher Busby

Jacobs University, Bremen, Germany

References

[1] Caufield, K. (1989), *Multiple Exposure: Chronicles of the Radiation Age* (London: Secker and Warburg).

[2] Gracheva LM and Korolev VG (1980) *Genetic effects of the decay of radionuclides in cells.* Edited by I.A.Zacharov. Calcutta: Oxonian Press PVT

[3] ICRP Publication 72 (1996) Age-dependent Doses to the Members of the Public from Intake of Radionuclides Part 5, Compilation of Ingestion and Inhalation Coefficients. Amsterdam: Elsevier

[4] ECRR2003 (2003) *2003 recommendations of the European Committee on Radiation Risk-The health effects of ionizing radiation at low dose--Regulator's edition.* Busby C.C editor with Bertell R, Yablokov A, Schmitz Feuerhake I and Scott Cato M (Brussels: ECRR-2003) Translations of the above into French Japanese Russian and Spanish (see www.euradcom.org for details)

[5] ECRR2010 (2010) *The 2010 Recommendations of the European Committee on Radiation Risk. The Health Effects of Ionizing Radiation at Low Doses and Low Dose Rates.* Eds. Busby C, Yablolov AV, Schmitz Feuerhake I, Bertell R and Scott Cato M Brussels: ECRR; Aberystwyth Green Audit

[6] CERRIE (2004) Report of the Committee Examining Radiation Risk from Internal Emitters (CERRIE) *Chilton, UK: National Radiological Protection Board*

[7] Busby CC, Bramhall R and Dorfman P (2004) *CERRIE Minority Report 2004: Minority Report of the UK Department of Health/ Department of Environment (DEFRA) Committee Examining Radiation Risk from Internal Emitters (CERRIE)* Aberystwyth: Sosiumi Press

[8] IRSN (2005) Les consequences sanitaire des contaminations internes chroniques par les radionucleides. Ed:-F.Paquet Rapport DRPH/2005-20 Fontenay aux Roses: Institut de Radioprotection et de Surete Nucliare

[9] UNSCEAR, (1977) *Sources and Effects of Ionising Radiation*, Report to the General Assembly, with annexes, (New York, United Nations).

[10] Busby, C. (1994), Increase in Cancer in Wales Unexplained, BMJ, 308: 268.

[11] Busby C.C (2002). 'High Risks at low doses.' Proceedings of 4th International Conference on the Health Effects of Low-level Radiation: Keble College Oxford, Sept 24 2002. (London: British Nuclear Energy Society).

[12] BEIR V (1990) BEIR (Committee on Biological Effects of Ionising Radiation), (1990) The Health Effects of Exposure to Low Levels of Ionising Radiation, BEIR V, (Washington: National Academy Press).

[13] Mothershill C and Seymour C (2012) Human and environmental health effects of low doses of radiation. In *Fukushima—What to expect. Proceedings of the 3rd International Conference of the European Committee on Radiation Risk, Lesvos, Greece May 5/6 2009* Eds Busby C, Busby J, Rietuma D and de Messieres M. Brussels, Belgium: ECRR

[14] Baverstock, K.F. & Charlton, D.E. 1988. *DNA Damage by Auger Emitters*. Taylor and Francis, London

[15] Hofer KG (1998) Biophysical aspects of Auger Processes- a review. *Acta Oncol.*35 798-96

[16] Ward, J.F., Limoli, P., Calabro-Jones, P. & Evans, W.F. 1988. Radiation vs.chemical damage to DNA. In: Nygard, O.F., Simic, M. & Cerutti, P. (eds.), Anticarcinogenesis and Radiation Protection. Plenum, New York.

[17] Busby Chris and Schnug Ewald (2008) Advanced biochemical and biophysical aspects of uranium contamination. In: (Eds) De Kok, L.J. and Schnug, E. *Loads and Fate of Fertilizer Derived Uranium.* Backhuys Publishers, Leiden, The Netherlands, ISBN/EAN 978-90-5782-193-6.

[18] Busby, C.C. (1995), *Wings of Death: Nuclear Pollution and Human Health* (Aberystwyth: Green Audit)

[19] Busby, C.C.(1998), 'Enhanced mutagenicity from internal sequentially decaying beta emitters from second event effects.' In 'Die Wirkung niedriger Strahlendosen- im kindes-und Jugendalter, in der Medizin, Umwelt ind technik, am Arbeitsplatz'. Proceedings of International Congress of the German Society for Radiation Protection. Eds: Koehnlein W and Nussbaum R. Muenster, 28 March 1998 (Bremen: Gesellschaft fur Strahlenschutz)

[20] Goodhead DT (1991) Biophysical features of radiation at low dose and low dose rate. In CB Seymour and C Mothershill (Eds) New Developments in fundamental and applied radiobiology London: Taylor and Francis.

[21] Apelgot S and DBuguesne M (1963) Energie dissipee par le Tritium dans les microorganisms. *Int.J.Radiation Biol.* 7 (1) 65-74

[22] Robertson JS and Hughes WL (1959) Intranuclear irradiation with Tritium labeled thymidine. In: Proc 1st National Biophysical Conference Columbus Ohio 278-283

[23] ICRU 35 (1984) Radiation Dosimetry: Electron Beams with energies between 1 and 50MEv. Bethesda MD USA: ICRU

[24] Burlakova E B; Goloshchapov A N, Zhizhina G P, Konradov A A, (2000) New aspects of regularities in the action of low doses of low level irradiation. In Low Doses of Ra-

diation—Are They Dangerous? Burlakova E B, Ed.; Nova Science Publishers: New York, NY, USA,

[25] Busby C.C. (2009) Very Low Dose Fetal Exposure to Chernobyl Contamination Resulted in Increases in Infant Leukemia in Europe and Raises Questions about Current Radiation Risk Models. International Journal of Environmental Research and Public Health.; 6(12):3105-3114. http://www.mdpi.com/1660-4601/6/12/3105

[26] Luning, K. G., Frolen, H., Nelson, A., and Ronnbaeck, C. (1963), `Genetic Effects of Strontium-90 Injected into Male Mice', Nature, No 4864 197: 304-5.

[27] Smirnova, E. I. and Lyaginska, A. M. (1969), `Heart Development of Sr-90 Injured Rats', in Y. I. Moskalev and Y. I. Izd (eds.), Radioaktiv Izotopy Organizs (Moscow: Medizina), 348

[28] Stokke, T., Oftedal, P. and Pappas, A. (1968), `Effects of Small Doses of Radioactive Strontium on the Rate Bone Marrow', Acta Radiologica, 7: 321-9.

[29] Loutit J F (1970), Malignancy from Radium. Brit.J.Cancer 24(2) 17-207

[30] Nielsen, P.E, Hiort, C., Soennischsen, S.O., Buchardt, O., Dahl, O. & Norden, B. 1992. DNA binding and photocleavage by Uranyl VI salts. J. Am. Chem. Soc. 114: 4967-4975.

[31] Huxley, H.E. & Zubay, G. 1961. Preferential staining of nucleic acid containing structures for electron microscopy. Biophys. Biochem. Cytol. 11: 273.

[32] Petkau A (1980) Radiation carcinogenesis from a membrane perspective *Acta Physiologica Scandinavica* suppl. 492. 81-90

[33] Tamplin AR and Cochran TB (1974) Radiation standards for hot particles. A report on the inadequacy of existing radiation standards related to exposures of man to insoluble particles of plutonium and other alpha emitting hot particles. Washington, USA: Natural Resources Defence Council

[34] Charles MW, Mill AJ and Darley P (2003) Carcinogenic risk from hot particles. J.Radiol.Prot. 23 5-28

[35] Popplewell, DS, Ham GJ, Dodd NJ, Shuttler SD (1988) 'Plutonium and Cs-137 in autopsy tissues in Great Britain' Sci. Tot. Environment 70 321-34

[36] Priest, N. D., O'Donnell, R.G., Mitchell, P. I., Strange, L., Fox, A., Henshaw, D. L., and Long, S. C. (1997), 'Variations in the concentration of plutonium, strontium-90 and total alpha emitters in human teeth collected within the British Isles', Science of the Total Environment, 201, 235-243.

[37] Beral V E. Roman, and M. Bobrow (eds.) (1993), Childhood Cancer and Nuclear Installations (London: British Medical Journal).

[38] Independent Advisory Group (1984), Investigation of the Possible Increased Incidence of Cancer in West Cumbria, The Black Report, (London: HMSO).

[39] Royal Society (2001) The Health Effect of Depleted Uranium Weapons Vol 1 London: Royal Society

[40] COMARE, (1996) The Incidence of Cancer and Leukaemia in Young People in the Vicinity of the Sellafield Site in West Cumbria: Further Studies and Update since the Report of the Black Advisory Group in 1984, COMARE 4th Report (Wetherby: Department of Health).

[41] Hohenemser C, Deicher M, Hofsass H, Lindner G, Recknagel E and Budnick J (1986) Agricultural impact of Chernobyl: a warning. *Nature* 321 817

[42] Apelgot S (1968) Effect letal de la disintegration des atomes radioactifs (3H, 14C, 32P) incorpres dans bacteria. In: Biological Effects of Transmutation and decay of incorporated Radioisotopes. Vienna: IAEA 147-163

[43] Apelgot S and Latarjet R (1962) Marquage d'un acide deoxyribonucleique bacterien par le radiophosphore, le radiocarbone at le tritium: comparison des effets letaux. Biochim et Biophys Acta 55(1) 40-55

[44] Andersen FA and Person S (1971) Incorporation of 14C labeled precursors into *e.coli* The lethal and mutagenic effects. *Radiation Res.* Abstr. 47 (1) p261

[45] Andersen FA and Person S (1973) Incorporation of 14C labeled precursors into *e.coli* The lethal and mutagenic effects. *Radiation Res.* Abstr. 21 (1) p4

[46] Pluchnnik G (1965) Mutagenyii protsess u khorelly pri assimilatsii radioaktivnoi uglekiskoty (mutagenic process in Chlorella during assimilation of radioactive carbon dioxide) *Genetika* 1(5) 19-25)

[47] Pluchennik G (1966) Comparative study of mutagenic effects of incorporated isotopes of the most important organogenic elements. Report 1. Mutagenic effects of incorporated 14C from the example of reversion to autotrophy of diploid yeasts that are homozygous in relation to mutations for adenine requirement. (Russian) *Genetika* 2(5) 117-124

[48] Totter IR, Zerle MR and Hollister H (1958) Hazards to man of Carbon-14. *Science* 128 1490-1495

[49] Pauling L (1958)Genetic and somatic effects of Carbon-14 *Science* 128 1183-1186

[50] Beal JM (1950) Chromosome aberrations in onion roots from plants grown in an atmosphere containing 14-CO_2. *Amer.J.Botany* 37 660-661

[51] Beal M and Scully NJ (1950) Chromosome aberrations in onion roots from plants grown in an atmosphere containing 14-CO_2. *Botan.Gas*112 232-235

[52] McQuade HA, Friedkin M and Atchison AA (1956) Radiation effects of thymidine 2-[14]C. 1. Uptake of thymidine 2-^{14}C and thymine 2-^{14}C in the onion root tip. 2. Chromosomal aberrations caused by thymidine 2-[14]C. and thymine 2-^{14}C in the onion root tip. *Expt.Cell.Res.* 11 (2) 249-264

[53] McQuade HA and Friedkin M (1960) Radiation effects of thymidine ^{3}H and thymidine ^{14}C. *Expt.Cell.Res.* 21(1) 118-125

[54] Kuzin AM (1962) Biological effects of increased concentrations of ^{14}C in the atmosphere (Russian) in *Radiation Genetics* Izd-vo AN SSSR Moscow 274-278

[55] Kuzin AM (1962) Effectiveness of biological action of ^{14}C when incorporated in live cells. In *Radionatsionnaya Genetika* Izd-vo AN SSSR Moscow 267-273

[56] Purdom CE (1965) Genetic effect of incorporated ^{14}C in *D.MelanogasterMutation Res.* 2(2) 156-167

[57] Kuzin AM, Glembotskii Ya M, Lapkin Yu A (1964) Mutagenic efficiency of incorporated Carbon 14. (Russian) *Radiobiologiya* 4(6) 804-809

[58] Jha AM, Dogra Y, Turner A and Millward GE (2005) Impact of low doses of tritium on the marine mussel *myrtilis edulis.* Genotoxic effects and tissue specific bioconcentrations. *Mutat.Res* 56(1) 47-57

[59] Cox R and Edwards A (2000) Commentary on the Second Event Theory of Busby *Int.J.Radiat.Biol* 76(1) 119-125

[60] Busby C.,(2000), 'Response to Commentary on the Second Event Theory by Busby' *International Journal of Radiation Biology* 76 (1) 123-125

[61] Gracheva LM and Shanshiashvili TA (1983) Genetic effects of decay of radionuclide products of fission of nuclear fuel II.Lethal and mutagenic effects on the mutation of cells of the yeast *saccharomyces cerevisiae* induced by Sr-90 and Sr-89. *Genetika* (Moscow) 9/4 532-5

[62] Miller RC and Hall EJ (1978) X-ray dose fractionation and oncogenic transformations in culture mouse embryo cells. *Nature* 272 58-60

[63] Borek C and Hall EJ (1974Rffects of split doses of X-rays on Neoplastic transformation of single cells *Nature* 252 499-501

[64] Speirs, F.W. 1949. The influence of energy absorption and electron range on dosage in irradiated bone. *Brit. J. Radiol.* 22: 521-533.

[65] Hainfeld, J.F., Slatkin, D.N. & Smilowitz, H.M. 2004. The use of gold nanoparticles to enhance radiotherapy in mice. *Phys. Med. Biol.* 49: N309-N315.

[66] Busby CC (2005) Does uranium contamination amplify natural background radiation dose to the DNA? *European J. Biology and Bioelectromagnetics.* 1 (2) 120-131

[67] Busby CC (2005) Depleted Uranium Weapons, metal particles and radiation dose. *European J. Biology and Bioelectromagnetics.* 1(1) 82-93

[68] Pattison J E, Hugtenburg R P, Green S, (2009) Enhancement of natural background gamma-radiation dose around uranium micro-particles in the human body. J.Royal http://rsif.royalsocietypublishing.org/content/early/2009/09/23/rsif.2009.0300.abstract

[69] Eakins, JS, Jansen J. Th. M. and Tanner R. J. (2011) A Monte Carlo analysis of possible cell dose enhancements effects by Uranium microparticles in photon fields Radiation Protection Dosimetry (2011), Vol. 143, No. 2–4, pp. 177–180 doi:10.1093/rpd/ncq398

[70] Regulla, D.F., Hieber, L.B. & Seidenbusch, M. 1998. Physical and biological interface dose effects in tissue due to X-ray induced release of secondary radiation from metallic gold surfaces. *Radiat. Res.* 150: 92-100

[71] Elsaesser A, Busby C, McKerr G and Howard CV (2007) Nanoparticles and radiation. EMBO Conference: Nanoparticles. October 2007 Madrid

[72] Harrison J and Day P (2008) Radiation doses and risks from internal emitters. *J Radiol.Prot.* 28 137-159

[73] Wick RR, Chmelevsky,D and Goessner W (1984) Risk to bone and haematopoetic tissue in ankylosing spondilitis patients. In W.Goessner, GB Gerber , U Hagen and A Luz. Eds *The radiobiology of radium and thorotrast.* Munich Germany: Urban and Schwartzenberg

[74] Mori T, Kumatori T, Kato Y, Hatakeyama S, Kamiyama R, Mori W, Irie H, Maruyama T and Iwata S (1983) Present status of medical study of thorotrast patients in Japan. Pp123-135 . In W.Goessner, GB Gerber , U Hagen and A Luz. Eds *The radiobiology of radium and thorotrast.* Munich Germany: Urban and Schwartzenberg

[75] ECRR2009 (2009) The health effects of exposure to Uranium. Brussels: ECRR (www.euradcom.org)

[76] Guseva Canu I, Laurier D, Caër-Lorho S, Samson E, Timarche M, Auriol B, Bérard P, Collomb P, Quesned B, Blanchardone E (2010) Characterisation of protracted low-level exposure to uranium in the workplace: A comparison of two approaches. *International Journal of Hygiene and Environmental Health* 213 (2010) 270–277

[77] Guseva Canu, Irina, Garsi, Jerome-Philippe, Cae°r-Lorho Sylvaine, Jacob SophieCollomb, Philippe, Acker Alain, Laurier Dominique (2012) Does uranium induce circulatory ? First results from a French cohort of uranium workers *Occup. Envir. Med.* OEM Online First, published on March 3, 2012 as 10.1136/oemed-2011-100495

[78] Guseva Canu I, Jacob S Cardis E, Wild P Cae°r –Lorho S, Auriol B, Garsi JP, Tirmarche M, Laurier D (2010) Uranium carcinogenicity in humans might depend on the physical and chemical nature of uranium and its isotopic composition: results from pilot epidemiological study of French nuclear workers. Cancer Causes Control DOI 10.1007/s10552-011-9833-5

[79] Urquhart T D, Black R T, Muirhead M T, et al., (1991) Case-control study of leukaemia and non-Hodgkins lymphoma in children in Caithness near the Dounreay nuclear installation. British Medical Journal; 302:687-692.

[80] Viel J-F, Poubel D, Carre A, (1995) Incidence of leukaemia in young people and the La Hague nuclear waste reprocessing plant: a sensitivity analysis. Statistics in Medicine, 14, 2459-2472.

[81] Busby C, and M. Scott Cato, (1997)`Death Rates from Leukemia are Higher than Expected in Areas around Nuclear Sites in Berkshire and Oxfordshire', *British Medical Journal*, 315 (1997): 309

[82] Busby C, and M. Scott Cato, (1997)`Death Rates from Leukemia are Higher than Expected in Areas around Nuclear Sites in Berkshire and Oxfordshire', British Medical Journal, 315 (1997): 309

[83] Bowie C, Ewings P D, (1988) Leukaemia incidence in Somerset with particular reference to Hinkley Point, Taunton: Somerset Health Authority.

[84] Kaatsch P, Spix C, Schulze-Rath R, Schmiedel S, Blettner M, (2008) Leukaemias in young children living in the vicinity of German nuclear power plants. Int J Cancer 122, pp. 721-726.

[85] Sermage-Faure Claire, Laurier Dominique, Goujon-Bellec Stéphanie, Chartier Michel, Guyot-Goubin, Aurélie, Rudant Jérémie, Hémon Denis, Clavel Jacqueline (2012) Childhood leukemia around French nuclear power plants—The geocap study, 2002–2007 International Journal of Cancer Volume 131 (5) E769–E780 DOI: 10.1002/ijc. 27425

[86] COMARE (Committee on Medical Aspects of Radiation in the Environment), (1986) The Implications of the New Data on the Releases from Sellafield in the 1950s for the Conclusions of the Report on the Investigation of a Possible Increased Incidence of Cancer in West Cumbria, COMARE 1st Report (London: HMSO).

[87] NRPB, (1995) Risks of leukaemia and other cancers in Seascale from all sources of ionising radiation NRPB R-276 (Chilton: NRPB).

[88] Wakeford R and Little MP (20030) Risk coefficients for childhood cancer after intrauterine irradiation. A review. Int.J.Rad.Biol. 79 293-309

[89] Michaelis J, Kaletsch U, Burkart W and Grosche B, (1997) Infant leukaemia after the Chernobyl Accident Nature 387, 246.

[90] Petridou E, Trichopoulos D, Dessypris N, Flytzani V, Haidas S, Kalmanti M, Koliouskas D, Kosmidis H, Piperolou F, Tzortzatou F, (1996) Infant Leukaemia after in utero exposure to radiation from Chernobyl, Nature, 382:25, 352.

[91] Gibson B E S, Eden O B, Barrett A, et al., (1988) Leukaemia in young children in Scotland, The Lancet, 630

[92] Busby, C. C. and Cato, M. S. (2000), 'Increases in leukemia in infants in Wales and Scotland following Chernobyl: evidence for errors in risk estimates' Energy and Environment 11(2) 127-139 92 Busby C.C. and Cato M.S. (2001) 'Increases in leukemia in infants in Wales and Scotland following Chernobyl: Evidence for errors in statutory risk estimates and dose response assumptions'. International Journal of Radiation Medicine 3 (1) 23

[93] Busby C.C. (2009) Very Low Dose Fetal Exposure to Chernobyl Contamination Resulted in Increases in Infant Leukemia in Europe and Raises Questions about Current

Radiation Risk Models. International Journal of Environmental Research and Public Health.; 6(12):3105-3114. http://www.mdpi.com/1660-4601/6/12/3105

[94] Mangano J, (1997) Childhood leukaemia in the US may have risen due to fallout from Chernobyl, British Medical Journal, 314: 1200.

[95] Tondel M, Hjalmarsson P, Hardell L, Carisson G, Axelson A, (2004) Increase in regional total cancer incidence in Northern Sweden. *J Epidem. Community Health.* 58 1011-1016.

[96] ICRP, (2007) *The 2007 recommendations of the International Commission on Radiological Protection.* ICRP 103 Orlando USA: Elsevier

[97] Scherb H and Voigt K (2010) The human sex odds ratio at birth after the atmospheric bomb tests, Chernobyl, and in the vicinity of nuclear facilities. Env.Sci.Pollut.Res. Int 18 (5) 697-707

[98] Sternglass E J, (1971) Environmental Radiation and Human Health, in Proceedings of the Sixth Berkeley Symposium on Mathematical Statistics and Probability, ed. J. Neyman (Berkeley, Calif.: University of California Press).

[99] Whyte R K, (1992) First Day Neonatal Mortality since 1935: A Re-examination of the Cross Hypothesis, *British Medical Journal*, 304: 343-6.

[100] Scherb H and Voigt K (2011) in Busby C, Busby J, Rietuma D and de Messieres M Eds. (2011) Fukushima: What to Expect. Proceedings of the 3rd International Conference of the European Committee on Radiation Risk May 5/6th Lesvos Greece. Brussels: ECRR; Aberystywth UK: GreenAudit

[101] Padmanabhan VT (2011) in Busby C, Busby J, Rietuma D and de Messieres M Eds. (2011) Fukushima: What to Expect. Proceedings of the 3rd International Conference of the European Committee on Radiation Risk May 5/6th Lesvos Greece. Brussels: ECRR; Aberystywth UK: GreenAudit

[102] Alaani Samira, Tafash Muhammed, Busby Christopher, Hamdan Malak and Blaurock-Busch Eleonore (2011) Uranium and other contaminants in hair from the parents of children with congenital anomalies in Fallujah, Iraq Conflict and Health 2011, 5:15 doi:10.1186/1752-1505-5-15

[103] Busby, Chris; Hamdan, Malak; Ariabi, Entesar. (2010) Cancer, Infant Mortality and Birth Sex-Ratio in Fallujah, Iraq 2005–2009. Int. J. Environ. Res. Public Health 7, no. 7: 2828-2837. doi:10.3390/ijerph7072828

[104] Alaani S, Al Fallouji M, Busby C and Hamdan M (2012) Pilot study of congenital rates at birth in Fallujah, Iraq, 2010 *J. Islam. Med. Assoc. N. Amer.* Sept 1st 2012

[105] Busby C, Yablokov A V (2006, 2009) *ECRR 2006. Chernobyl 20 years On. The health Effects of the Chernoby lAccident* Aberystwyth: Green Audit

[106] Yablokov A V, Nesterenko V B, Nesterenko A V, (2009) Chernobyl: Consequences of the Catastrophe for people and the environment. Annals of the New York Academy of Sciences. Vol 1181 Massachusetts USA: Blackwell

[107] Busby C, Busby J, Rietuma D and de Messieres M Eds. (2011) Fukushima: What to Expect. Proceedings of the 3rd International Conference of the European Committee on Radiation Risk May 5/6th Lesvos Greece. Brussels: ECRR; Aberystwyth UK: Green-Audit

[108] Malko M V, (1998) Chernobyl accident: the crisis of the international radiation community in Imanaka T: Research activities about the radiological consequences of the Chernobyl NPS accident and social activities to assist the sufferers of the accident. (Kyoto University: Research Reactor Institute).

[109] Dubrova Y E, Nesterov V N, Jeffreys A J et al., (1997) Further evidence for elevated human minisatellite mutation rate in Belarus eight years after the Chernobyl accident. Mutation Research 381 267-278.

[110] Ellegren H, Lindgren G, Primmer C R, Moeller A P, (1997), Fitness loss and Germline mutations in Barn Swallows breeding in Chernobyl, *Nature* 389/9, 583-4.

[111] Møller AP, Bonisoli-Alquati A, Rudolfsen G, Mousseau TA (2012) Elevated Mortality among Birds in Chernobyl as Judged from Skewed Age and Sex Ratios. PLoS ONE 7(4): e35223. doi:10.1371/journal.pone.0035223

Chapter 8

The Molecular Epidemiology of
DNA Repair Polymorphisms in Carcinogenesis

Paul W. Brandt-Rauf, Yongliang Li,
Changmin Long and Regina Monaco

Additional information is available at the end of the chapter

1. Introduction

There are well-established examples of highly penetrant mutations in genes that are directly involved in carcinogenesis and result in a high risk of cancer in the individuals who carry these mutations. Some of the best examples include syndromes of defective DNA repair, such as xeroderma pigmentosum [1]. However, these examples tend to be very rare and thus contribute minimally to the overall burden of cancer risk. Nevertheless, it has long been suspected that less penetrant susceptibility may be produced by much more common variants in the same cancer-related genes, for example, in the form of single nucleotide polymorphisms (SNPs), that presumably would be less disruptive and therefore produce more subtle effects on the function of the encoded proteins but which could contribute greatly to overall cancer attributable risk in populations due to their widespread occurrence [1]. Because several of these common polymorphisms occur in DNA repair proteins, many epidemiologic studies have examined their relationship to cancer risk [2-4].

These studies have looked at all different types of cancer, many different at-risk populations, several different DNA repair pathways, and a variety of polymorphisms at different sites [5-18]. The results to date at best have been inconsistent, conflicting and confusing with many examples of positive, negative or null associations between particular polymorphisms and particular cancers, even in multiple large meta-analyses of the data. For example, a very recent large, rigorous and systematic review of the literature on the involvement of DNA repair polymorphisms in human cancer reached the conclusion that because of the inconsistencies in the literature "none of the cancer genome-wide association studies (GWAs) published so far showed highly statistically significant associations for any of the common DNA repair gene variants" and "clarification of the discrepancies in the literature is needed." [4] It was suggested that one way to proceed would be that "gene/environment and gene/lifestyle interactions for carcinogenic mechanisms involving DNA repair should be investigated more systematically and with less classification error." [4] However, even in studies of populations with exposures to known environmental carcinogens and the cancers most closely associated with those exposures, the results of DNA repair polymorphism studies have not always been clear-cut; these inconsistencies may also be the result of poor exposure classification, multiple confounders and/or poor understanding of the exact mechanisms of DNA damage and/or repair [19-23]. In other words, what is needed is to study model systems where there are clear linkages between the exposure to the carcinogenic risk factor and the specific DNA damage that it produces with the DNA repair mechanisms that would correct those particular defects.

In environmental carcinogenesis studies of DNA repair polymorphisms, the majority of the work has focused on base excision repair (BER) or nucleotide excision repair (NER) pathways, since these are thought to play dominant roles in the repair of damage from exogenous carcinogens, including chemical carcinogens. In both of these pathways, numerous polymorphisms in numerous proteins that make up the DNA repair machinery have been examined. However, much of the focus has been on the particular proteins in the respective pathways that contain the most common polymorphic variants, in particular the x-ray cross complementing-1 (XRCC1) protein in BER and the xeroderma pigmentosum-D (XPD) protein in NER [24-38].

This is also understandable because of the critical roles that each of these proteins play in their respective pathways. For example, in BER the particular type of damage produced by exposure to a chemical carcinogen is usually recognized and removed by a specific DNA glycoslase. The BER apparatus includes numerous other proteins that complete the repair at the resultant abasic site once the damage is removed: apurinic/apyrimidinic endonuclease (APE1), poly(ADP-ribose) polymerase-1 (PARP-1), poly(ADP-ribose) polymerase-2 (PARP-2), DNA polymerase β (Pol β) and DNA ligase IIIα (Lig III). AP endonuclease is responsible for cleaving the phosphodiester bond at the abasic site created by the glycosylase. PARP-1 and to a lesser extent PARP-2 participate in the repair process by catalyzing ribosylation of a number of DNA-bound proteins, thereby decreasing the affinity of these proteins for DNA, and allowing the repair machinery to access the damaged site. Pol β, the polymerase involved in short patch repair, provides two essential activities, deoxyribophosphodiesterase activity which releases the 5' sugar phosphate group, and gap filling synthesis, where one nucleotide is added to the 3' OH. Finally, Lig III seals the nick in an ATP-dependent manner [39, 40]. The XRCC1 protein is critical to this process since it acts as a scaffold protein in this pathway and appears to enhance the activity of the other BER proteins. Although XRCC1 has not been demonstrated to contain enzymatic activity of its own, it is thus necessary for coordinating and regulating the early and late stages of BER through its protein interaction modules [41, 42].

XRCC1 is known to contain three common polymorphic sites that might be expected to have an effect on XRCC1 structure and function because they occur in or near important protein domains [11]. For example, the polymorphism at amino acid residue 194, which results in the substitution of a tryptophan for the normal arginine, occurs in the XRCC1 N-terminal domain from amino acid residues 1-195 that has been observed to mediate its interaction with the palm-thumb domain of Pol β [43]. A second polymorphism at amino acid residue 280, which results in the substitution of a histidine for the normal arginine, occurs in the region between the N-terminal domain and the BRCA1 carboxy terminal (BRCT1) domain of the protein and close to the nuclear localization signal site and thus could affect the relationship between these two critical domains and/or the protein's localization ability [44]. The third and most common polymorphism in XRCC1 occurs at amino acid residue 399, resulting in the substitution of a glutamine for the normal arginine, within the highly conserved BRCT1 domain from amino acid residues 315-403, which has been associated with the functioning of PARP1, PARP2 and APE1 [45].

Like BER, NER occurs in a series of steps: damage recognition, unwinding and demarcation of the DNA, excision of the single-stranded fragment containing the damaged site, and DNA re-synthesis. NER is accomplished primarily through the action of proteins of the xeroderma pigmentosum family of genes which are categorized into 7 different groups (A-G). XPC and XPE proteins are involved in recognition of different types of DNA damage. XPB and XPD are DNA helicases that function as subunits of the transcription factor IIH complex (TFIIH) to promote DNA bubble formation at the damaged site by unwinding the DNA as XPA complexes with replication protein A (RPA) for demarcation. XPF and XPG are structure-specific endonucleases for excision of the damaged site. Finally, replicative DNA polymerase and DNA ligase I complete the repair [46, 47]. XPD is one of the major players in NER and is essential for life [48, 49].

XPD is also known to contain at least two common polymorphic sites, namely at amino acid residues 312 (aspartic acid->asparagine) and 751 (lysine->glutamine) [50]. The 751 site is assumed to be particularly important for XPD function since it occurs in the C-terminal domain of the protein which has been suggested to interact with the p44 helicase activator protein of the TFIIH complex [51]; also, it is been shown that an XPD mutation that results in the loss of the final 17 C-terminal amino acids, including residue 751, results in the clinical disease phenotype of trichothiodystrophy [52].

In summary, an ideal system for investigating the role of DNA repair polymorphisms in carcinogenesis might be an exposure to a known chemical carcinogen that produces specific types of DNA damage that are repaired by the BER and/or NER pathways where the effects of common polymorphisms in XRCC1 and XPD on the damage and repair could be studied.

2. A model for the study of the epidemiology of dna repair polymorphisms in carcinogenesis

Such a potential model system for the study of the role of DNA repair polymorphisms in chemical carcinogenesis is provided by the known carcinogen vinyl chloride (VC) because considerable detail is available concerning the molecular biology of its pathogenic pathway which allows for careful study of the role of DNA damage and repair in the carcinogenic process in exposed human populations through the application of molecular epidemiologic approaches (Figure 1).

As noted, VC is a well-established animal and human carcinogen. It is most strongly associated with liver cancer, in particular the rare, sentinel neoplasm of angiosarcoma of the liver (ASL), a malignant tumor of the endothelial cells of the liver [53]. However, VC has also been identified as a cause of hepatocellular carcinoma (HCC), the corresponding malignant tumor of the parenchymal cells of the liver [54]. In addition, it has been associated with other malignancies, e.g., lung and brain, although these associations remain much more controversial. The most significant exposures to VC occur in the petrochemical and plastics industries because VC is used in the manufacture of polyvinyl chloride, one of most high-volume plastics in the world. For example, it is estimated that worldwide more than 2,200,000 workers are probably occupationally exposed to VC. General population exposures also occur primarily through the air and water. For example, elevated levels of VC have been found not only in the air near VC manufacturing and processing facilities but also in the vicinity of many hazardous waste sites and municipal landfills, either due to the direct disposal of VC or from the microbial degradation of other chlorinated solvents to form VC. In some cases, dangerously high levels have been detected in the air at some of these landfills [53]. General population exposures may also occur from tobacco smoke, drinking water from PVC pipe, and consumption of food and beverages from PVC packaging and bottles, although probably at much lower levels.

VC is a gas so the most significant exposures are respiratory. Following inhalation, absorption is rapid in humans and most subsequent metabolism occurs in the liver [53]. Phase I metabolism is primarily via the cytochrome P-450 isoenzyme 2E1 (CYP2E1) to generate the reactive intermediates chloroethylene oxide (CEO) and chloracetaldehyde (CAA) which are further metabolized in phase II reactions by glutathione-S-transferases (GSTs) and aldehyde dehydrogenase 2 (ALDH2) to end products for ultimate excretion. However, CEO and CAA can readily interact with cellular macromolecules, including DNA, to produce promutagenic effects. VC biotransformation to CEO probably occurs principally in hepatocytes, but the epoxide can also reach and react with adjacent sinusoidal lining cells, so that mutagenic effects can occur in parenchymal liver cells and non-parenchymal endothelial cells, providing a logical rationale for the association between VC exposure and ASL as well as HCC [55]. The major VC-associated liver DNA adduct is 7-(2-oxoethyl)guanine, comprising up to 98% of all adducts formed. However, this adduct is eliminated from the DNA with a very short half-life, principally by chemical depurination, and is not considered to be promutagenic. On the other hand, three etheno DNA adducts are also formed in much less abundance, but they are known to be promutagenic. These are: N^2,3-ethenoguanine (εG) ; 1,N^6-ethenoadenine (εA) ; and 3,N^4-ethenocytosine (εC) [56].

The promutagenic properties of etheno-DNA adducts that are not fully repaired by one or another of the DNA repair pathways have been well documented in experimental systems *in vitro*, as well as *in vivo* in bacterial and mammalian cells. The εA adduct generates A->T, A->G and A->C base changes; the εG adduct generates G->A base changes; and the εC adduct generates C->A and C->T base changes [55]. These experimental results are consistent with the tumor mutational spectra identified in exposed animals and humans in oncogenes and tumor suppressor genes. Of particular interest have been the A->T transversions at codons 179, 249 and 255 of the *TP53* tumor suppressor gene generated by εA adducts and the G->A transitions at codon 13 of the K-*ras* oncogene generated by εG adducts, because of their frequent occurrence in human ASLs from VC-exposed individuals but not in sporadic ASLs in individuals without VC exposure. In addition, other results suggest that these VC-associated mutations, particularly the codon 13 K-*ras* mutation, may be a relatively early event in VC carcinogenesis, and thus the occurrence of these mutations may be useful biomarkers of cancer risk in exposed individuals, as discussed below.

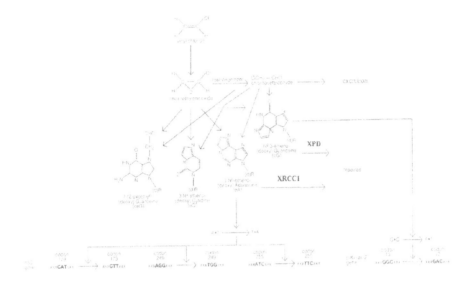

Figure 1. Proposed mechanism of VC-induced DNA damage and repair as a model system for the study of the effects of polymorphisms in BER and NER pathways.

The G->A transition at codon 13 of K-*ras* results in the substitution of an aspartic acid for the normal glycine at amino acid residue 13 in the encoded p21 protein product. This substitution is believed to be oncogenic, having been identified in other human tumors as well. The oncogenic mechanism of action of this substitution is thought to be through the production of a conformational change in p21 which may be responsible for altering its intrinsic GTPase activity, thus affecting signal transduction within the cell leading to uncontrolled growth and division [57]. Similarly, the A->T transversions at various codons of *p53* produce their corresponding amino acid substitutions in the encoded p53 protein product, all changes that have been shown to cause the protein to adopt its so-called "malignant" conformation with a concomitant loss of its normal tumor suppressor activity [57]. These protein changes provide a useful indicator of the pathogenic consequences of the occurrence of the corresponding mutations, as well as convenient intermediate biomarkers of VC effect to study the molecular epidemiology of VC carcinogenesis in exposed human populations, including the effects of polymorphisms in the relevant DNA repair pathways.

It has been shown that the mutant *ras*-p21 protein containing aspartic acid for glycine at amino acid residue 13 can be distinguished from the wild-type protein and other mutant *ras*-p21 proteins immunologically with a mouse monoclonal antibody specific for this protein. For cells in culture that contain the mutant *ras* gene, it is possible to use this monoclonal antibody to detect mutant *ras*-p21 expression in the cells by immunocyto-chemistry and in the extracellular supernatant by immunoblotting. In analogous situations *in vivo*, mutant Asp 13 *ras*-p21 can be detected in tumor tissue by immunohistochemistry and in the serum by immunoblotting of VC-exposed workers with ASLs known to contain the mutant *ras* gene but not in the serum of VC-exposed workers with ASLs that do not contain the mutation or in unexposed controls [57-59].

An analogous, although slightly more complicated situation occurs with p53. As noted, all of the VC-induced mutations in the *p53* gene have been shown to cause a similar conformational change in the encoded p53 protein that results in the exposure of a common epitope, which is normally not immunologically detectable in the wild-type protein. Thus, these mutant p53 proteins can be distinguished from wild-type p53 immunologically with a mouse monoclonal antibody that binds to this mutant-specific epitope. For cells in culture that contain the mutant *p53* genes, it is possible to use this monoclonal antibody to detect mutant p53 protein expression in the cells by immunocytochemistry and in the extracellular supernatant by immuno-blotting or by enzyme-linked immunosorbent assay (ELISA). In the analogous situation *in vivo*, mutant p53 can be detected in the tumor tissue by immunohistochemistry and in the serum by immunoblotting or ELISA of VC-exposed workers with ASLs known to contain the mutant p53 genes but not of VC-exposed workers with ASLs that do not contain the mutations or in unexposed controls. In some cases of mutant p53-positive tumors, it is known that individuals can also develop an antibody response to the mutant p53 which can obscure the detection of the mutant p53 protein itself. However, it is also possible to detect these auto-antibodies to mutant p53 using an ELISA. Thus, the detection in serum of mutant p53 protein and/or an antibody response to mutant p53 protein can be used together to best identify individuals who have a *p53* mutation in their tumors [57, 60, 61].

Based on the above evidence, it seems that these serum biomarkers for mutant *ras*-p21 and mutant p53 accurately reflect the occurrence of the corresponding DNA damage in the target tissue of VC-exposed workers. In addition, these biomarkers have been identified not only in VC-exposed workers with ASLs but also in VC-exposed workers with non-malignant (but potentially pre-malignant) angiomatous lesions and in VC-exposed workers without any apparent neoplastic disease [57, 62-64]. In a large cohort of French VC workers, the presence of these biomarkers was found to occur with a highly statistically significant dose-response relationship with regard to estimated, cumulative VC exposure, supporting the claim that the generation of the biomarkers was indeed the result of the exposure [65]. Similar results with these biomarkers have been noted in several other VC workers cohorts around the world [66-71]. To date in these various studies, at least five VC-exposed biomarker-positive workers without ASL have developed subsequent liver lesions presumed to be ASLs, also suggesting that these biomarkers may have predictive value for the subsequent occurrence of cancer.

However, at any given level of VC exposure, some workers will have none, one or both mutant biomarkers. One possible explanation for this inter-individual variability is genetic differences in the proteins that metabolize VC or repair the DNA damage it produces. Although polymorphisms in the proteins involved in metabolizing VC have been shown to have an effect, polymorphisms in DNA repair proteins have been found to be even more significant.

There are several potential mechanisms by which VC-induced adducts could be repaired before they have a chance to cause mutations. As noted above, the oxoethyl adduct is removed rapidly by chemical depurination. The potential repair of the etheno adducts, however, is more complicated and involves the BER and NER pathways.

For example, the $1,N^6$-εA adducts are recognized and removed by 3-methyl adenine DNA glycosylase which is part of the BER pathway [55]. Likewise, the $3,N^4$-ethenocytosine adducts are also repaired with high efficiency by BER via the thymine DNA glycosylase. Therefore, polymorphisms in the BER pathway that could decrease DNA repair efficiency, particularly the polymorphisms in XRCC1, might be expected to result in an increase in εA and εC adduct levels at any given level of exposure in VC-exposed individuals with a resultant increase in the VC-associated mutant biomarkers, particularly the mutant p53 biomarker. In contrast, the $N^2,3$-ethenoguanine adducts have been shown to be not very efficiently repaired by BER [56, 72]. Thus, if they are repaired, it is likely to be by a different DNA repair pathway such as NER. Therefore, polymorphisms in the NER pathway that could decrease DNA repair efficiency, particularly the polymorphisms in XPD, might be expected to result in an increase in εG adduct levels at any given level of exposure in VC-exposed individuals with a resultant increase in the VC-associated mutant biomarkers, particularly the mutant ras-p21 biomarker.

In fact in the aforementioned French VC worker cohort, we have been able to identify the effect of the XRCC1 polymorphisms on the occurrence of the mutant p53 biomarker, but not the mutant ras-p21 biomarker [73-75]. The difference in effect on the two biomarkers is expected, since, as noted the εA adducts that result in the mutant p53 biomarker are repaired efficiently by BER but the εG adducts that result in the mutant ras–p21 biomarker are not, so changes in XRCC1 might affect the former but should not affect the latter. Among the three XRCC1 polymorphisms, the most significant effect on the mutant p53 biomarker was attributable to the residue 399 polymorphism. In this case, individuals who were homozygous variant Gln-Gln at 399 had a statistically significant 1.9-fold risk of occurrence of the mutant p53 biomarker compared to homozygous Arg-Arg wild-type individuals, even after controlling for potential confounders including cumulative VC exposure, and the gene-environment interaction between the polymorphism and VC exposure appeared to be potentially supra-multiplicative [75]. Studies in other VC worker populations have found similar effects of the XRCC1 polymorphisms, particularly the 399 polymorphism, on the mutant p53 biomarker, as well as other biomarkers of DNA damage [76-79].

This is also consistent with various experimental results examining this model system. For example, molecular modeling of the BRCT1 domains of the normal and polymorphic forms of XRCC1 demonstrates that the 399 substitution produces significant conformational changes in this domain, including the loss of secondary structural features such as α-helices that can be critical for mediating protein-protein interactions that would allow XRCC1 to coordinate

BER [80]. Also, studies of lymphoblasts from individuals of different genotypes exposed *in vitro* to the reactive metabolites of VC showed that cells with the XRCC1 399 homozygous variant Gln-Gln genotype had an approximate 4-fold decrease in efficiency of repair of εA DNA adducts compared to cells with the homozygous wild-type Arg-Arg genotype [74, 81], resulting in an approximate 1.8-fold increase in mutation frequency in the polymorphic cells compared to the wild-type cells as determined by the hypoxanthine-guanine phosphoribo-syltransferase (HPRT) assay [82]. Based on mutational spectrum studies in CAA-exposed human cell lines [83], the resultant increase in * A DNA adducts would especially result in an increase in A->T transversions consistent with those found in the tumors of VC-exposed workers, as noted above.

Furthermore, in the French VC worker cohort, we have been able to identify the effect of the XPD polymorphisms on the occurrence of both mutant biomarkers, although the most marked and statistically significant effect was on the mutant *ras*-p21 biomarker, as expected [75]. In this case, individuals who were homozygous variant at either residue 312 or 751 had a statistically significant 2.6-3.0-fold increased risk of occurrence of the mutant *ras*–p21 bio-marker compared to homozygous wild-type individuals, even after controlling for potential confounders including cumulative VC exposure. Furthermore, in the case of the residue 751 polymorphism, the gene-environment interaction between the polymorphism and VC exposure, as well as the gene-gene interaction between the XPD and CYP2E1 polymorphisms (which could increase VC metabolism to its promutagenic reactive metabolites and thus also increase etheno-DNA adducts at any given level of VC exposure with a resultant increase in the mutant biomarkers) appeared to be potentially multiplicative [75]. Once again, studies in other VC worker populations have found similar effects of the XPD polymorphisms on other biomarkers of DNA damage [77].

This is also consistent with various experimental results in this model system. For example, molecular modeling of the normal and polymorphic forms of XPD demonstrates that these substitutions produce discrete local conformational changes in the protein which affect its overall structure and could affect its function [82, 84], and, in particular, are projected to interfere with its protein-protein interactions and binding to other components of the TFIIH complex (Figure 2; adapted from Gibbons et al. [85]). Also, studies of lymphoblasts from individuals of different genotypes exposed *in vitro* to the reactive metabolites of VC showed that cells with the XPD 751 homozygous variant Gln-Gln genotype had an approximate 5-fold decrease in efficiency of repair of εG DNA adducts compared to cells with the homozygous wild-type Lys-Lys genotype [82], resulting in an approximate 4.8-fold increase in mutation frequency in the polymorphic cells compared to the wild-type cells as determined by the HPRT assay, even though there is no difference in the level of expression of the XPD protein among cells that are homozygous wild-type, heterozygous or homozygous polymorphic at this codon (Figure 3). Once again, based on mutational spectrum studies in CAA-exposed human cell lines [83], the resultant increase in * G DNA adducts would especially result in an increase in G->A transitions consistent with those found in the tumors of VC-exposed workers.

A thorough understanding of the molecular biology and molecular epidemiology of VC carcinogenesis can provide the basis for new molecular approaches to the prevention of VC-induced cancers and potentially other cancers related to DNA-damaging agents. For example, one approach to secondary prevention could be based on "personalized prevention" derived from knowledge of the status of individual's DNA repair capability. Although little is currently known about methods for altering DNA repair activity, there is some evidence to suggest augmenting DNA repair may be possible. Several *in vitro* studies have shown that DNA repair processes can be increased by selenium-based compounds in response to radiation or chemically induced DNA damage [86]. More recently, a study in mice has suggested that selenocystine administration, although it did not protect against immediate DNA damage following ionizing radiation exposure, was nevertheless protective because it enhanced the rate of repair of the induced DNA damage [87]. In cohorts exposed to DNA damaging agents, determination of the dose of selenium compounds to provide an optimum effect on DNA repair could be based on the genetic status of the exposed individuals in terms of the presence of polymorphisms in key components of the repair apparatus, and the success of such interventions could be effectively monitored by following mutant biomarkers of DNA damage.

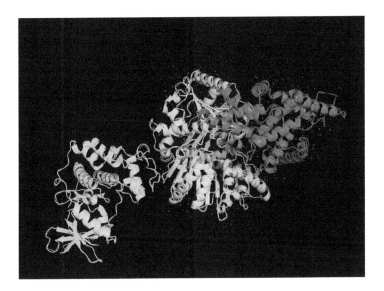

Figure 2. Protein backbone structures showing the proposed interaction effect of XPD (blue), cyclin H (pink) and cdk7 (green) in the TFIIH complex.

3. Conclusion

VC provides an instructive model for the study of the role of DNA repair polymorphisms in chemical carcinogenesis. A detailed understanding of the molecular biology of VC carcinogenesis has provided new ways of studying the molecular epidemiology of

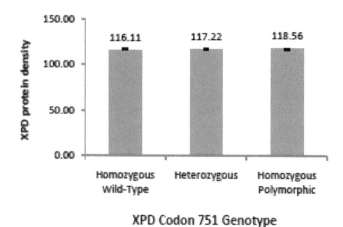

Figure 3. Levels of expression of the XPD protein in lymphocyte cell lines that are homozygous wild-type, heterozygous or homozygous polymorphic at codon 751.

VC carcinogenesis in exposed humans, which in turn may provide the basis for new approaches to the prevention and treatment of VC-related cancer. This model could also have much broader implications, since other potential carcinogenic exposures share some of the same molecular biologic pathways of damage and repair as VC similar molecular epidemiologic biomarkers could be useful for monitoring their carcinogenic process and the effect of altered susceptibility due to changes in DNA repair capability. Such studies in additional model systems would further help to define the exact significance of DNA repair polymorphisms in the development of human cancers.

Acknowledgements

This work was supported in part by grants to PWB-R from NIOSH, R01-OH04192 and R01-OH07590.

Author details

Paul W. Brandt-Rauf*, Yongliang Li, Changmin Long and Regina Monaco

*Address all correspondence to: pwb1@uic.edu

Division of Environmental and Occupational Health Sciences, School of Public Health, University of Illinois at Chicago, Chicago, USA

References

[1] Au WW (2006) Heritable Susceptibility Factors for the Development of Cancer. J. radiat. res. 47:B13-B17.

[2] Berwick M, Vineis P (2000) Markers of DNA Repair and Susceptibility to Cancer in Humans: An Epidemiologic Review. J. natl. cancer inst. 92:874-897.

[3] Goode EL, Ulrich CM, Potter JD (2002) Polymorphisms in DNA Repair Genes and Associations with Cancer Risk. Cancer epidemiol biomarkers prev. 11:1513-1530.

[4] Ricceri F, Matullo G, Vineis P (2012) Is There Evidence of Involvement of DNA Repair Polymorphisms in Human Cancer? Mutat. res. 736:117-121.

[5] Jiricny J, Nystrom-Lahti M (2000) Mismatch Repair Defects in Cancer. Curr. opin. genet. dev. 10:157-161.

[6] Hung RJ, Hall J, Brennan P, Boffetta P (2005) Genetic Polymorphisms in the Base Excision Repair Pathway and Cancer Risk: A HuGE Review. Am. j. epidemiol. 162:925-942.

[7] Weiss JM, Goode EL, Ladiges WC, Ulrich CM (2005) Polymorphic Variation in hOGG1 and Risk of Cancer: A review of the Functional and Epidemiologic Literature. Mol. carcinog. 42:127-141.

[8] Clarkson SG, Wood RD (2005) Polymorphisms in the Human XPD (ERCC2) Gene, DNA Repair Capacity and Cancer Susceptibility: An Appraisal. DNA repair 4:1068-1074.

[9] Hu Z, Ma H, Chen F, Wei Q, Shen H (2005) XRCC1 Polymorphisms and Cancer Risk: A Meta-Analysis of 38 Case-Control Studies. Cancer epidemiol. biomarkers prev. 14:1810-1818.

[10] Manuguerra M, Saletta F, Karagas MR, Berwick M, Vegila F, Vineis P, Matullo G (2006) XRCC3 and XPD/ERCC2 Single Nucleotide Polymorphisms and the Risk of Cancer: A HuGE Review. Am. j. epidemiol. 164:297-302.

[11] Tudek B (2007) Base Excision Repair Modulation as a Risk Factor for Human Cancers. Mol. aspects med. 28:258-275.

[12] Bugni JM, Han J, Tsai M, Hunter DJ, Samson LD (2007) Genetic Association and Functional Studies of Major Polymorphic Variants of MGMT. DNA repair 6:1116-1126.

[13] Qiu L, Wang Z, Shi X, Wang Z (2008) Associations between XPC Polymorphisms and Risk of Cancers: A Meta-Analysis. Eur. j. cancer 44:2241-2253.

[14] Wang F, Chang D, Hu F, Sui H, Han B, Li D, Zhao Y (2008) DNA Repair Gene XPD Polymorphisms and Cancer Risk: A Meta-Analysis Based on 56 Case-Control Studies. Cancer epidemiol. biomarkers prev. 17:507-517.

[15] Jiang J, Zhang X, Yang H, Wang W (2009) Polymorphisms of DNA Repair Genes: ADPRT, XRCC1, and XPD and Cancer Risk in Genetic Epidemiology. In: Verma M,

editor. Methods in Molecular Biology, Cancer Epidemiology, vol. 471. Totowa: Humana Press. pp. 305-333.

[16] Wilson DM, Kim D, Berquist BR, Sigurdson AJ (2011) Variation in Base Excision Repair Capacity. Mutat. res. 711:100-112.

[17] Ding D, Zhang Y, Yu H, Guo Y, Jiang L, He X, Ma W, Zheng W (2012) Genetic Variation of XPA Gene and Risk of Cancer: A Systematic Review and Pooled Analysis. Int. j. cancer 131:488-496.

[18] Gossage L, Perry C, Abbotts R, Madhusudan S (2012) Base Excision Repair Factors are Promising Prognostic and Predictive Markers in Cancer. Curr. mol. pharmacol. 5:115-124.

[19] Hulla JE, Miller MS, Taylor JA, Hein DW, Furlong CE, Omiecinski CJ, Kunkel TA (1999) Symposium Overview: The Role of Genetic Polymorphism and Repair Deficiencies in Environmental Disease. Toxicol. sci. 47:135-143.

[20] de Boer JG (2002) Polymorphisms in DNA Repair and Environmental Interactions. Mutat. res. 509:201-210.

[21] Belitsky GA, Yakubovskaya MG (2008) Genetic Polymorphism and Variability of Chemical Carcinogenesis. Biochemistry 73:543-554.

[22] Klaunig JE, Wang Z, Pu X, Zhou S (2011) Oxidative Stress and Oxidative Damage in Chemical Carcinogenesis. Toxicol. appl. pharmacol. 254:86-99.

[23] Simonelli V, Mazzei F, D'Errico M, Dogliotti E (2012) Gene Susceptibility to Oxidative Damage: From Single Nucleotide Polymorphisms to Function. Mutat. res. 736:104-116.

[24] Stern MC, Johnson LR, Bell DA, Taylor JA (2002) XPD Codon 751 Polymorphism, Metabolism Genes, Smoking, and Bladder Cancer Risk. Cancer epidemiol. biomarkers prev. 11:1004-1011.

[25] Shen M, Hung RJ, Brennan P, Malaveille C, Donato F, Placidi D, Carta A, Hautefeuille A, Boffetta P, Porru S (2003) Polymorphisms of the DNA Repair Genes XRCC1, XRCC3, XPD, Interaction with Environmental Exposures, and Bladder Cancer Risk in a Case-Control Study in Northern Italy. Cancer epidemiol. biomarkers prev. 12:1234-1240.

[26] Gao WM, Romkes M, Day RD, Siegfried JM, Luketich JD, Mady HH, Melhem MF, Keohavong P (2003) Association of the DNA Repair Gene XPD Asp312Asn Polymorphism with p53 Gene Mutations in Tobacco-Related Non-Small Cell Lung Cancer. Carcinogenesis 24:1671-1676.

[27] Terry MB, Gammon MD, Zhang FF, Eng SM, Sagiv SK, Paykin AB, Wang Q, Hayes S, Teitelbaum SL, Neugut AI, Santella RM (2004) Polymorphism in the DNA Repair Gene XPD, Polycyclic Aromatic Hydrocarbon-DNA Adducts, Cigarette Smoking, and Breast Cancer Risk. Cancer epidemiol. biomarkers prev. 13:2053-2058.

[28] Han J, Hankinson SE, Colditz GA, Hunter DJ (2004) Genetic Variation in XRCC1, Sun Exposure, and Risk of Skin Cancer. Br. j. cancer 91:1604-1609.

[29] Neumann AS, Sturgis EM, Wei Q (2005) Nucleotide Excision Repair as a Marker for Susceptibility to Tobacco-Related Cancers: A Review of Molecular Epidemiological Studies. Mol. carcinog. 42:65-92.

[30] Han J, Colditz GA, Liu JS, Hunter DJ (2005) Genetic Variation in XPD, Sun Exposure, and Risk of Skin Cancer. Cancer epidemiol. biomarkers prev. 14:1539-1544.

[31] Leng S, Cheng J, Zhang L, Niu Y, Dai Y, Pan Z, Li B, He F, Zheng Y (2005) The Association of XRCC1 Haplotypes and Chromosomal Damage Levels in Peripheral Blood Lymphocytes Among Coke-Oven Workers. Cancer epidemiol. biomarkers prev. 14: 1295-1301.

[32] Zhao XH, Jia G, Liu YQ, Liu SW, Yan L, Jin Y, Liu N (2006) Association Between Polymorphisms of DNA Repair Gene XRCC1 and DNA Damage in Asbestos-Exposed Workers. Biomed. environ. sci. 19:232-238.

[33] Long XD, Ma Y, Wei YP, Deng ZL (2006) The Polymorphisms of GSTM1, GSTT1, HYL1*2, and XRCC1, and Aflatoxin B1-Related Hepatocellular Carcinoma in Guangxi Population, China. Hepatol. res. 36:48-55.

[34] Neri M, Ugolini D, Dianzani I, Gemignani F, Landi S, Cesario A, Magnani C, Mutti L, Puntoni R, Bonassi S (2008) Genetic Susceptibility to Malignant Pleural Mesothelioma and Other Asbestos-Associated Diseases. Mutat. res. 659:126-136.

[35] Long XD, Ma Y, Zhou YF, Yao JG, Ban FZ, Huang YZ, Huang BC (2009) XPD Codon 312 and 751 Polymorphisms, and AFB1 Exposure, and Hepatocellular Carcinoma Risk. BMC cancer 9:400.

[36] Sterpone S, Cozzi R (2010) Influence of XRCC1 Genetic Polymorphisms on Ionizing Radiation-Induced DNA Damage and Repair. J. nucleic acids

[37] Toumpanakis D, Theocharis SE (2011) DNA Repair Systems in Malignant Mesothelioma. Cancer lett. 312:143-149.

[38] Zhang XH, Zhang X, Zhang L, Chen Q, Yang Z, Yu J, Fu H, Zhu YM (2012) XRCC1 Arg399Gln Was Associated with Repair Capacity for DNA Damage Induced by Occupational Chromium Exposure. BMC res. notes 5:263.

[39] Fortini P, Pascucci B, Parlanti E, D'Errico M, Simonelli V, Dogliotti E (2003) The Base Excision Repair: Mechanisms and Its Relevance for Cancer Susceptibility. Biochemie 85:1053-1071.

[40] Baute J, Depicker A (2008) Base Excision Repair and Its Role in Maintaining Genome Stability. Crit. rev. biochem mol. biol. 43:239-276.

[41] Thompson LH, West MG (2000) XRCC1 Keeps DNA from Getting Stranded. Mutat. res. 459:1-18.

[42] Caldecott KW (2003) XRCC1 and DNA Strand Break Repair. DNA repair 2:955-969.

[43] Marintchev A, Mullen MA, Maciejewski MW, Pan B, Gryk MR, Mullen GP (1999) Solution Structure of the Single-Strand Break Repair Protein XRCC1 N-Terminal Domain. Nat. struct. biol. 6: 884-893.

[44] Marintchev A, Robertson A, Dimitriadis EK, Prasad R, Wilson SH, Mullen GP (2000) Domain Specific Interaction in the XRCC1 Polymerase Beta Complex. Nucleic acids res. 28:2049-2059.

[45] Vidal AE, Boiteux S, Hickson ID, Radicella JP (2001) XRCC1 Coordinates the Initial and Late Stages of DNA Abasic Site Repair through Protein-Protein Interactions. EMBO j. 20:6530-6539.

[46] Nouspikel T (2009) DNA Repair in Mammalian Cells: Nucleotide Excision Repair: Variations in Versatility. Cell. mol. life sci. 66:994-1009.

[47] Bergoglio V, Magnaldo T (2006) Nucleotide Excision Repair and Related Human Diseases. Genome dyn. 1:35-52.

[48] Oksenych V, Coin F (2010) The Long Unwinding Road: XPB and XPD Helicases in Damaged DNA Opening. Cell cycle 9:90-96.

[49] Lehmann AR (2008) XPD Structure Reveals Its Secrets. DNA repair 7:1912-1915.

[50] Benhamou S, Sarasin A (2002) ERCC2/XPD Gene Polymorphisms and Cancer Risk. Mutagenesis 17:463-469.

[51] Bienstock RJ, Skovaga M. Mandavilli BS, Van Houten B (2003) Structural and Functional Characterization of the Human DNA Repair Helicase XPD by Comparative Molecular Modeling and Site-Directed Mutagenesis of the Bacterial Repair Protein UvrB. J. biol. chem.. 278:5309-5316.

[52] Botta E, Nardo T, Broughton BC, Marinoni S, Lehmann AR, Stefanini M (1998) Analysis of Mutations in the XPD Gene in Italian Patients with Trichothiodystrophy: Site of Mutation Correlates with Repair Deficiency, but Gene Dosage Appears to Determine Clinical Severity. J. hum. genet. 63:1036-1048.

[53] Agency for Toxic Substances and Disease Registry (2006) Toxicological Profile for Vinyl Chloride. Atlanta: US Department of Health and Human Services.

[54] Grosse Y, Baan R, Straif K, Secretan B, El Ghissassi F, Bouvard V, Altieri A, Cogliano V (2007) Carcinogenicity of 1,3-Butadiene, Ethylene Oxide, Vinyl Chloride, Vinyl Fluoride, and Vinyl Bromide. Lancet oncol. 8:679-680.

[55] Dogliotti E (2006) Molecular Mechanisms of Carcinogenesis by Vinyl Chloride. Ann. 1st super sanita 42:163-169.

[56] Cohen SM, Storer RD, Criswell KA, Doerrer NG, Dellarco VL, Pegg DG, Wojcinski ZW, Malarkey DE, Jacobs AC, Klaunig JE, Swenberg JA, Cook JC (2009) Hemangiosarcoma in Rodents: Mode-of-Action Evaluation and Human Relevance. Toxicol. sci. 111:4-18.

[57] Li Y, Asherova M, Marion MJ, Brandt-Rauf PW (1998) Mutant Oncoprotein Biomarkers in Chemical Carcinogenesis. In: Mendelsohn ML, Mohr LC, Peeters JP, editors.

Biomarkers – Medical and Workplace Applications. Washington: Joseph Henry Press. pp. 345-353.

[58] DeVivo I, Marion MJ, Smith SJ, Carney WP, Brandt-Rauf PW (1994) Mutant c-Ki-ras p21 Protein in Chemical Carcinogenesis in Humans Exposed to Vinyl Chloride. Cancer causes control 5:273-278.

[59] Li Y, Marion MJ, Asherova M, Coulibaly D, Smith SJ, Do T, Carney WP, Brandt-Rauf PW (1998) Mutant p21ras in Vinyl Chloride Exposed Workers. Biomarkers 3:433-439.

[60] Brandt-Rauf PW, Chen JM, Marion MJ, Smith SJ, Luo JC, Carney W, Pincus MR (1996) Conformational Effects in the p53 Protein of Mutations Induced during Chemical Carcinogenesis: Molecular Dynamic and Immunologic Analyses. J. protein chem. 15:367-375.

[61] Smith SJ, Li Y, Whitney R, Marion MJ, Partilo S, Carney WP, Brandt-Rauf PW (1998) Molecular Epidemiology of p53 Protein Mutations in Workers Exposed to Vinyl Chloride. Am. j. epidemiol. 147:302-308.

[62] Brandt-Rauf PW, Luo JC, Cheng TJ, Du CL, Wang JD, Marion MJ (2000) Mutant Oncoprotein Biomarkers of Vinyl Chloride Exposure: Application to Risk Assessment. In: Anderson D, Karakaya AE, Sram RJ, editors. Human Monitoring after Environmental and Occupational Exposure to Chemical and Physical Agents. Amsterdam: IOS Press. pp. 243-248.

[63] Brandt-Rauf PW, Luo JC, Cheng TJ, Du CL, Wang JD, Rosal R, Do T, Marion MJ (2002) Molecular Biomarkers and Epidemiologic Risk Assessment. Hum. ecol. risk assess. 8:1295-1301.

[64] Marion MJ, DeVivo I, Smith S, Luo JC, Brandt-Rauf PW (1996) The Molecular Epidemiology of Occupational Carcinogenesis in Vinyl Chloride Exposed Workers. Int. arch. occup. environ. health 68:394-398.

[65] Schindler J, Li Y, Marion MJ, Paroly A, Brandt-Rauf PW (2007) The Effect of Genetic Polymorphisms in the Vinyl Chloride Metabolic Pathway on Mutagenic Risk. J. hum. genet. 52:448-455.

[66] Trivers GE, Cawley HL, DeBenedetti VM, Hollstein M, Marion MJ, Bennett ML, Hoover ML, Prives CC, Tamburro CC, Harris CC (1995) Anti-p53 Antibodies in Sera of Workers Occupationally Exposed to Vinyl Chloride. J. natl. cancer inst. 87:1400-1407.

[67] Luo JC, Liu HT, Cheng TJ, Du CL, Wang JD (1998) Plasma Asp13-Ki-ras Oncoprotein Expression in Vinyl Chloride Monomer Workers in Taiwan. J. occup. environ. med. 40:1053-1058.

[68] Luo JC, Liu HT, Cheng TJ, Du CL, Wang JD (1999) Plasma p53 Protein and Anti-p53 Antibody Expression in Vinyl Chloride Monomer Workers in Taiwan. J. occup. environ. med. 41:521-526.

[69] Luo JC, Cheng TJ, Du CL, Wang JD (2003) Molecular Epidemiology of Plasma Onco-proteins in Vinyl Chloride Monomer Workers in Taiwan. Cancer detect. prev. 27:94-101.

[70] Mocci F, De Biasio AL, Nettuno M (2003) Anti-p53 Antibodies as Markers of Carcinogenesis in Exposures to Vinyl Chloride. G. ital. med. lav. ergon. 25:S21-S23.

[71] Mocci F, Nettuno M (2006) Plasma Mutant-p53 Protein and Anti-p53 Antibody as a Marker: An Experience in Vinyl Chloride Workers in Italy. J. occup. environ. med. 48:158-164.

[72] Dosanjh MK, Chenna A, Kim E, Fraenkel-Conrat H, Samson L, Singer B (1994) All Four Known Cyclic Adducts Formed in DNA by the Vinyl Chloride Metabolite Chloroacetaldehyde are Released by a Human DNA Glycosylase. Proc. natl. acad. sci. usa 91:1024-1028.

[73] Li Y, Marion MJ, Rundle A, Brandt-Rauf PW (2003) A common Polymorphism in XRCC1 as a Biomarker of Susceptibility for Chemically Induced Genetic Damage. Biomarkers 8:408-414.

[74] Li Y, Marion MJ, Zipprich J, Freyer G, Santella RM, Kanki C, Brandt-Rauf PW (2006) The Role of XRCC1 Polymorphisms in Base Excision Repair of Etheno-DNA Adducts in French Vinyl Chloride Workers. Int. j. occup. med. environ. health 19:45-52.

[75] Li Y, Marion MJ, Zipprich J, Santella RM, Freyer G, Brandt-Rauf PW (2009) Gene-Environment Interactions between DNA Repair Polymorphisms and Exposure to the Carcinogen Vinyl Chloride. Biomarkers 14:148-155.

[76] Wong RH, Du CL, Wang JD, Chan CC, Luo JC, Cheng TJ (2002) XRCC1 and CYP2E1 Polymorphisms as Susceptibility Factors of Plasma Mutant p53 Protein and Anti-p53 Antibody Expression in Vinyl Chloride Monomer-Exposed Polyvinyl Chloride Workers. Cancer epidemiol. biomarkers prev. 11:475-482.

[77] Zhu SM, Xia ZL, Wang AH, Ren XF, Jiao J, Zhao NQ, Qian J, Jin L, Christiani DC (2008) Polymorphisms and Haplotypes of DNA Repair and Xenobiotic Metabolism Genes and Risk of DNA Damage in Chinese Vinyl Chloride Monomer (VCM)-Exposed Workers. Toxicol. let. 178:88-94.

[78] Ji F, Wang W, Xia ZL, Zheng YJ, Qiu YL, Wu F, Miao WB, Jin RF, Qian J, Jin L, Zhu YL, Christiani DC (2010) Prevalence and Persistence of Chromosomal Damage and Susceptible Genotypes of Metabolic and DNA Repair Genes in Chinese Vinyl Chloride-Exposed Workers. Carcinogenesis 31:648-653.

[79] Wang Q, Ji F, Sun Y, Qiu YL, Wang W, Wu F, Miao WB, Li Y, Brandt-Rauf PW, Xia ZL (2010) Genetic Polymorphisms of XRCC1, HOGG1 and MGMT and Micronucleus Occurrence in Chinese Vinyl Chloride-Exposed Workers. Carcinogenesis 31:1068-1073.

[80] Monaco R, Rosal R, Dolan MA, Pincus MR, Brandt-Rauf PW (2007) Conformational Effects of a Common Codon 399 Polymorphism on the BRCT1 Domain of the X-Ray Cross-Complementing-1 Protein. Protein j. 26:541-546.

[81] Li Y, Long C, Lin G, Marion MJ, Freyer G. Santella RM, Brandt-Rauf PW (2009) Effects of the XRCC1 Codon 399 Polymorphism on the Repair of Vinyl Chloride Metabolite-Induced DNA Damage. J. carcinogen. 8:108-112.

[82] Brandt-Rauf PW, Li Y, Monaco R, Kovvali G, Marion MJ (2012) Plastics and Carcinogenesis: The Example of Vinyl Chloride. J. carcinogen. 11:50-60.

[83] Matsuda T, Yagi T, Kawanishi M, Matsui S, Takebe H (1995) Molecular Analysis of Mutations Induced by 2-Chloroacetaldehyde, the Ultimate Carcinogenic Form of Vinyl Chloride, in Human Cells Using Shuttle Vectors. Carcinogenesis 16:2389-2394.

[84] Monaco R, Rosal R, Dolan MA, Pincus MR, Freyer G, Brandt-Rauf PW (2009) Conformational Effects of a Common Codon 751 Polymorphism on the C-Terminal Domain of the Xeroderma Pigmentosum D Protein. J carcinogen. 8:93-97.

[85] Gibbons BJ, Brignole EJ, Azubel M, Murakami K, Voss NR, Bushnell DA, Asturias FJ, Kornberg RD (2012) Subunit Architecture of General Transcription Factor TFIIH. Proc. natl. acad. sci. usa 109:1949-1954.

[86] Valdiglesias V, Pasaro E, Mendez J, Laffon B (2010) In Vitro Evaluation of Selenium Genotoxic, Cytotoxic, and Protective Effects: A Review. Arch. toxicol. 84:337-351.

[87] Kunwar A, Jayakumar S, Bhilwade HN, Bag PP, Bhatt H, Chaubey RC, Priyadarsini KI (2011) Protective Effects of Selenocystine Against γ-Radiation-Induced Genotoxicity in Swiss Albino Mice. Radiat. environ. biophys. 50:271-280.

Chapter 9

Radiosensitization Strategies Through Modification of DNA Double-Strand Break Repair

Yoshihisa Matsumoto, Shoji Imamichi,
Mikoto Fukuchi, Sicheng Liu, Wanotayan Rujira,
Shingo Kuniyoshi, Kazuki Yoshida,
Yasuhiro Mae and Mukesh Kumar Sharma

Additional information is available at the end of the chapter

1. Introduction

DNA double-strand break (DSB) is considered most critical type of DNA damage. In eukar-yote, DSB is repaired mainly through non-homologous end-joining (NHEJ) and homologous recombination (HR). Our understanding on the molecular mechanisms of these DNA repair mechanisms has been greatly deepened in the last two decades.

In NHEJ, DSB is first recognized by Ku protein (Fig.1 (1)), heterodimer consisting of Ku70 and Ku86 (also known as Ku80), which in turn recruits DNA-PK catalytic subunit (DNA-PKcs) (Fig.1 (2)). The comprex consisting of Ku70, Ku86 and DNA-PKcs is termed DNA-dependent protein kinase (DNA-PK). When the DSB are not readily ligatable, processing takes place prior to ligation (Fig.1 (3)). Processing might involve a number of enzymes depending on the shape of each DNA end and compatibility of two ends to be ligated: Artemis nuclease, DNA polymerase μ/λ, polynucleotide kinase/phosphatase (PNKP), Aprataxin (APTX) and Apra-taxin and PNKP-like factor (APLF, also known as PALF, C2orf13 or Xip1). DSBs are finally joined by DNA ligase IV, which is in tight association with XRCC4 (Fig.1 (4)). XRCC4-like factor (XLF, also known as Cernunnos), is essential at this step, especially when two ends are not compatible.

In HR, a complex consisting of Mre11, Rad50 and Nbs1, termed MRN complex, is thought to play two important roles in the initial stage (Fig.1 (1')): recruitment of ATM (Fig.1 (2'')) and resection of one of the strands (Fig.1 (2')). ATM is a protein kinase structurally similar to DNA-PKcs. Although ATM is thought to phosphorylate a great number of proteins as revealed by phosphoproteomic analyses, the phosphorylation of histone H2AX at Ser139 is thought one of the most important events, triggering signal transduction cascade involving mediator protein like MDC1 and ubiquitin ligases like RNF8 and RNF168. As Mre11 bears 5'-3' exonuclease activity, MRN resects one of the DNA strands to generate single-stranded DNA (ssDNA), which serves as a probe for the search for homology. Replication protein A (RPA) binds to ssDNA (Fig.1 (2')) and facilitate the formation of Rad51 filament in cooperation with BRCA2, PALB2, Rad52 and Rad51 paralogues (Fig.1 (3')). RPA also recruites ATRIP, which in turn recruits ATR, another protein kinase structurally related to DNA-PK and ATM (Fig.1 (3'')). ATR phosphorylate checkpoint kinase Chk1 to initiate signal transduction pathway leading to cell cycle checkpoints. Rad51 promotes strand exchange between homologous sequences (Fig.1 (4')). Template-dependent strand synthesis is proceeded by replication machinery including PCNA and DNA polymerase δ and ε (Fig.1 (5')). Finally, the junction of two DNA molecules (Holliday's junction) are resolved by nucleases Mus81-Eme1, ERCC1-XPF or SLX1-SLX4 (Fig.1 (6')). Alternatively, synthesized strand aneals with opposite end of DSB, detaches from the temprate strand, followed by synthesis and ligation of complementary strand (synthesis-dependent strand anealing; SDSA, not shown here).

Here, we will overview approaches to radiosensitization through the modification of DSB repair enzymes.

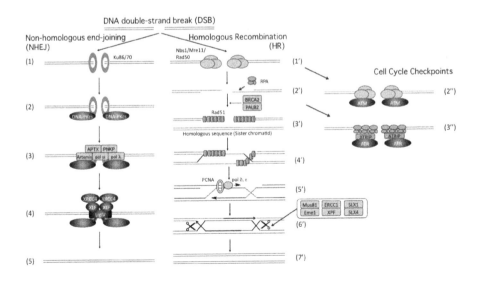

Figure 1. DNA double-strand break repair mechanisms.

2. DNA-PK, ATM and ATR kinases as targets for radiosensitizer

2.1. DNA-PK

DNA-PK was initially found in the extracts of HeLa cell, rabbit reticulocyte, Xenopus egg and sea urchin egg (Walker et al., 1985) and was purified from Hela cell nuclei as a 300-350 kDa protein, which is now called DNA-PKcs (Carter et al., 1990; Lees-Miller et al., 1990). Later it was found that Ku is an essential component of DNA-PK (Dvir et al., 1992, 1993; Gottlieb and Jackson, 1993). Furthermore, it was also shown that DNA-PK requires binding of DNA-PKcs to DNA ends via Ku to be activated, suggesting its possible role in sensoring DSBs (Gottlieb and Jackson, 1993). Ku86 was shown to be equivalent to XRCC5 (X-ray repair cross comple-menting) gene product, which is missing in X-ray sensitive rodent cell lines including xrs-5, -6, XR-V9B and XR-V15B (Taccioli et al., 1994; Smider et al., 1994). Subsequently, DNA-PKcs was found to correspond to XRCC7, which is deficient in $scid$ mouse as well as several radiosensitive cultured human and rodent cell lines (Kirchgessner et al., 1995; Blunt et al., 1995; Peterson et al., 1995; Lees-Miller et al., 1995). DNA-PK is abundant in human cells and its activity can be measured using synthetic peptides derived from p53 (Lees-Miller et al., 1992), enabling extensive studies on its biochemical properties even before molecular cloning of DNA-PKcs.

First reported selective inhibitor of DNA-PK is OK-1035, 3-cyano-5-(4-pyridyl)-6-hydrazono-methyl- 2-pyridone (Fig.2 A), which was found by screening of more than 10,000 microbial extracts and synthetic compounds (Take et al., 1995). IC_{50} (50% inhibitory concentration) on DNA-PK was 8 μM, which was more than 50-fold lower than that on other seven kinases examined, although it was reported to be much higher, $i.e.$, 100 μM, in others' study (Stockley et al., 2001). OK-1035 was shown to suppress adriamycin-induced p21 expression in cultured human carcinoma cell at concentrations 500 - 2000 μM (Take et al., 1996) and also to retard the repair of DSB measured by neutral single cell gel electrophoresis (comet) assay (Kruszewski et al., 1998).

Sequence of DNA-PKcs revealed its similarity to phosphatidylinositol 3-kinase (PI3K) (Hartley et al., 1995). This study also that fungal metabolite wortmannin (Fig.2 B), which had been known as an inhibitor of PI3K, could inhibit DNA-PK (Hartley et al., 1995). IC_{50} of wortmannin is reported to be 0.016 μM and 0.12 μM (Sarkaria et al., 1998; Izzard et al., 1999). It was also shown that wortmannin binds covalently to DNA-PKcs and functions as non-competitive, irreversible inhibitor of DNA-PK (Sarkaria et al., 1998; Izzard et al., 1999). Expectedly, a number of studies have demonstrated radiosensitizing effects of wortmannin but there is a concern whether the observed radiosensitization was really due to inhibition of DNA-PK. In this regard, some studies showed that radiosensitization by these compounds could be observed even in DNA-PKcs-deficient cells (Rosenzweig, et al., 1997; Hosoi et al., 1998), indicating that radiosensitization by these compounds was not solely due to inhibition of DNA-PK. In addition to ATM discussed next, PI3K-Akt pathway, which might be even more sensitive to wortmannin, might be important to sustain cell survival after irradiation. On the other hand, there are also studies showing that radiosensitization was not observed in DNA-PKcs deficient cells (Chernikova, et al., 1999; Hashimoto, et al., 2003). These studies argue that, even if wortmannin affect PI3K or other kinase more potently than DNA-PK, the radiosensitizing effect might be mainly due to inhibition of DNA-PK.

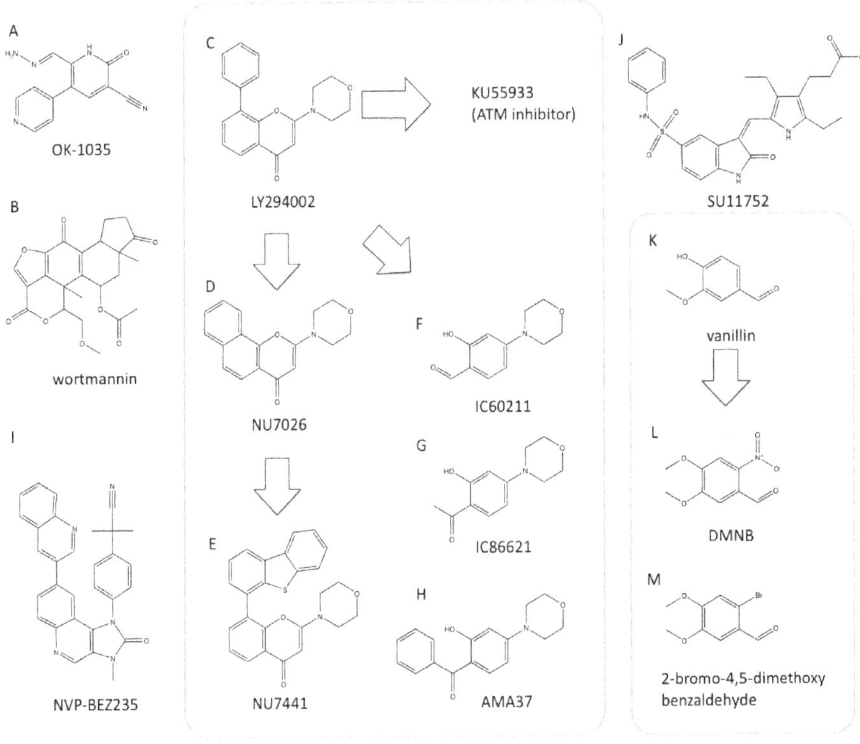

Figure 2. Strucuture of DNA-PK inhibitors.

Another PI3K inhibitor LY294002, 2-(4-morpholinyl)-8-phenyl-4H-1-benzopyran-4-one (Fig.2 C) was also shown to inhibit DNA-PK. In contrast to wortmannin, LY294002 competes with ATP (Izzard et al., 1999). IC_{50} of LY294002 is reported to be 6 µM (Izzard et al., 1999). LY294002 was used as a leading compound to explore more potent and selective inhibitors of DNA-PK. NU7026, 2-(morpholin-4-yl)-benzo[h]chromen4-one (Fig.2 D), was found as selective inhibitor of DNA-PK (Veuger et al., 2003). IC_{50} of NU7026 was 0.23 µM for DNA-PK, 13 µM for PI3K and >100 µM for ATM and ATR (Veuger et al., 2003). NU7026 sensitized cultured cells toward radiation in a manner dependent on DNA-PK (Veuger et al., 2003). Synthesis and screening of chromen-4-one library resulted in identification of NU7441 (Fig.2 E), 8-dibenzothiophen-4-yl-2-morpholin-4-yl-chromen-4-one (Leahy et al., 2004; Hardcastle et al., 2005). IC_{50} of NU7441 was 0.014 µM for DNA-PK, 5.0 µM for PI3K and >100 µM for ATM and ATR (Leahy et al., 2004). NU7441 sensitized cultured cells toward radiation and etoposide in a manner dependent on DNA-PK at 0.5 µM (Zhao et al., 2006). Screening of the derivatives of LY294002 also lead to the identification of other selective inhibitors of DNA-PK; IC60211 (Fig.2 F, 2-Hydroxy-4-morpholin-4-yl-benzaldehyde, IC_{50}: 0.43 µM), IC86621 (Fig.2 G, 1-(2-Hydroxy-4-morpholin-4-yl-phenyl) -ethanone, IC_{50}: 0.12 - 0.17 µM), AMA37 (Fig.2 H, 1-(2-Hydroxy- 4-morpholin-4-yl-phenyl)-phenyl-methanone, IC_{50}: 0.27 µM) (Kashishian et al., 2003; Knight et al., 2004).

It was recently reported that NVP-BEZ235 (Fig.2 I, 2-methyl-2-(4-(3-methyl-2-oxo-8-(quino-lin-3-yl)-2,3-dihydro-1H-imidazo[4,5-c]quinolin-1-yl)phenyl)propanenitrile), which had been initially identified as a dual inhibitor for PI3K and mammalian target of rapamycin (mTOR) (Maira et al., 2008), inhibited DNA-PK, ATM and ATR and sensitizes cells to ionizing radiation (Toledo et al., 2011; Mukherjee et al., 2012). NVP-BEZ235 sensitized the cultured cells to radiation and inhibited DSB repair, as shown by persistence of 53BP1 foci, to a greater extent than NU7026 and KU55933 (Mukherjee et al., 2012). NVP-BEZ235 sensitized ATM-deficient cells, i.e., fibroblast from ataxia telangiectasia patient, and also DNA-PKcs-deficient human glioma cell M059J (Mukherjee et al., 2012), which could be due to dual inhibition of DNA-PK and ATM. Moreover, inhibition of *in cellulo* phosphorylation mediated by DNA-PK and ATM was achieved at low concentration, i.e., 0.1 - 0.5 µM, while the similar extent of inhibition was achieved at 10 µM (Mukherjee et al., 2012).

Screening of a three-substituted indoline-2-one library lead to identification of SU11752 (Fig. 2 J) as selective DNA-PK inhibitor (IC_{50}: 0.13 µM) (Ismail, et al., 2004). Vanillin, 4-hydroxy-3-methoxybenzoaldehyde (Fig.2 K), was found to inhibit DNA-PK albeit at a relatively high concentration, i.e., IC_{50} 1500 µM (Durant and Karran, 2003). Screening of library of vanillin derivatives lead to finding of more potent inhibitors, 4,5-dimethoxy-2-nitrobenzaldehyde (DMNB, Fig.2 L) and 2-bromo-4,5- dimethoxybenzaldehyde (Fig.2 M), whose IC_{50} were 15 µM and 30 µM, respectively (Durant and Karran, 2003).

DNA-PK can be inhibited by homopolymeric phosphorythioate oligonucleotides, suramin and heparin (Hosoi et al., 2002). Inhibitory activities of homopolymeric phosphorothioate oligo-nucleotides on DNA-PK were independent of base composition but were dependent on length. IC_{50} decreased as length increased: 0.975 µM for 12 mer and 0.013 µM for 36 mer (Hosoi et al., 2002). IC_{50} of suramin and heparin were 1.7 µM and 0.27 µg ml^{-1}, respectively (Hosoi et al., 2002). Suramin sensitized cultured human cancer cell toward ionizing radiation but not to ultraviolet radiation (Hosoi et al., 2004). Furthermore, suramin did not affect the radiation sensitivity of *scid* cells, which are deficient in DNA-PK, indicating that radiosensitizating effects of suramin were mediated through inhibition of DNA-PK (Hosoi et al., 2004).

Single chain antibody variable fragment (scFv) is another approach to achieve specific inhibition of DNA-PK. ScFv was initially generated from existing murine monoclonal antibody 18-2, expressed in *E. coli* and introduced into the cell by microinjection (Li et al., 2003). The epitope of scFv 18-2 was mapped within 2001-2025 region, which is outside of kinase domain and thus ScFv 18-2 inhibited DNA-PK activity only modestly (Li et al., 2003). Nevertheless, microinjection of scFv 18-2 resulted in the inhibition of NHEJ, indicated by persistence of γ-H2AX foci and sensitized cells toward ionizing radiation (Li et al., 2003). However, the use of scFv as clinical radiosensitization might be difficult without a method to deliver it efficiently into the cell nucleus. ScFv 18-2 conjugated with nuclear localization signal was developed

(Xiong et al., 2009). In more recent study, scFv was conjugated with folate and introduced into the cell nucleus via folate receptor-mediated endocytosis and exhibited radiosensitization in terms of clonogenic survival (Xiong et al., 2012). Another study screened a phage-displayed library of humanized scFv and identified a new antibody against DNA-PKcs, anti-DPK3-scFv (Du et al., 2010). Transfection of cDNA of anti-DPK3-scFv into human cancer cells resulted in increased radiosensitivity with decreased repair capability (Du et al., 2010). It also sensitized transplanted tumor on mice toward radiation (Du et al., 2010).

2.2. ATM

ATM, ataxia-telangiectasia mutated, was identified as the gene responsible for the genetic disorder ataxia telangiectasia, showing similarity to PI3K (Savitsky et al., 1995). Subsequently similarity between ATM and DNA-PKcs, suggesting ATM might also be a protein kinase rather than a lipid kinase (Hartley et al., 1995). ATM was shown to be a protein kinase, which is activated by DNA damage and phosphorylates p53 at Ser15 (Banin et al., 1998; Canman et al., 1998).

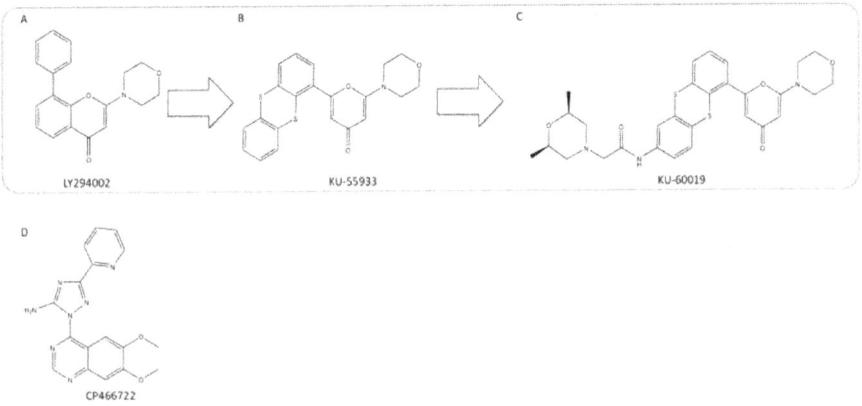

Figure 3. Structure of ATM inhibitors.

Like DNA-PK, ATM was also shown to be inhibited by wortmannin with IC_{50} of 0.15 μM (Sarkaria et al., 1998). It was also shown that administration of wortmannin to cultured cell phenocopies the defect of ataxia telangiectasia cell, *e.g.*, defective accumulation of p53 (Price and Youmell, 1996) and radioresistant DNA synthesis, which is thought to reflect defective G1/S- or S-phase checkpoint (Hosoi et al., 1998; Sarkaria et al., 1998). Caffeine, which was known to abrogate cell cycle checkpoint, was shown to inhibit ATM and ATR (Sarkaria et al., 1999; Hall-Jakson et al., 1999). IC_{50} for ATM and ATR was 200 μM and 1,100 μM, respectively (Sarkaria et al., 1999).

Selective inhibitors were found from the small molecule library of LY294002 derivatives (Hickson, et al., 2004; Hollick et al., 2007). Among them KU-55933, 2-morpholin-4-yl-6-

thialanthren-1-yl-pyran-4-one (Fig.3 B) showed inhibition of ATM with IC_{50} of 0.013 μM (Hickson, et al., 2004). IC_{50} values for other PI3K-related kinases were greater than 1.8 μM, which is approximately 200-fold higher than that for ATM (Hickson, et al., 2004). As in the case of DNA-PK inhibitors, morpholine group is important for inhibitory activity, as KU-58050, in which morpholine group was replaced by piperidine group was much less effective: IC_{50} was 300 μM (Hickson, et al., 2004). KU-55933 inhibited *in cellulo* phosphorylation of ATM substrates, *e.g.*, p53 at Ser15 and histone H2AX at Ser139, 10 μM induced by ionizing radiation, but not that induced by ultraviolet irradiation (Hickson, et al., 2004). Even at lower concentration, *i.e.*, 0.3 μM, the inhibition of p53 phosphorylation was significant, although there was trace amount of residual phosphorylation (Hickson, et al., 2004). It was also shown that KU-55933 sensitized cultured cell to ionizing radiation and to radiomimetic compounds, *e.g.*, etoposide and doxorubicin but did not alter the sensitivity of fibroblast from ataxia telangiectasia patients to ionizing radiation (Hickson, et al., 2004). Futhermore, KU-55933 was found to suppress HIV infection (Lau et al., 2005).

Modification of KU-55933 lead to identification of KU-60019 (Fig.3 C), 2-((2R, 6S)-2, 6-Dimethyl-morpholin-4-yl)-N-[5-(6-morpholin-4-   yl-4-oxo-4H-pyran-2-yl)-9H-thioxanthen-2-yl]-acetamideas a more potent inhibitor of ATM (Golding et al., 2009). IC_{50} of KU-60019 for ATM was 0.0063 μM, whereas IC_{50} values for DNA-PKcs and ATR were 1.7 μM and >10 μM, respectively (Golding et al., 2009). KU-60019 mostly abolished ionizing radiation-induced phosphorylation of p53 at Ser15 and Chk2 at Thr68 at 1 to 3 μM, whereas > 10 μM concentration of KU-55933 was required to obtain similar extent of inhibition (Golding et al., 2009). KU-60019 at 1 μM showed similar extent of radiosensitization to KU-55933 at 10 μM (Golding et al., 2009).

Independent screening of chemical library lead to identification of CP466722, 2-(6,7-dimethoxyquinazolin-4-yl)-5- (pyridin-2-yl)-2H-1,2,4-triazole-3-amine, as a novel inhibitor of ATM (Rainey, et al., 2008). CPP466722 inhibited in cellolo phoshorylation of ATM at Ser1981, SMC1 at Ser957 and Chk2 at Thr68 but not affected the phosphorylation events, which are thought to be mediated through other PI3K-related kinases (Rainey, et al., 2008). CPP466722 sensitized cultured cells to radiation to a similar extent to KU-55933 (Rainey, et al., 2008).

CGK733 was reported to be a dual inhibitor of ATM and ATR, but the report was retracted thereafter because of fabrication. Even after the retraction, CGK733 was marketed as an inhibitor of ATM and ATR and several studies used CGK733 to show the involvement of ATM and/or ATR in response to DNA damage caused by a variety of agents. On the other hand, however, there is a report that this compound did not affect ATM and ATR kinase as shown, respectively, by ionizing radiation-induced phosphorylation of ATM at Ser1981 and Chk2 at Thr 68 and by ultraviolet radiation-induced phosphorylation of Chk1 at Ser317 (Choi et al., 2011, and references therein).

2.3. ATR

ATR was initially identified as a molecule structurally related to human ATM and yeast Rad3 (Cimprich et al, 1996; Keegan et al., 1996). ATR was then shown to be a protein kinase, which is capable of phosphorylating itself and p53 at Ser15 (Canman et al., 1998). ATR is thought to be a sensor of single-stranded DNA (ssDNA), binding to RPA (Replication Protein A) via ATRIP (ATR-interacting protein) (Zou and Elledge, 2003).

Despite of its structural similarity to DNA-PKcs, ATM and PI3K, ATR appeared refractory to wortmannin inhibition: IC_{50} of wortmannin for ATR was 1.8 µM, which was 10- to 100-fold higher than that for DNA-PKcs and ATM (Sarkaria et al., 1998). Selective inhibitors of ATR emerged recently.

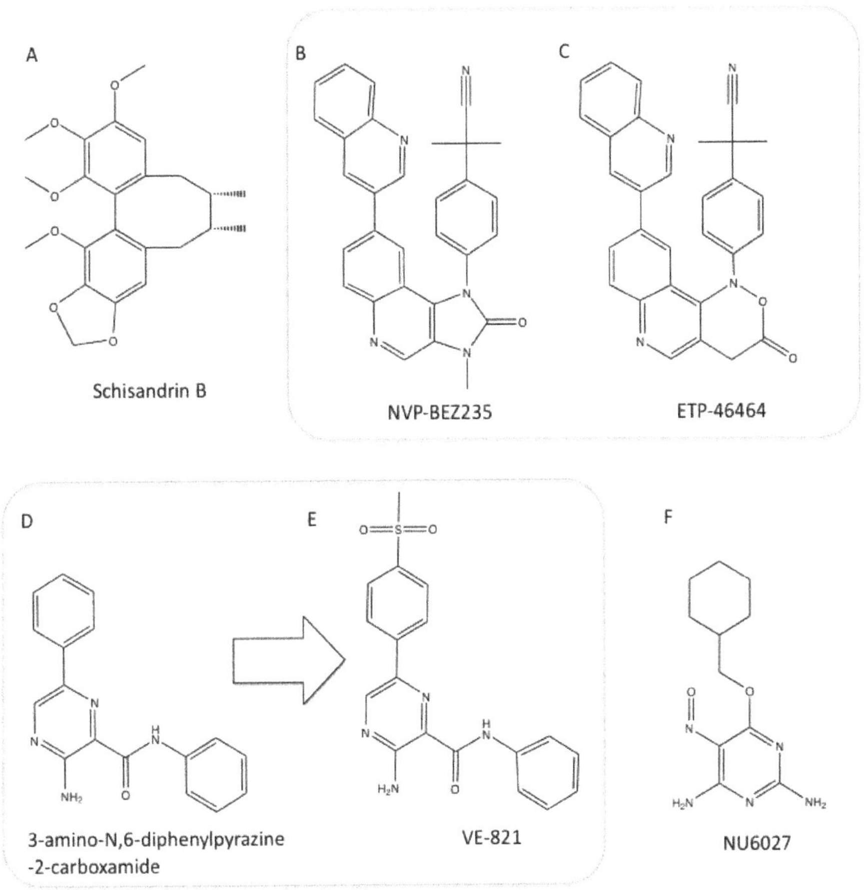

Figure 4. Structure of ATR inhibitors

Schisandrin B is an active ingredient of *Fructus schisandrae*, which has been used in traditional Chinese medicine to treat hepatitis and myocardial disorders (Fig.4 A). Schisandrin
B was found to inhibit ATR (Nishida et al., 2009). IC_{50} of Schisandrin for ATR and ATM
were, respectively, 7.25 μM and 1,700 μM and DNA-PK, PI3K and mTOR were not inhibited up to ~100 μM (Nishida et al., 2009). Schisandrin B sensitized cultured human
cells to ultraviolet radiation and ionizing radiation at concentrations 1 - 30 μM (Nishida
et al., 2009). Sensitization was not observed in cells from Seckel patient, who harbor mutation in ATR gene (Nishida et al., 2009), showing that the sensitizing effect is mediated
through ATR.

Library of 623 compounds, which had exhibited some inhibitory effects on PI3K, was screened
for their effects on *in cellulo* phosphorylaiton of H2AX stimulated by ATR-activating domain
of TopBP1 (Toledo et al., 2011). This screening identified NVP-BEZ235 (Fig.4 B) and ETP-46464
(Fig.4 C) (Toledo et al., 2011). Whereas NVP-BEZ235 also inhibited DNA-PK and ATM (see
above), ETP-464 did not affect DNA-PK and ATM (Toledo et al., 2011). These compounds
mostly inhibited the phosphorylation *in cellulo* of H2AX and other ATR substrates, *e.g.*, Chk1,
even at 0.1 - 0.5 μM (Toledo et al., 2011).

High throughput screening of ATR by *in vitro* kinase assay identified 3-amino-N,6-diphe-
nylpyrazine-2-carboxamide (Charrier et al., 2011). IC_{50} of this compounds for ATR was
0.62 μM, whereas that for ATM and DNA-PK was > 8 μM (Charrier et al., 2011). Then
the derivatives of this compound were synthesized and subjected to test for ATR inhibition. VE-821, 3-amino-6-(4-(methylsulfonyl)phenyl)-N-phenylpyrazine-2-carboxamide, was
found as most potent and selective inhibitor of ATR (Charrier et al., 2011). IC_{50} of VE-821
for ATR was 0.026 μM, whereas that for ATM and DNA-PK was > 8 μM and 4.4 μM, respectively (Charrier et al., 2011).

NU6027, 2,6-diamino-4-cyclohexyl-methyloxy-5-nitroso-pyrimidine, was initially developed
as an inhibitor of cyclin- dependent kinases (CDKs) (Arris et al, 2000). NU6027 was recently
found, however, to inhibit ATR more potently than CDK2 (Peasland et al., 2011). NU6027
inhibited *in cellulo* phosphorylation of Chk1 at Ser345 with IC_{50} of 6.7 μM, whereas autophosphorylation of DNA-PKcs at Ser2056 and ATM at Ser1981 were not affected at 10 μM (Peasland
et al., 2011). NU6027 sensitized cultured cells to hydroxyurea and cisplatin, but this effect was
not observed in ATR-knocked down cells, showing that sensitization was mediated through
ATR (Peasland et al., 2011).

It might be added that p53-deficient cells, than p53-proficient cells, exhibited greater extent of sensitization toward ionizing radiation and other DNA damaging agents by ATR
inhibitors NVP-BEZ235, ETP-46464 (Toledo et al., 2011), VE-821 (Reaper et al., 2011) and
NU6027 (Peasland et al., 2011). This could be due to simultaneous inactivation of two
checkpoint pathways mediated through ATM and ATR, respectively, the former of which
involves p53. As most of cancer cells lose p53 function, inhibition of ATR might be a
promising approach to achieve selective killing of cancer cells, minimizing the effects to
surrounding normal cells.

3. Other DSB repair enzymes as targets for radiosensitizer

3.1. MRN complex

Mirin, Z-5-(4-hydroxybenzylidene)-2-imino-1,3-thiazolidin-4-one (Fig.5), was identified in a screen for smalll molecules inhibiting MRN-ATM pathway (Dupre et al., 2009). Restriction enzyme-digested plasmid was added to cell-free extract prepared from *Xenopus laevis* egg in 96-well format and the phosphorylation of H2AX-mimicking peptide was quantified. Approximately 10,000 compounds, which had exhibited inhibition of p53 activity or inteference with mitosis and spindle dynamics, were subjected to screen. Mirin inhibited H2AX phosphorylation in *Xenopus laevis* egg cell free extract with an IC_{50} of 66 μM and also autophosphorylation of ATM at Ser1981 in human cells within 25 - 100 μM range (Dupre et al., 2009). Mirin inhibited nuclease activity of Mre11, but did not affect DNA binding or DNA tethering activity of MRN complex (Dupre et al., 2009). Mirin also abrogated G2/M checkpoint, reduced homologous recombination and showed radiosensitizing effects in cultured human cells within 25 - 100 μM range (Dupre et al., 2009).

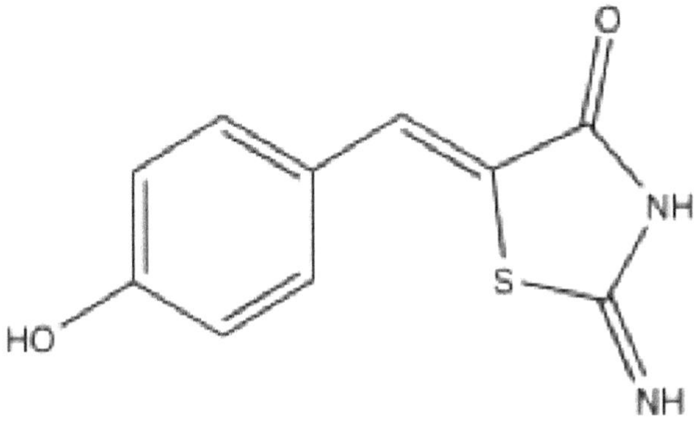

Figure 5. Structure of Mirin.

3.2. DNA ligase IV

Inhibitors of DNA ligases were searched in a database of 1.5 million commercially available low molecular weight chemicals by computer-aided drug design approach based on crystal structure of DNA ligase I (Chen et al., 2008). In this approach, L82 ((E)-2-((2-(2-((3,5-dibromo-4-methylphenyl)amino)ethyl)hydrazono)methyl)-4-nitrophenol, Fig.6 A), inhibiting DNA ligase I, L67 ((E)-4-chloro-5-(2-(4-hydroxy-3-nitrobenzylidene)hydrazinyl)pyridazin-3(2H)-one, Fig.6 B), inhibiting DNA ligases I and III, and L189 ((E)-6-amino-5-(benzylideneamino)-2-mercaptopyrimidin-4-ol, Fig.6 C), inhibiting DNA ligases I, III and IV, were identified. None of them inhibited the activity of T4 ligase (Chen et al., 2008). Kinetic alalysis indicated that, whereas L82 is non-competitive inhibitor, L67 and L189 competes with DNA substrate (Chen et al., 2008). L67 sensitized cultured human cancer cells to methylmethansulfonate at 3 μM (Chen et al., 2008). Similarly, L189 sensitized cultured human cancer cells to ionizing radiation at 20 μM (Chen et al., 2008). It might be noted that the sensitizing effects of L67 and L189 were not observed in non-cancer cells, suggesting is selective effects on cancer cells (Chen et al., 2008). These compounds can be a leading compounds for the development of more potent and/or more selective inhibitors of DNA ligases.

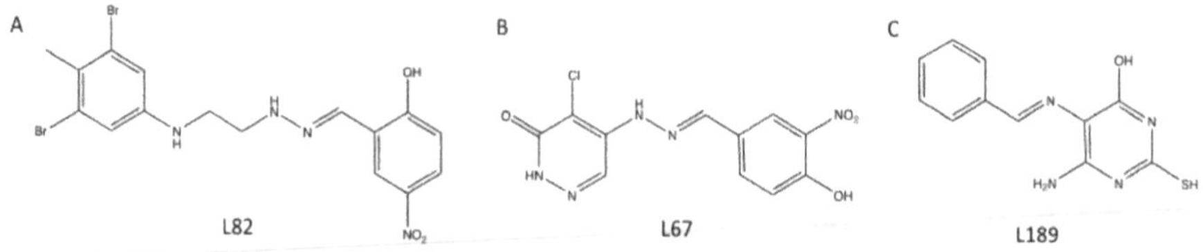

Figure 6. Strucutre of DNA ligase inhibitors.

3.3. DPYD as a new target

Gimeracil, 5-chloro-2,4-dihydroxypyridine (Fig.7), is an inhibitor of dihydropyrimidine dehydrogenease (DPYD) and used as a component of oral anti-cancer medicine S-1, in order to suppress degradation of 5-fluorouracil. The results of clinical trial of concurrent chemora-diotherapy using S-1 suggested possible radiosensitizing effect of S-1. Gimeracil increased radiosensitivity of cultured human cancer cells of various origin within 200 – 5,000 μM, being maximal within 1,000 – 5,000 μM range (Takagi et al., 2010). Cell lines deficient for DNA-PKcs or Ku86 were sensitized by gimeracil to radiation even to a greater extent than respective control cells (Takagi et al., 2010). On the other hand, radiosensitiztion was not observed in cell lines deficient for XRCC3, NBS1 or FANCD2 (Takagi et al., 2010). These observations collectively suggested that gimeracil exert radiosensitizing effects through inhibition of HR-mediated DSB repair. Gimeracil reduced the frequency of homologous recombination of chromosomal substrate including the restriction site of I-SceI by approximately 15% (Takagi et al., 2010). Gimeracil reduced the formation of ionizing radiation-induced foci of Rad51 and RPA but increased that of Nbs1, Mre11, Rad50 and FancD2 (Sakata et al., 2011). This observation suggested that gimeracil might have inhibited the step after strand resection by Mre11-Rad50-Nbs1 complex but before the loading of RPA and Rad51 onto single-stranded DNA. Although the role of DPYD in HR has not been described, treatment with siRNA for DPYD sensitized cells to ionizing radiation to a similar extent to gimeracil and also diminished the radiosensitization by gimeracil (Sakata et al., 2011). These results collectively indicate that gimeracil exerts radiosensitizing effects through inhibition of DPYD, which might have a novel role in HR.

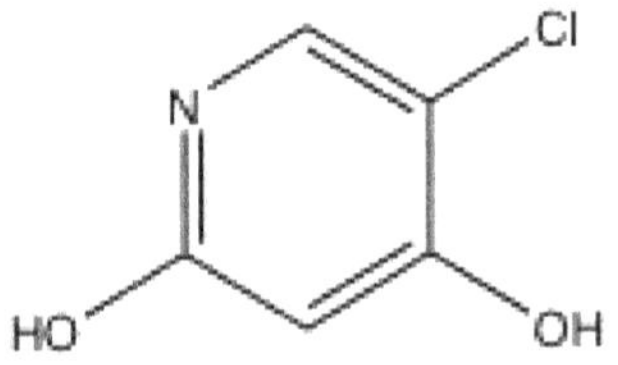

Figure 7. Structure of gimeracil.

4. Radiosensitization by hyperthermia

Hyperthermia, heating parts of body at 40 - 45 °C, has been used to treat cancer mostly combined with ionizing radiation. Hyperthemia is known to sensitize cells to ionizing radiation, inhibiting the repair of DNA damages including DSBs, but the molecular mechanism of radiosensitization by hyperthermia has remaind to be clarified.

Effects of hyperthermia on DNA polymerases α and β have been studied for a long time. These studies suggested that DNA polymerase β was sensitive to hyperthermia and its inactivation was correlated to radiosensitization as well as to cell killing (Spiro et al., 1982). Later it was reported that DNA polymerase β knocked out cells or overexpressed cells exhibited radio-sensitization by hyperthermia indifferent from control cells (Raaphorst et al., 2004). Elucidation of DSB repair mechanisms through NHEJ and HR provided clues to the mechanisms of radiosensitization by hyperthermia.

Among essential factors in NHEJ, Ku is shown to be affected by hypethermia. Purified DNA-PK lost its activity upon incubation at 44°C for 5 - 30 min (Matsumoto et al., 1997). When DNA-PKcs and Ku were heated separately, heating of Ku, but not DNA-PKcs, lead to decrease in DNA-PK activity, suggesting that Ku, rather than DNA-PKcs, is heat sensitive component (Matsumoto et al., 1997). Inactivation of DNA-PK activity by hyperthermia was observed also *in cellulo*, *i.e.*, when culture cells were heated at 44 – 47°C (Burgman et al., 1997; Ihara et al., 1999; Umeda et al., 2003). It might be noted, however, that the extent of the loss of DNA-PK was greatly different between mouse, hamster and human cell, being greatest in mouse and least in human (Umeda et al. 2003). In murin cells, significant loss of DNA-PK activity was observed at lower tempertures, *i.e.*, 41°C or 42°C (our unpublished observations). DNA-PK activity could be restored by mixing the lysate of heated cells with the lysate of DNA-PKcs-deficient cells, but not with Ku86-deficient cells, indicating that *in cellulo* inactivation of DNA-PK by hyperthermia might be also due to the property of Ku rather than DNA-PKcs (Ihara et al., 1999). Moreover, Ku was identified as constitutive heat shock element-binding factor, CHBF, whose activity was lost by hyperthermia, allowing the binding of HSF1 (Kim et al., 1995). DNA binding activity of Ku correlated with extent of radiosensitization by hyperthermia (Burgman et al., 1997). Reduced solubility of Ku in aquaous buffer after hyperthermia was also reported, which might reflect aggregation (Beck and Dynlacht, 2001). However, the hypothesis that radiosensitization by hyperthermia is due to inactivation of Ku or DNA-PK has been challenged by genetic studies, showing that cells deficient for Ku or DNA-PKcs could be radiosensitized by hyperthermia to a similar extent or even to a greater extent than control cells (Kampinga et al., 1993; Raaphorst et al., 1993; Woudstra, et al, 1999, Raaphorst et al., 2004), although there are studies, in contrast, showing no or reduced radiosensitization in Ku- or DNA-PKcs-deficient cells (Iliakis and Seaner, 1990; O'Hara et al., 1995). Moreover, chicken lymphocyte DT40 derivative lacking Ku70 and Rad54, therefore, deficient in both of NHEJ and HR, still showed radiosensitization by hyperthermia (Raaphorst et al., 2004; Yin et al., 2004).

There is also accumulating studies on the effects of hyperthermia on MRN complex. It was initially found that Mre11, Rad50 and Nbs1 exported from nucleus to cytoplasm upon hyperthemia at 42.5°C or 45.5°C (Zhu et al., 2001; Seno and Dynlacht, 2004). This nuclear export of MRN complex increased when cells were irradiated prior to hyperther- mia (Zhu et al., 2001; Seno and Dynlacht, 2004). Similar phonomenon was observed in mild hyperthermia at 41.1°C (Xu et al., 2002). However, in a recent study, inhibition of nuclear export of MRN complex by leptomycin B did not diminish radiosensitization by hyperthermia at 45.5°C for 10 min (Dynlacht et al., 2011). It was also shown, neverthe- less, that ATLD cells, which have mutated in Mre11, did not show radiosensitization by hyperthermia at 41.5°C for 2 hrs or at 45.5°C for 10 min (Dynlacht et al., 2011). On the other hand, radiosensitization by hyperthermia was observed in NBS cells and Rad50- knocked down cells (Dynlacht et al., 2011). Exonuclease activity of Mre11 was decreased to ~10% by 42.5°C treatment for 15 min (Dynlacht et al., 2011). These results collectively indicate Mre11 as target for radiosensitization by hyperthermia.

Hyperthermia is shown to affect BRCA1 and BRCA2. Heating cultured human cancer cells at 42°C for 1 - 2 hrs or more decreased the amount of BRCA1 (Ma et al., 2003). It might be caused by protein degradation, but various inhibitors of proteases, so far as tested, failed to suppress the decrease of BRCA1 (Ma et al., 2003). Alternatively, it might be caused by protein aggrega- tion and reduced solubility in aqueous buffers. It was also shown that BRCA1 deficient cells were sensitive to hyperthermia (Ma et al., 2003). Recent study reported the degradation of BRCA2 induced by mild hyperthermia at 41°C to 42.5°C (Krawczyk et al., 2011). Rad54- deficient ES cells and cells treated with XRCC3 siRNA were not radiosensitized by mild hyperthermia (Krawczyk et al., 2011). Furthermore, mild hyperthermia showed synthetic lethality with PARP-1 inhibitor oraparib, like BRCA2 deficient cancer cells (Krawczyk et al., 2011). These data collectively indicated BRCA2 as a major target of mild hyperthermia.

Obviously, hyperthermia inactivates many enzymes and induces aggregation of many proteins. In this regard, hyperthermia is not specific on certain enzyme, unlike inhibitors described above. However, susceptiblity to inactivation by hyperthermia might be greatly different among proteins. The extent of radiosensitization by hyperthermia can be greatly infuluenced by many factors, *e.g.*, cell type, genetic background, physiological conditions, heating temperature, duration of heating, sequence of heating and radiation and the in- terval between them. Further studies would be required to examine the effects of hyper- thermia on various repair enzymes and and its relationship to radiosentizing effects under various conditions.

5. Concluding remarks and future perspectives

Because of great advances in our understanding of the moleuclar mechanisms of DSB repair in past two decades, extensive studies have been done to achieve radiosensitization by modification of DSB repair molecules. Especially, a number of inhibitors have been developed for DNA-PK, ATM and ATR protein kinases. We saw here that preceeding studies on DNA-PK and on PI3K greatly facilitated the studies on ATM and ATR. It might be underscored that LY294002, preexisting inhibitor of PI3K, served as a leading compound and enabled the finding of potent and specific inhibitors like NU7441 and KU-55933.

Studies toward the clinical application of these compounds are underway. Preclinical studies of pharmacokinetics and metabolism in mice were conducted for NU7026 and NU7441. In the case of NU7026, the radiosensitizing effect on cultured cancer cell was marginal upon the treatment at 10 μM for 2 hrs and could be increased by extending the treatment time up to 24 hrs (Nutley et al., 2005). On the other hand, however, NU7026 underwent rapid plasma clearance in mice, presumably because of oxidation and ring opening of morpholino group (Nutley et al., 2005). It was estimated that NU7026 should be administered four times per day at 100 mg/kg intraperitoneally in order to obtain radiosensitization (Nutley et al., 2005). In the case of NU7441, the radiosensitizating effect on cultured cancer cell could be obtained by treatment at 1 μM for 1 hr (Zhao et al., 2006). The concentration of NU7441 required for radiosensitization could be maintained within tumor tissues for more than 4 hrs at nontoxic dose (Zhao et al., 2006). The administration of etoposide and NU7441 to mice bearing human tumor xenografts synergistically delayed tumor growth, indicating the chemosensitizing effect of NU7441 *in vivo* (Zhao et al., 2006). Studies are still going on to obtain compounds with better characteristics, *e.g.*, higher aquaous solubility (Cano et al., 2010).

Search for inhibitors of enzymes other than protein kinases has been difficult due to the absence of assay system suitable for highthroughput screening. However, inhibitors of other enzymes, *i.e.*, Mre11 nuclease and DNA ligase IV have been developed, although few at present. Now these compuonds are obtained, more potent and specific inhibitors can be obtained by molecular evolution as in the case of DNA-PK, ATM and ATR protein kinases. Additionally, search for other inhibitors will be greatly facilitated by an aid of computer-based structural prediction and drug designing.

In addition to use of each chemicals alone, use of two or more chemicals together to inhibit two pathways of DSB repair or one of them with other repair mechanisms, which is called synthetic lethality approach, will be promising. Successful example is shown in the treatment of cancers arisen in the carriers of BRCA2 mutation with PARP-1 inhibitors. When PARP-1, which is essential for single-strand break (SSB) repair, is inactivated, SSB is converted to DSB, which requires BRCA2 to be repaired. As BRCA2 mutation is heterozygotic, normal cells retain BRCA2 function. On the other hand cancer cells have lost BRCA2 function and, therefore, shows extreme sensitivity to increased sensitivity to converted DSBs. This is instructive also to find a means to discriminate cancer cells and normal cells. As described in the previous section, although hyperthemia is not an approach to target a certain molecule specifically, it did show synthetic lethal effects with PARP-1 inhibitor. These examples underscores the importance of the choice of agents based on the thorough consideration of biological charac-teristics and genetic background of each cancer and patient. In addition to continuing persuit for the new radiosensitizing agents, extensive studies would be necessary regarding combi-natorial approach and personalized medicine.

Note

We apologize for not citing many important literatures because of space limitation.

Acknowledgements

Our study was supported in part by Grant-in-Aid for Scientific Research from the Ministry of Education, Culture, Sport, Science and Technology of Japan to YM. MKS is supported by Takeda Science Foundation, Japan Society for Promotion of Sciences and Tokyo Biochemistry Research Foundation.

Author details

Yoshihisa Matsumoto, Shoji Imamichi, Mikoto Fukuchi, Sicheng Liu, Wanotayan Rujira, Shingo Kuniyoshi, Kazuki Yoshida, Yasuhiro Mae and Mukesh Kumar Sharma

Research Laboratory for Nuclear Reactors, Tokyo Institute of Technology, Tokyo, Japan

References

[1] Arris, C.E., Boyle, F.T., Calvert, A.H., Curtin, N.J., Endicott, J.A., Garman, E.F.,Gibson, A.E., Golding, B.T., Grant, S., Griffin, R.J., Jewsbury, P., Johnson, L.N., Lawrie, A.M., Newell, D.R., Noble, M.E., Sausville, E.A., Schultz, R. & Yu, W. (2000) Identification of novel purine and pyrimidine cyclin-dependent kinase inhibitors with distinct molec-ular interactions and tumor cell growth inhibition profiles. *J. Med. Chem.*, 43: 2797-2804.

[2] Banin, S., Moyal, L., Shieh, S.-Y. Taya, Y., Anderson, C. W., Chessa, L., Smorodinsky, N. I., Prives, C., Reiss, Y., Shiloh, Y., & Ziv, Y. (1998) Enhanced phosphorylation of p53 by ATM in response to DNA damage. *Science*, 281:1674-1677.

[3] Beck, B.D. & Dynlacht, J.R. (2001) Heat-induced aggregation of XRCC5 (Ku80) in nontolerant and thermotolerant cells. *Radiat. Res.*, 156: 767-774.

[4] Blunt, T.; Finnie,N.; Taccioli, G.; Smith, G.; Demengeot, J.; Gottlieb, T.; Mizuta, R.; Varghese, A.; Alt, F.; Jeggo, P. & Jackson, S.P. (1995) Defective DNA-dependent protein

kinase activity is linked to V(D)J recombination and DNA repair defects associated with the murine scid mutation. *Cell*, 80: 813-823.

[5]　Burgman, P., Ouyang, H., Peterson, S., Chen, D.J. & Li, G.C. (1997) Heat inactivation of Ku autoantigen: possible role in hyperthermic radiosensitization. *Cancer Res.*, 57: 2847-2850.

[6]　Canman, C. E., Lim, D.-S., Cimprich, K. A., Taya, Y., Tamai, K., Sakaguchi, K., Appella, E., Kastan, M. & Siliciano, J. D. (1998) Activation of the ATM kinase by ionizing radiation and phosphorylation of p53. *Science*, 281: 1677-1679.

[7]　Cano, C., Barbeau, O.R., Bailey, C., Cockcroft, X.-L., Curtin, N.J., Guggan, H., Frigerio, M., Golding, B.T., Hardcastle, I.R., Hummersone, M.G., Knights, C., Menear, K.A., Newell, D.R., Richardson, C.J., Smith, G.C.M., Spittle, B. & Griffin, R.J. (2010) DNA-dependent protein kinase (DNA-PK) inhibitors. Synthesis and biological activity of quinolin-4-one and pyridopyrimidin-4-one surrogates for the chromen-4-one chemotype. *J. Med. Chem.*, 53: 8498-8507.

[8]　Carter, T., Vancurova, I., Sun, I., Lou, W. & DeLeon, S. (1990) A DNA-activated protein kinase from HeLa cell nuclei. *Mol. Cell. Biol.*, 10: 6460-6471.

[9]　Charrier, J.D., Durrant, S.J., Golec, J.M.C., Kay, D.P., Knegtel, R.M.A., MacCormick, S., Mortimore, M., O'Donnell, M.E., Pinder, J.L., Reaper, P.M., Rutherford, A.P., Wang, P.S.H., Young, S.C. & Pollard, J.R. (2011) Discovery of potent and selective inhibitors of ataxia telangiectasia mutated and Rad3 related (ATR) protein kinase as potential anticancer agents. *J. Med. Chem.*, 54: 2320-2330.

[10]　Chen, X., Zhong, S., Zhu, X., Dziegielewska, B., Ellenberger, T., Wilson, G.M., MacKerell, A.D. & Tomkinson, A.E. (2008) Rational design of human DNA ligase inhibitors that target cellular DNA replication and repair. *Cancer Res.*, 68: 3169-3177.

[11]　Chernikova, S.B., Wells, R.L. & Elkind, M.M. (1999) Wortmannin sensitizes mammalian cells to radiation by inhibiting the DNA-dependent protein kinase-mediated rejoining of double-strand breaks. *Radiat. Res.*, 151: 159-166.

[12]　Choi, S., Toledo, L.I., Fernandez-Capetillo, O. & Bakkenist, C.J. (2011) CGK733 does not inhibit ATM or ATR kinase activity in H460 human lung cancer cells. *DNA Repair*, 10: 1000-1001.

[13]　Cimprich, K.A., Shin, T.B., Keith, C.T. & Shreiber, S.L. (1996) cDNA cloning and gene mapping of a candidate human cell cycle checkpoint protein. *Proc. Natl. Acad. Sci. USA*, 93: 2850-2855.

[14]　Dvir, A., Peterson, S.R., Knuth, M.W., Lu, H. & Dynan, W.S. (1992) Ku autoantigen is the regulatory component of a template-associated protein kinase that phosphorylates RNA polymerase II. *Proc. Natl. Acad. Sci. USA*, 89: 11920 -11924

[15]　Dvir, A., Stein, L.Y., Calore, B.L. & Dynan,W.S. (1993) Purification and characterization of a template-associated protein kinase that phosphorylates RNA polymerase II. *J. Biol. Chem.*, 268: 10440-10447.

[16] Du, L., Zhou, L.-J., Pan, X.-J., Wang, Y.-X., Xu, Q.-Z., Yang, Z.-H., Wang, Y., Liu, X.-D., Zhu, M.-X. & Zhou, P.-K. (2010) Radiosensitization and growth inhibition of cancer cells mediated by an scFv antibody gene against DNA-Pkcs in vitro and in vivo. *Radiat. Oncol.*, 5: 70.

[17] Dupre, A., Boyer-Chatenet, Sattler, R.M., Modi, A.P., Lee, J.-H., Nicolette, M.L., Kopelovih, L., Jasin, M., Baer, R., Paull, T.T. & Gautier, J. (2009) A forward chemical genetic screen reveals an inhibitor of the Mre11-Rad50-Nbs1 complex. *Nat. Chem. Biol.*, 4: 119-125. (Corringendum Vol. 5, pp.1)

[18] Durant, S. & Karran, P. (2003) Vanillins – a novel family of DNA-PK inhibitors. *Nucleic Acids Res.*, 31: 5501-5512.

[19] Dynlacht, J.R., Batuello, C.N., Lopez, J.T., Kim, K.K. & Turchi, J.J. (2011) Identification of Mre11 as a target for heat radiosensitization. *Radiat. Res.*, 176: 323-332.

[20] Golding, S.E., Rosenberg, E., Valerie, N., Hussaini, I., Frigerio, M., Cockcroft, X.F., Chong, W.Y., Hummersone, M., Rigoreau, L., Menear, K.A., O'Connor, M., Povirk, L., van Meter, T. & Valerie, K. (2009) Improved ATM kinase inhibitor KU-60019 radiosen-sitizes glioma cells, compromises insulin, AKT and ERK prosurvival signaling, and inhibits migration and invasion. *Mol. Cancer Ther.*, 8: 2894-2902.

[21] Gottlieb, T.M. & Jackson,S.P. (1993) The DNA-dependent protein kinase: requirement for DNA ends and association with Ku antigen. *Cell*, 72:.131-142.

[22] Griffin, R.J., Fontana, G., Golding, B.T., Guiard, S., Hardcastle, I.R., Leahy, J.J.J., Martin, N., Richardson, C., Rigoreau, L., Stockley, M. & Smith, G.C.M. (2005) Selective benzo-pyranone and pyramido[2,1-a]isoquinolin-4-one inhibitors of DNA-dependent protein kinase: synthesis, structure-activity studies, and radiosensitization of a human tumor cell line in vitro. *J. Med. Chem.*, 48: 569-585.

[23] Hall-Jackson, C.A., Cross, D.A.E., Morrice, N. & Smythe, C. (1999) ATR is a caffeine-sensitive, DNA-activated protein kinase with a substrate specificity distinct from DNA-PK. *Oncogene*, 18: 6707-6713.

[24] Hartley, K.; Gell, D.; Smith, C.; Zhang, H.; Divecha, N.; Connelly, M.; Admon, A.; Lees-Miller, S.; Anderson, C. & Jackson, S. (1995) DNA-dependent protein kinase catalytic subunit: a relative of phosphatidylinositol 3-kinase and the ataxia telangiectasia gene product. *Cell*, 82: 849-856.

[25] Hardcastle, I.R., Cockcroft, X., Curtin, N.J., El-Murr, M. D., Leahy, J.J.J., Stockley, M., Golding, B.T., Rigoreau, L., Richardson, C., Smith, G.C.M. & Griffin, R.J. (2005) Discovery of potent chromen-4-one inhibitors of the DNA-dependent protein kinase (DNA-PK) using a small-molecule library approach. *J. Med. Chem.*, 48: 7829-7846.

[26] Hashimoto, M., Rao, S., Tokuno, O., Yamamoto, K., Takata, M., Takeda, S. & Utsumi, H. (2003) DNA-PK: the major target for wortmannin-mediated radiosensitization by the inhibition of DSB repair via NHEJ pathway. *J. Radiat. Res.*, 44: 151-159.

[27] Hickson, I., Zhao, Y., Richardoson, C.J., Green, S.J., Martin, N.M.B., Orr, A.I., Reaper, P.M., Jackson, S.P., Curtin, N.J. & Smith, G.C.M. (2004) Identification and characterization of a novel and specific inhibitor of the ataxia-telangiectasia mutated kinase ATM. *Cancer Res.*, 64: 9152-9159.

[28] Hollick, J.J., Rigoreau, L.J.M., Cano-Soumillac, C., Cockcroft, X., Curtin, N.J., Frigerio, M., Golding, B.T., Guiard, S., Hardcastle, I.R., Hickson, I., Hummersone, M.G., Menear, K.A., Martin, N.M.B., Matthews, I., Newell, D.R., Ord, R., Richardson, C.J., Smith, G.C.M. & Griffin, R.J. (2007) Pyranone, thiopyranone, and pyridone inhibitors of phosphatidylinositol 3-kinase related kinases. Structure-activity relationships for DNA-dependent protein kinase inhibition, and identification of the first potent and selective inhibitor of the ataxia telangiectasia mutated kinase. *J. Med. Chem.*, 50: 1958-1972.

[29] Hosoi, Y., Matsumoto, Y., Enomoto, A., Morita, A., Green, J., Nakagawa, K., Naruse, K. & Suzuki, N. Suramin sensitizes cells to ionizing radiation by inactivating DNA-dependent protein kinase. *Radiat. Res.* 162: 308-314 (2004).

[30] Hosoi, Y., Matsumoto, Y., Tomita, M., Enomoto, A., Morita, A., Sakai, K. Umeda, N., Zhao, H.-J., Nakagawa, K., Ono, T. & Suzuki, N. (2002) Phosphorothioate oligonucleotides, suramin and heparin inhibit DNA-dependent protein kinase activity. *Brit. J. Cancer*, 86: 1143-1149.

[31] Hosoi, Y., Miyachi, H., Matsumoto, Y., Ikehata, H., Komura, J., Ishii, K., Zhao, H.J., Yoshida, M., Takai, Y., Yamada, S., Suzuki, N. & Ono, T. (1998) A phosphatidylinositol 3-kinase inhibitor wortmannin induces radioresistant DNA synthesis and sensitizes cells to bleomycin and ionizing radiation. *Int. J. Cancer*, 78: 642-647.

[32] Ihara, M., Suwa, A., Komatsu, K., Shimasaki, T., Okaichi, K., Hendrickson, E.A. & Okumura, Y. (1999) Heat sensitivity of double-stranded DNA-dependent protein kinase (DNA-PK) activity. *Int. J. Radiat. Biol.*, 75: 253-258.

[33] Iliakis, G. & Seaner, R. (1990) A DNA double-strand break repair-deficient mutant of CHO cells shows reduced radiosensitization after exposure to hyperthermic temperatures in the plateau phase of growth. *Int. J. Hyperthermia*, 6: 801-812.

[34] Ismail, I.H., Marternsson, S., Moshinsky, D., Rice, A., Tang, C., Howlett, A., McMahon, G. & Hammarste, O. (2004) SU11752 inhibits the DNA-dependent protein kinase and DNA double-strand break repair resulting in ionizing radiation sensitization. *Oncogene*, 23: 873-882.

[35] Izzard, R.A., Jackson, S.P. & Smith, G.C.M. (1999) Competitive and noncompetitive inhibition of the DNA-dependent protein kinase. *Cancer Res.*, 59: 2581-2586.

[36] Kamping, H.H., Kanon, B., Konings, A.W.T., Stackhouse, M.A. & Bedford, J.S. (1993) Thermal radiosensitization in heat- and radiation-sensitive mutants of CHO cells. *Int. J. Radiat. Biol.*, 64: 225-230.

[37] Kashishian, A., Douangpanya, H., Clark, D., Schlachter, S.T., Eary, C.T., Schiro, J.G., Huang, H., Burgess, L.E., Kesicki, E.A. & Halbrook, J. (2003) DNA-dependent protein kinase inhibitors as drug candidates for the treatment of cancer. *Mol.Cancer Ther.*, 2: 1257-1264.

[38] Keegan, K.S., Holzman, D.A., Plug, A.W., Christenson, E.R., Brainerd, E.E., Flaggs, G., Bentley, N.J., Taylor, E.M., Meyn, M.S., Moss, S.B., Carr, A.M., Ashley, T. & Hoekstra, M.F. (1996) The Atr and Atm protein kinases associate with different sites along meiotically paring chromosomes. *Genes Dev.*, 10: 2423-2437.

[39] Kim, D., Ouyang, H., Yang, S.H., Nussenzweig, A., Burgman, P. & Li, G.C. (1995) A constitutive heat shock element-binding factor is immunologically identical to the Ku autoantigen. *J. Biol. Chem.*, 270: 15277-15284.

[40] Kirchgessner, C.; Patil, C.; Evans, J.; Cuomo, C.; Fried, L.; Carter, T.; Oettinger, M. & Brown, M. (1995) DNA-dependent kinase (p350) as a candidate gene for the murine SCID defect. *Science*, 267: 1178-1183.

[41] Knight, Z,A., Chiang, G.G., Alaimo, P.J., Kenski, D.M., Ho, C.B., Coan, K., Abraham, R.T. & Shokat, K.M. (2004) Isoform-specific phosphoinositide 3-kinase inhibitors from an arylmorpholine scaffold. *Bioorg. Med. Chem.*, 12: 4749-4759.

[42] Krawczyk, P.M., Eppink, B., Essers, J., Stap, J., Rodermond, H., Odijk, H., Zelensky, A., van Bree, C., Stalpers, L.J., Buist, M.R., Soullie, T., Rens, J., Verhagen, H.J., O'Connor, M.J., Franken N.A., ten Hagen, T.L., Kanaar, R. & Aten, J.A. (2011) Mild hyperthermia inhibits homologous recombination, induces BRCA2 degradation, and sensitizes cancer cells to poly (ADP-ribose) polymerase-1 inhibition. *Proc. Natl. Acad. Sci. USA*, 108: 9851-9856.

[43] Kruszweski, M., Wojewódzka, M., Iwanen´ko, T, Szumiel, I. & Okuyama, A. (1998) Differential inhibitory effect of OK-1035 on DNA repair in L5178Y murine lymphoma sublines with functional or defective repair of double strand breaks. *Mutat. Res.*, 409: 31-36.

[44] Lau, A., SwinBank, K.M., Ahmed, P.S., Taylor, D.L., Jackson, S.P., Smith, G.C.M. & O'Connor, M.J. (2005) Suppression of HIV-1 infection by a small molecule inhibitor of the ATM kinase. *Nat. Cell Biol.*, 7: 493-500.

[45] Leahy, J.J.J., Golding, B.T., Griffin, R.J., Hardcastle, I.R., Richardson, C., Rigoreau, L. & Smith, G.C.M. (2004) Identification of a highly potent and selective DNA-dependent protein kinase (DNA-PK) inhibitor (NU7441) by screening of chromenone libraries. *Bioorg. Med. Chem. Lett.*, 14: 6083-6087.

[46] Lees-Miller, S.P., Chen,Y.-R. & Anderson,C.W. (1990) Human cells contain a DNA-activated protein kinase that phosphorylates simian virus 40 T antigen, mouse p53, and the human Ku autoantigen. *Mol. Cell. Biol.*, 10: 6472-6481.

[47] Lees-Miller, S., Godbout, R., Chan, D., Weinfeld, M., Day III, R., Barron, G. & Allalunis-Turner, J. (1995). Absence of p350 subunit of DNA-activated protein kinase from a radiosensitive human cell line. *Science*, 267: 1183-1185.

[48] Lees-Miller, S., Sakaguchi, K., Ullrich, S., Appella, E. & Anderson, C. (1992) Human DNA-activated protein kinase phosphorylates serines 15 and 37 in the amino-terminal transactivation domain of human p53. *Mol. Cell. Biol.*, 12: 5041-5049.

[49] Li, S., Takeda, Y., Wragg, S., Barrett, J., Phillips, A. & Dynan, W.S. (2003) Modification of the ionizing radiation response in living cells by an scFv against the DNA-dependent protein kinase. *Nucleic Acids Res.*, 31: 5848-5857.

[50] Ma, Y.X., Fan, S., Yuan, R., Meng, Q., Gao, M., Goldberg, I.D., Fuqua, S.A., Pestell, R.G. & Rosen, E.M. (2003) Role of BRCA1 in heat shock response. *Oncogene*, 22: 10-27.

[51] Maira, S.M., Stauffer, F., Brueggen, J., Furet, P., Schnell, C., Fritsch, C., Brachmann, S., Chene, P., De Pover, A., Schoemaker, K., Fabbro, D., Gabriel, D., Simonen, M., Murphy, L., Finan, P., Sellers, W. & Garcia-Echeverria, C. (2008) Identification and characterization of NVP-BEZ235, a new orally available dual phosphatidylinositol 3-kinase/mammalian target of rapamycin inhibitor with potent in vivo antitumor activity. *Mol. Cancer Ther.*, 7: 1851-1863.

[52] Matsumoto, Y., Suzuki, N., Sakai, K., Morimatsu, A., Hirano, K. & Murofushi, H. (1997) A possible mechanism for hyperthermic radiosensitization mediated through hyperthermic lability of Ku subunits in DNA-dependent protein kinase. *Biochem. Biophys. Res. Commun.*, 234: 568-572.

[53] Mukherjee, B., Tomimatsu, N., Amancherla, K., Camacho, C.V., Pichamoorthy, N. & Burma, S. (2012) The dual PI3K/mTOR inhibitor NVP-BEZ235 is a potent inhibitor of ATM- and DNAPKcs-mediated DNA damage responses. *Neoplasia*, 14: 34-43.

[54] Nishida, H., Tatewaki, N., Nakajima, Y., Magara, T., Ko, K.M., Hamamori, Y. & Konishi, T. (2009) Inhibition of ATR protein kinase activity by schisandrin B in DNA damage response. *Nucleic Acids Res.*, 37: 5678-5689.

[55] Nutley, B.P., Smith, N.F., Hayes, A., Kelland, L.R., Brunton, L., Golding, B.T., Smith, G.C.M., Martin, N.M.B., Workman, P., & Raynaud, F.I. (2005) Preclinical pharmacokinetics and metabolism of a novel prototype DNA-PK inhibitor NU7026. *Brit. J. Cancer*, 93: 1011-1018.

[56] O'Hara, M.D., Pollard, M.D., Wheatley, G., Regine, W.F., Mohiuddin, M. & Leeper, D. B. (1995) Thermal response and hyperthermic radiosensitization of scid mouse bone marrow CFU-C. *Int. J. Radiat. Oncol. Biol. Phys.*, 31: 905-910.

[57] Peasland, A., Wang, L.-Z., Rowling, E., Kyle, S., Chen, T., Hopkins, A., Cliby, W.A., Sarkaria, J., Beale, G., Edmondson, R.J. & Curtin, N.J. (2011) Identification and evaluation of a potent novel ATR inhibitor NU6027, in breast and ovarian cancer cell lines. *Brit. J. Cancer*, 105: 372-381.

[58] Peterson, S., Kurimasa, A., Oshimura, M., Dynan, W., Bradbury, E. & Chen, D. (1995) Loss of the catalytic subunit of the DNA-dependent protein kinase in DNA double-strand-break-repair mutant mammalian cells. *Proc. Natl. Acad. Sci. USA*, 92: 3171-3174.

[59] Price, B.D. & Youmell, M.B. (1996) The phosphatidylinositol 3-kinase inhibitor wortmannin sensitizes murine fibroblast and human tumor cells to radiation and blocks induction of *p53* following DNA damage. *Cancer Res.*, 56: 246-250.

[60] Raaphorst, G.P., Maude-Leblanc, J. & Li, L. (2004) Evaluation of recombination repair pathways in thermal radiosensitization. *Radiat. Res.*, 161: 215-218.

[61] Raaphorst, G.P., Thakar, M. & Ng, C.E. (1993) Thermal radiosensitization in two pairs of CHO wild-type and radiation-sensitive mutant cell lines. *Int. J. Hyperthrmia*, 9: 383-391.

[62] Raaphorst, G.P., Yang, D.P. & Niedbala, G. (2004) Is DNA polymerase beta important in thermal radiosensitization? *Int. J. Hyperthermia*, 20: 140-143.

[63] Rainey, M.D., Charlton, M.E., Stanton, R.V. & Kastan, M.B. (2008) Transient inhibition of ATM kinase is sufficient to enhance cellular sensitivity to ionizing radiation. *Cancer Res.*, 68: 7466-7474.

[64] Reaper, P.M., Griffiths, M.R., Long, J.M., Charrier, J.-D., MacCormick, S., Charlton, P.A., Golec, J.M.C. & Pollard, J.R. (2011) Selective killing of ATM- or p53-deficient cancer cells through inhibition of ATR. *Nat. Chem. Biol.*, 7: 428-430.

[65] Rosenzweig, K.E., Youmell, M.B., Palayoor, S.T. & Price, B.D. (1997). Radiosensitization of human tumor cells by the phosphatidylinositol3-kinase inhibitors wortmannin and LY294002 correlates with inhibition of DNA-dependent protein kinase and prolonged G2-M delay. *Clin. Cancer Res.*, 3: 1149-1156.

[66] Sakata, K., Someya, M., Matsumoto, Y., Tauchi, H., Kai, M., Toyota, M., Takagi, M., Hareyama, M. & Fukushima, M. (2011) Gimeracil, an inhibitor of dihydropyrimidine dehydrogenase, inhibits the early step in homologous recombination. *Cancer Sci.*, 102: 1712-1716.

[67] Sarkaria, J.N., Busby, E.C., Tibbetts, R.S., Roos, P., Taya, Y., Karnitz, L.M. & Abraham, R.T. (1999) Inhibition of ATM and ATR kinase activities by the radiosnsitizing agent, caffeine. *Cancer Res.*, 59: 4375-4382.

[68] Sarkaria, J.N., Tibbetts, R.S., Busby, E.C., Kennedy, A.P., Hill, D.E. & Abraham, R.T. (1998) Inhibition of phosphoinositide 3-kinase related kinases by the radiosensitizing agent wortmannin. *Cancer Res.*, 58: 4375-4382.

[69] Savitsky, K.; Bar-Shira, A.; Gilad, S.; Rotman, G.; Ziv, Y.; Vanagaite, L.; Tagle, D.A.; Smith, S.; Uziel, T.; Sfez, S.; Ashkenazi, M.; Pecker, I.; Frydman, M.; Harnik, R.; Patanjali, S.R.; Simmons, A.; Clines, G.A.; Sartiel, A.; Jaspers, N.G.J.; Taylor, A.M.R.; Arlett, C.F.; Miki, T.; Weissmn, S.M.; Lovett, M.; Collins, F.S. & Shiloh,Y. (1995) A single ataxia telangiectasia gene with a product similar to PI-3 kinase. *Science*, 268: 1749-1753.

[70] Seno, J.D. & Dynlacht, J.R. (2004) Intracellular redistribution and modification of proteins of Mre11/Rad50/Nbs1 DNA repair complex following irradiation and heat-shock. *J. Cell. Physiol.*, 199: 157-170.

[71] Smider, V., Rathmell, W.K., Lieber,M.R. & Chu,G. (1994) Restoration of X-ray resistance and V(D)J recombination in mutant cells by Ku cDNA. *Science*, 266: 288-291.

[72] Spiro, I.J., Denman, D.L. & Dewey, W.C. (1982) Effect of hyperthermia on CHO DNA plymerase alpha and beta. *Radiat. Res.*, 89: 134-149.

[73] Stockley, M., Clegg, W., Fontana G., Golding, B.T., Martin, N., Rigoreau, L.J., Smith, G.C., & Griffin, R.J. (2001) Synthesis, crystal structure determination, and biological properties of the DNA-dependent protein kinase (DNA-PK) inhibitor 3-cyano-6-hydrazonomethyl-5-(4-pyridyl)prid-[1H]-2-one (OK-1035). *Bioorg. Med. Chem. Lett.*, 11: 2837 -2841.

[74] Taccioli, G.E., Gottlieb, T.M., Blunt, T., Priestley, A., Demengeot, J., Mizuta, R., Lehmann, A.R., Alt, F.W., Jackson, S.P. & Jeggo, P.A. (1994) Ku80: product of the *XRCC5* gene and its role in DNA repair and V(D)J recombination. *Science*, 265: 1442-1445.

[75] Takagi, M., Sakata, K., Someya, M., Tauchi, H., Iijima, K., Matsumoto, Y., Torigoe, T., Takahashi, A., Hareyama, M. & Fukushima, M. (2010) Gimeracil sensitizes cells to radiation via inhibition of homologous recombination. *Radiother. Oncol.*, 96: 259-266.

[76] Take, Y., Kumano, M., Hamano, Y., Fukatsu, H., Teraoka, H., Nishimura, S. & Oku-mura, A. (1995) OK-1035, a selective inhibitor of DNA-dependent protein kinase. *Biochem. Biophys. Res. Commun.*, 215: 41-47.

[77] Take, Y.; Kumano, M.; Teraoka, H.; Nishimura, S. & Okuyama, A. (1996) DNA-dependent protein kinase inhibitor (OK-1035) suppresses p21 expression in HCT116 cells containing wild-type p53 induced by adriamycin. *Biochem. Biophys. Res. Commun.*, 221: 207-212.

[78] Toledo, L.I., Murga, M., Zur, R., Soria, R., Rodriguez, A., Martinez, S., Oyarzabal, J., Pastor, J., Bischoff, J.R. & Fernandez- Capetillo., O. (2011) A cell-based screen identifies ATR inhibitors with synthetic lethal properties for cancer-associated mutations. *Nat. Struct. Mol. Biol.*, 18: 721-727.

[79] Veuger, S.J., Curtin, N.J., Richardson, C.J., Smith, G.C.M. & Durkacz, B.W. (2003) Radiosensitization and DNA Repair Inhibition by the Combined Use of Novel Inhibitors of DNA-dependent Protein Kinase and Poly(ADP-Ribose) Polymerase-1. *Cancer Res.*, 63: 6008-6015.

[80] Walker, A.I., Hunt, T., Jackson, R.J. & Anderson, C.W. (1985) Double-stranded DNA induces the phosphorylation of several proteins including the 90 000 mol. wt. heat-shock protein in animal cell extracts. *EMBO J.*, 4: 139-145.

[81] Woudstra, E.C., Konings, A.W.T., Jeggo, P.A. & Kampinga, H.H. (1999) Role of DNA-PK subunits in radiosensitization by hyperthermia. *Radiat. Res.*, 152: 214-218.

[82] Xiong, H., Lee, R.J., Haura, E.B., Edwards, J.G., Dynan, W.S. & Li, S. (2012) Intranuclear delivery of a novel antibody-derived radiosensitizer targeting the DNA-dependent protein kinase catalytic subunit. *Int. J. Radiat. Oncol. Biol. Phys.*, 83: 1023-1030.

[83] Xiong, H., Li, S., Yang, Z., Burgess, R.B., & Dynan, W.S. (2009) E. coli expression of a soluble, active single-chain antibody variable fragment containing a nuclear localization signal. *Protein Expr. Purif.*, 66: 172-180.

[84] Xu, M., Myerson, R.J., Straube, W.L., Moros, E.G., Lagroye, I., Wang, L.L., Lee, J.T. & Roti Roti, J.L. (2002) Radiosensitization of heat resistant human tumour cells by 1 hour at 41.1°C and its effect on DNA repair. *Int. J. Hyperthermia*, 18: 385-403.

[85] Yin, H.-L., Suzuki, Y., Matsumoto, Y, Tomita, M, Furusawa, Y, Enomoto, A, Morita, A, Aoki, M, Yatagai, F, Suzuki, T, Hosoi, Y, Ohtomo, K & Suzuki N. (2004) Radiosensitization by hyperthermia in chicken B lymphocyte cell line DT40 and its derivatives lacking non-homologous end-joining and/or homologous recombination pathways of DNA double-strand break repair. Radiat Res 162: 433-441.

[86] Zhao, Y., Thomas, H.D., Matey, M.A., Cowell, I.G., Rihardson, C.J., Griffin, R.J., Calvert, A.H., Newell, D.R., Smith, G.C.M. & Curtin, N.J. (2006) Preclinical evaluation of a potent novel DNA-dependent protein kinase inhibitor NU7441. *Cancer Res.*, 66: 5354-5362.

[87] Zhu, W.-G., Seno, J.D., Beck, B.D. & Dynlacht, J.R. (2001) Translocation of MRE11 from the nucleus to the cytoplasm as a mechanism of radiosensitization by heat. *Radiat. Res.*, 156: 95-102.

[88] Zou, L & Ellege, S.J. (2003) Sensing DNA damage through ATRIP recognition of RPA-ssDNA complex. *Science*, 300: 1542-1548.

Permissions

The contributors of this book come from diverse backgrounds, making this book a truly international effort. This book will bring forth new frontiers with its revolutionizing research information and detailed analysis of the nascent developments around the world.

We would like to thank Clark C. Chen, M.D., Ph.D., for lending his expertise to make the book truly unique. He has played a crucial role in the development of this book. Without his invaluable contribution this book wouldn't have been possible. He has made vital efforts to compile up to date information on the varied aspects of this subject to make this book a valuable addition to the collection of many professionals and students.

This book was conceptualized with the vision of imparting up-to-date information and advanced data in this field. To ensure the same, a matchless editorial board was set up. Every individual on the board went through rigorous rounds of assessment to prove their worth. After which they invested a large part of their time researching and compiling the most relevant data for our readers. Conferences and sessions were held from time to time between the editorial board and the contributing authors to present the data in the most comprehensible form. The editorial team has worked tirelessly to provide valuable and valid information to help people across the globe.

Every chapter published in this book has been scrutinized by our experts. Their significance has been extensively debated. The topics covered herein carry significant findings which will fuel the growth of the discipline. They may even be implemented as practical applications or may be referred to as a beginning point for another development. Chapters in this book were first published by InTech; hereby published with permission under the Creative Commons Attribution License or equivalent.

The editorial board has been involved in producing this book since its inception. They have spent rigorous hours researching and exploring the diverse topics which have resulted in the successful publishing of this book. They have passed on their knowledge of decades through this book. To expedite this challenging task, the publisher supported the team at every step. A small team of assistant editors was also appointed to further simplify the editing procedure and attain best results for the readers.

Our editorial team has been hand-picked from every corner of the world. Their multi-ethnicity adds dynamic inputs to the discussions which result in innovative outcomes. These outcomes are then further discussed with the researchers and contributors who give their valuable feedback and opinion regarding the same. The feedback is then collaborated with the researches and they are edited in a comprehensive manner to aid the understanding of the subject.

Apart from the editorial board, the designing team has also invested a significant amount of their time in understanding the subject and creating the most relevant covers. They scrutinized every image to scout for the most suitable representation of the subject and create an appropriate cover for the book.

The publishing team has been involved in this book since its early stages. They were actively engaged in every process, be it collecting the data, connecting with the contributors or procuring relevant information. The team has been an ardent support to the editorial, designing and production team. Their endless efforts to recruit the best for this project, has resulted in the accomplishment of this book. They are a veteran in the field of academics and their pool of knowledge is as vast as their experience in printing. Their expertise and guidance has proved useful at every step. Their uncompromising quality standards have made this book an exceptional effort. Their encouragement from time to time has been an inspiration for everyone.

The publisher and the editorial board hope that this book will prove to be a valuable piece of knowledge for researchers, students, practitioners and scholars across the globe.

List of Contributors

Qiang Xia, Feng Xue, Jian-Jun Zhang, Bo Zhai and Xi-Dai Long
Department of Liver Surgery, the Affiliated Renji Hospital, Shanghai Jiao Tong University School of Medicine, Shanghai, P.R. China

Xiao-Ying Huang, Chao Wang and Zhao-Quan Huang
Department of Pathology, the Affiliated Hospital, Youjiang Medical College for Nationalities, Baise, P.R. China

De-Chun Kong
Department of Physiology, the Basic Medicine, Shanghai Jiao Tong University School of Medicine, Shanghai, P.R. China

Carol Bernstein
Research Service Line, Southern Arizona Veterans Affairs Health Care System, Tucson, AZ, USA

Anil R. Prasad
Department of Pathology, University of Arizona, Tucson, AZ, USA

Valentine Nfonsam
Department of Surgery, University of Arizona, Tucson, AZ, USA

Harris Bernstein
Department of Cellular and Molecular Medicine, University of Arizona, Tucson, AZ, USA

António S. Rodrigues, Bruno Costa Gomes, Célia Martins, Marta Gromicho and José Rueff
CIGMH – Department of Genetics, Faculty of Medical Sciences, Universidade Nova de Lisboa, Lisboa, Portugal

Nuno G. Oliveira and Patrícia S. Guerreiro
Research Institute for Medicines and Pharmaceutical Sciences (iMed.UL), UL, Faculty of Pharmacy, Universidade de Lisboa, Lisboa, Portugal

Vivek Mohan and Srinivasan Madhusudan
Translational DNA Repair Group, Academic Unit of Oncology, School of Molecular Medical Sciences, University of Nottingham, Nottingham University Hospitals, Nottingham NG 1PB, UK

Axelle Renodon-Cornière, Pierre Weigel, Magali Le Breton and Fabrice Fleury
Unité UFIP, CNRS FRE 3478, University of Nantes, France

K. Barakat
Department of Physics, University of Alberta, Canada
Department of Engineering Mathematics and Physics, Fayoum University, Fayoum, Egypt

J. Tuszynski
Department of Physics, University of Alberta, Canada
Department of Oncology, University of Alberta, Canada

Christopher Busby
Jacobs University, Bremen, Germany

Paul W. Brandt-Rauf, Yongliang Li, Changmin Long and Regina Monaco
Division of Environmental and Occupational Health Sciences, School of Public Health, University of Illinois at Chicago, Chicago, USA

Yoshihisa Matsumoto, Shoji Imamichi, Mikoto Fukuchi, Sicheng Liu, Wanotayan Rujira, Shingo Kuniyoshi, Kazuki Yoshida, Yasuhiro Mae and Mukesh Kumar Sharma
Research Laboratory for Nuclear Reactors, Tokyo Institute of Technology, Tokyo, Japan

Printed in the USA
CPSIA information can be obtained
at www.ICGtesting.com
JSHW011458221024
72173JS00005B/1124